$S=EX^2$

Pere Estupinyà

S=EX²

The Science of Sex

Copernicus Books is a brand of Springer

Pere Estupinyà
Tortosa, Spain

Translated by
Mara Faye Lethem

The original version of this book was revised. An erratum to this book can be found at DOI 10.1007/978-3-319-31726-7_19

ISBN 978-3-319-31725-0   ISBN 978-3-319-31726-7 (eBook)
DOI 10.1007/978-3-319-31726-7

Library of Congress Control Number: 2016949746

© Springer International Publishing Switzerland 2016
This work is subject to copyright. All rights are reserved by the Publisher, whether the whole or part of the material is concerned, specifically the rights of translation, reprinting, reuse of illustrations, recitation, broadcasting, reproduction on microfilms or in any other physical way, and transmission or information storage and retrieval, electronic adaptation, computer software, or by similar or dissimilar methodology now known or hereafter developed.
The use of general descriptive names, registered names, trademarks, service marks, etc. in this publication does not imply, even in the absence of a specific statement, that such names are exempt from the relevant protective laws and regulations and therefore free for general use.
The publisher, the authors and the editors are safe to assume that the advice and information in this book are believed to be true and accurate at the date of publication. Neither the publisher nor the authors or the editors give a warranty, express or implied, with respect to the material contained herein or for any errors or omissions that may have been made.

S=ex^2. La Ciencia del Sexo by Pere Estupinyà Copyright © Debate 2013
S=ex^2. La Ciència del Sexe by Pere Estupinyà Copyright © Rosa dels vents 2013
Copyright © Springer Cham Heidelberg New York Dordrecht London 2016
Springer International Publishing AG is part of Springer Science+Business Media
All Rights Reserved.

Cover illustration: Txema Sanz + Mikel Urmeneta copyright Kukuxumusu

Printed on acid-free paper

This Copernicus imprint is published by Springer Nature
The registered company is Springer International Publishing AG Switzerland

To Fazia, who sends every single one of my hormonal levels off the charts, even those yet to be discovered.

Contents

1	**Sex in Our Cells**	1
	Searching for the Hormones of Desire	5
	Few Differences Between the Male and the Female	8
	The Chemistry of Our Sexual Behavior	15
2	**Sex in Our Genitals**	21
	Sympathetic and Parasympathetic Nerves in Sexual Arousal	23
	The Erection of the Penis and the Clitoris	24
	Erection Problems and Premature Ejaculation Due to Stress	29
	Desire and Arousal Are Not the Same	33
	Orgasm Based on the Distance Between the Clitoris and the Vagina	36
3	**Sex in Our Brains**	41
	Science Is More Interesting than Sex	41
	My Orgasm as Revealed by fMRI	46
	Want-Like-Learn, and the Empire of the Senses	52
4	**Sex in Our Minds**	59
	Sex Is an Irrational Act	59
	Measuring Sexual Arousal at the Kinsey Institute	62
	The Lack of Agreement Between the Female Mind and the Female Genitals	65
	Surveys and Statistics on Sexuality	69
	A Brief History of Scientific Research on Sex	73

5 Sex in Our Beds — 81
Even Scientists Have Trouble Finding the G-Spot — 82
The Two Types of Female Ejaculation — 86
The Genetic Component to Female
Multi-orgasmic Capability — 89
I Was Multi-orgasmic and Didn't Know It — 93
The "Coolidge Effect" and My Envy of Men Without
a Refractory Period — 95
Masturbation and Its Disadvantages As Compared to Intercourse — 100
Vibrators, Lubricants and Aphrodisiacs to Increase
Sexual Pleasure — 105
The Treacherous Effects of Alcohol in Arousal and Orgasm — 109
Motivations for Anal Sex — 112
Penis Size Does Matter, but Clitoral Size Doesn't — 115

6 Sex in the Doctor's Office — 123
Expectations and Other Male Sexual Dysfunctions — 125
Female Concerns. The Key Is Satisfaction, Not Desire — 129
The Microorganisms That Cohabitate in or Invade Our Genitals — 137
Sexually Transmitted Diseases — 141

7 Sex in Nature — 145
Why Do Ducks Have Penises and Roosters Don't? — 147
The Origin of Sex in Bacteria, Amoebas and Sea Sponges — 148
Hermaphroditic Potatoes and Animal Sex Changes — 150
Sexual Dimorphism: You've Only Ever Eaten Female Monkfish — 152

8 Sex in Evolution — 155
The Trap of Hidden Ovulation in Women — 156
Monogamy Is Natural, Faithfulness Isn't — 158
"Bonobos' Way of Life," Are You More of a Bonobo
or a Chimpanzee? — 160

9 Sex in Bars — 165
Others' Beauty Depends on Ours — 166
The Power of the Unconscious in Physical Attraction — 168
The Internet Has Only Revolutionized the First Few Steps
in the Dating Process — 173

	Non-verbal Signs of Seduction	175
	The Magic of Kissing	179
	Sex Without Commitment and "Hookup Culture"	182
10	**Having an Orgasm with the Power of the Mind**	189
	Hyperventilation to Activate the Sympathetic Nervous System	192
	Meditation and Yoga Increase Sexual Pleasure	194
11	**Pornography: From Distortion to Education**	199
	Women Prefer Lesbian Porn Over Gay	201
	Porn Can Exacerbate Some Problems, but Doesn't Cause Them	203
12	**Let's Do It Tonight, Dear, I Have a Headache**	207
	Sexus Sanus in Corpore Sano	209
	Corpus Sanum in Sexu Sano	212
	Sex in Old Age	213
13	**Sex in a Wheelchair, for Love and Pleasure**	219
	Neurosurgery to Regain Genital Sensitivity	223
14	**Science in Sexual Orientation**	227
	Homosexual Fluidity: Behavior Is Not the Same as Orientation	231
	Yes, You Can Be Born Gay	235
	What's Damaging Is Homophobia, Not Homosexuality	243
	Does Male Bisexuality Exist?	244
	Learning from Asexuals	246
15	**Learning from S&M Clubs**	251
	When Pain Produces Pleasure and Takes Away Another Pain	257
	Fetishists from Head to Toe	262
	Sexual Fantasies: Inhibiting Sins in Our Thoughts Leads to More Sins of Word and Deed	267
16	**Disorders of Obsession, Impulsivity, and Lack of Control**	275
	Hypersexuality Is Not an Addiction	276
	Paraphilias: When Science Articles Are Stranger than Fiction	282
	Involuntary Orgasms During Rape	287

17	**Sexual Identities Beyond XX and XY**	291
	Intersexuality: When Chromosomes and Genitals Don't Match Up	294
	Transsexuality: The Mind Is in Control	297
	Sex Change Operations, and the Phantom Penis	303
18	**Marrying Social and Sexual Monogamy in Swingers' Clubs**	313
	Polyamory with Emotional Monogamy	317
	Partner Issues When Desire Wanes	319
	Genes Don't Justify Infidelity	324
	Love Addicts	326

Erratum E1

Epilogue: Sex and Science Don't End at Orgasm 329

Acknowledgments 339

About the Author 341

Bibliography 343

Introduction

When neuroscientist Barry Komisaruk asked me to participate in one of his studies on the physiology of sexual response, I immediately accepted. It was January of 2012 and I was doing research for this book on the science of sex. Volunteering in an experiment at Rutgers University seemed like a great opportunity to get to know this sort of research from the inside. I was ready to leap in without the slightest hesitation. Later, when Barry explained that my task would be to stimulate myself manually while an MRI scanner measured activity in different parts of my brain as I got myself aroused and reached orgasm, I told him I needed to think about it. Oh man! The image that came into my mind was pretty terrifying. A few days later, I sent Barry an email apologizing and saying: "Barry, I'm sorry, but it's too embarrassing. And, honestly, I don't know if I would be able to complete the objective in those conditions." He insisted that the experiment would take place in complete privacy, that the only thing that the team would see would be my brain on the computer screen, and that I shouldn't worry about being nervous, and that even without climax, part of the data would still be just as useful. He added that I would be paid two hundred dollars, but I wasn't sure whether, under those conditions, that was an incentive or a setback.

In Chapter 3, I'll tell you if I ended up becoming the first man in history to have an orgasm under a magnetic resonance scanner. But first I want to take a moment here to reflect on the sudden reaction I had after declining: "Embarrassed? What was I really embarrassed about?" After all, while researching my last book, *The Brain Snatcher*, I was happy to participate in a Harvard study to see if a brain scanner—identical to the one Komisaruk would use—could detect my lies. I also let them electrically stimulate a part of my frontal lobe at the National Institutes of Health to find out whether I would learn a motor task more quickly. I got dizzy spinning around in a short-radius centrifuge at the Massachusetts Institute of Technology (MIT) to analyze how my body reacted to the absence of gravity. And I've gone into all kinds of laboratories, including military research labs, with much more questionable ethics than Barry's. My objective has always been to get to know science from as much of an insider's perspective as possible. And every time I've had a chance to observe or actively participate in experiments, I've done it. So where did this sudden reticence come from? I had the opportunity to collaborate with a top researcher studying the relationship between the nervous system and sexual response, and I was turning it down out of "modesty"? Strange. Especially since I've always thought of myself as an open-minded person who approached sex as the most natural thing in the world. Besides, when a researcher from Barry's team explained to me, weeks earlier, how they'd stimulated different parts of her genitals to see which nerves and parts of her brain were involved in each type of arousal, I was surprised to hear that she could reach orgasm in 15 seconds, but at no point did I judge her participation in the study as something unseemly or in poor taste. It had all seemed fine, and very interesting. Until it was my turn and I saw for myself how deep our sexual prejudices run.

Who knows whether the root of my embarrassment was a biological instinct or the result of cultural influence. Sex is an irrational act and, as such, coolly predicting our reactions in the face of new, emotionally intense situations is arduous and complicated. It is not easy for the human mind to reconcile reason and emotion.

In fact, with our friends and family members, and even with our partners, as absurd as it may seem, sometimes we have a lot of trouble talking about sexuality without a hint of shame hindering our words and making us look away.

That worried me, since my intention was to write a book about the science of sex that could be recommended by a father to his son and by a grandson to his grandmother. I'm convinced that sexuality is the perfect topic to illustrate the workings of our hormonal and nervous systems, the physiology of our brains, and the scientific analysis of our minds and social behavior. That's why I wanted to write about sex just the way I do about science, making it accessible to everyone. So I've tried to avoid using crude or excessively explicit language that could make some readers uncomfortable. Of course, I won't draw on obscure euphemisms when describing the different nerve endings that lead to the clitoris, vagina, or cervix. I'll also admit that I like to spark readers' playfulness and imagination, that I won't be able to avoid some sarcasm when talking about the circumstances that cause premature ejaculation in guys who act all macho, and that I'll resort to failsafe humor when I find myself blushing in front of the computer. I am firmly convinced that medical education and practice really need to work on communicating about desires, fantasies, doubts, and problems in the sexual realm, and that this deficiency limits our ability to enjoy one of the activities that brings us the most well-being and satisfaction.

Take a look at the study carried out in 2010 by Harvard psychologist Daniel Gilbert: in 2,250 cell phones they installed an app that asked men and women at random times what they were doing and how happy they felt. With 0 being the lowest and 100 the highest, the subjective happiness of working came in at last place with a 61. Reading, watching TV, taking care of the kids, and listening to the news were activities that all rated around 65. Shopping was in tenth place at 68. Then taking a walk, praying, or meditating. Eating came in seventh. Listening to music was fifth, chatting was third with 74, exercising took second place with 77, and yes, you guessed it, having sexual relations took top ranking with 92. The conclusion was obvious: sex is the activity that makes us most happy, at least temporarily.

It's true that the study was limited to one sector of the population and that this general tendency doesn't reflect the enormous diversity within humans' highly varied sexual conduct and culture. But I'd like to make an important point here: some of you believe that sex is so heterogeneous and depends on so many factors that it can't be studied scientifically. I completely disagree. It would be absurd to suggest that sexual behavior can only be understood via science or that one piece of data on hormones is more relevant than the hundreds gathered by anthropologists. Of course not. But the scientific method does have a lot to contribute to the academic study of sex, particularly due to its ability to separate a whole into various parts, isolate the diverse factors that influence our behavior, and systematically analyze and provide solid information that, when put together with data from other disciplines, contributes to creating a more accurate global vision than one based on unproven opinions.

We are in the multidisciplinary age. It would be outmoded academic extremism to try to give complete answers from merely the biological or sociological perspective. The paradigm of sexual investigation that is in vogue, and which we will use in this book, is biopsychosociological. That's not just some trumped-up big word, it's the one used by many sexologists to define how biology, psychology, and sociology should work together as a team in which each field, using their best tools, contributes to the scientific understanding of human sexuality.

But that's enough of that straitlaced tone! Obviously this book endeavors to be accurate and informative, but also to be entertaining, surprising, and suggestive. And in order to do that I will stick to my maxim of being incredibly scrupulous when selecting the most rigorous scientific information available, and presenting it in the most accessible and fun language possible. I don't call my father sir and I don't put science on a pedestal. And besides, when you're someone's friend—and I consider science my friend—you can criticize them, joke around, and make light of things. Your friends know how to tell the difference between a joke and an insult.

I probably won't always use the nomenclature preferred by researchers; I will mercilessly edit their preambles to talk more about the topics I think you will be interested in, yet always with the deepest respect and never trivializing their message.

If you take a look at the bibliography you will see that for most of the topics, I not only cite isolated scientific articles, but I also consult expert peer-reviewed studies published in high-profile magazines.

I should admit from the beginning that the science of sex is still in its infancy and is rife with gaps. In many theoretical and practical aspects, I have learned much more from therapists, experts, and people from various subcultures than I have from the academic world. Not only did I visit laboratories and analyze rats, hormones, and statistics, I also participated in a tantric sex workshop, met with porn actors and actresses, visited S&M and swingers' clubs, operating rooms, and doctors' offices, and spoke with people who told me about all sorts of experiences, from the most fetishistic to the most mundane. All in the name of science, of course! My own sexual life has expanded and improved because of it, and my hope is that the same thing happens to you when reading this book.

Once I had gathered so many interesting testimonials, I decided to use these personal stories as a resource that researchers usually call a case report. Here we will meet asexuals who are perfectly happy with their sexless lives, young women worried about not reaching climax, others who have multiple orgasms, people with physical disabilities who want to continue giving their partners pleasure, intersexuals whose gender identity doesn't coincide with their chromosomal sex, and bisexuals who, contrary to what some scientists defend, clearly state no preference for men or women. We will hear stories of withering erections, and orgasms in strange circumstances. They will all be real cases that, accompanied by statistics, will be our starting point for fascinating subjects such as the relationship between pain and pleasure, the evolutionary determining factors of our behavior, the arousing effects of jealousy, or the benefits of yoga in solving sexual problems.

I am sure that some of these stories will make you say "well, that's not the case for me." Of course, that's a result of the diversity we mentioned earlier. But don't reject the statistical data. It is true that the expression "in general" is deceptive, and that normality only exists as a statistical average. But you can generalize in science; what you should never do is individualize. If they tell us that "men have higher sexual drive than women," we should interpret it as a piece of data along the lines of "boys are taller than girls" or "smokers have a higher risk of lung cancer." Science doesn't look for patterns in order to deny diversity or make us fit into a stereotype, but because they are the clues along the trail to confirming differences, trying to explore their origin, and discovering how nature functions on the most intimate levels, from the pituitary gland, cellular oxidation, or the role of testosterone in sexual desire. No one negates the exceptions, but recognizing trends is very useful, hence the revolution supposed by Alfred Kinsey's work, which we will obviously discuss in this book.

I work with hypotheses, not ideas. We trust scientific data more than intuition biased by personal experiences, but we would be very naive if we didn't take into account that researchers' interpretations are conditioned by their social context. It is very good to maintain our own independent opinions, but again the ideal is that they aren't hermetic and can be adapted based on the best available information. So you won't find lectures here, or indoctrination of any kind, but you will find some suggestions and practical advice, all of which are approved by the therapists, psychologists, and sexologists I've spoken with, and based on a complete respect for diversity and free thought.

I like to vindicate sex as something fun. Not only is it a wonderful exchange of love and pleasure between couples, but its mix of diversity and taboos also leads to extremely suggestive conversations, private games, and provocations. If we can manage to take the drama out of it and shed the outsized importance that society gives it, sex is really a world filled with curious things to discover—from an intimate perspective, but also from an intellectual one.

I hold curiosity in the highest regard, for me it's the fuel of knowledge. Surely you can live without knowing about the mechanism through which sexual arousal triggers an erection in the penis and the clitoris. But if you don't know it and someone offers to explain it to you, I can't imagine you not being intrigued to find out, or the answer not leading to more questions. It is that curiosity, and realizing that no one had ever explained anything medical or scientific about sex to me, that motivated me to write this book. And it is that curiosity that will lead us to talk about the phenomenon of the phantom penis after amputation, the possible relationship between clitoral position and a higher frequency of orgasms during penetration, historical anecdotes, and the neurophysiological explanation of the Coolidge effect, which proves that the time it takes a male rat to have a new erection after ejaculation is shorter if he is given a new female partner.

I confess that there is a huge hidden challenge in this book: to explain things about sex to you that you've never heard before. Sex is overrated;

everyone talks about sexuality to the point that it becomes so tiresome and repetitive that it seems there's nothing new to say about it. Until you talk with the most interesting professional community on the planet, whose work is nothing less than discovering the unknown. By definition, if you approach a scientist in any discipline and ask them what they are investigating, they will begin to talk to you about the mysteries of the universe, about life, the mind, your cells, the past, and the future. Honestly, scientific research is the most fascinating adventure in the world. And when you combine that with sex, the sky's the limit!

Without science we would have never known about the existence of extra-solar planets or about neutron stars, or that there are microscopic viruses that cause illness or that the Earth roars because of movements in the tectonic plates. Nor which gene on the Y chromosome starts the series of signals that makes a fetus male, or whether pederasts have more activity in one particular part of the brain. You could spend hours debating whether female ejaculate is more similar to semen or to urine, but a chromatographer can quickly clear that up.

Throughout the book we'll get a little more serious when analyzing medical aspects and psychosexual disorders. I visited clinics and therapists with different philosophies, and I have to say that while there are many fabulous sexologists, I found few professionals who approach sex from a complete, integral perspective. I've met doctors who prescribe testosterone cream with shocking frequency and psychotherapists who insist that every problem has an exclusively mental basis. A book's rigidity and inability to give immediate answers don't make it the ideal format to discuss these rapidly shifting topics, but I will definitely defend this biopsychosociological vision in which the physical, psychological, and social aspects must all be considered together before any diagnosis. In no way do I claim to replace the advice of doctors and therapists, and I recommend you don't hesitate to see a professional when you think you need more information.

Something similar happens with the sociological aspects. I want to make clear that I support the equality of men and women, the definitive acceptance of homosexuality on all levels, a committed celebration of diversity, and I encourage a positive vision of sexuality, and absolute respect for the limits that each person wants to establish with their partner and according to their convictions. But I am not here to defend any cause and, besides, the "isms" rub me a little bit the wrong way. The only thing I can say at this point is that evidence-based science supports a more open vision of sex and makes some of the more conservative moral statements about the negative effects of masturbation, pornography consumption, the use of sex toys, or the obsolete concept of "abnormality" look silly. And that last one goes for everybody. Another important message is that the key is to balance desire and satisfaction, not "the more different kinds of sex, the better." In fact, I've observed more happiness in asexuals than in confused polyamorous couples.

As I wrap up this introduction, I want to warn you that the first few chapters are more biological and may make for denser reading. In them I will try to explain basic aspects of physiology and research methodologies that I consider important for understanding the following chapters. But if you feel your interest waning, check the index and go straight to the section that most interests you. We will start with hormones, followed by the nervous system, the brain, and the study of human behavior. From there, we will deal with medical aspects and sexual disorders. We will explore sex in nature and our evolutionary past, we will be enlightened by tips from some fantastic therapists, and we will end up having sexual experiences in all sorts of highly peculiar environments. Again, I hope that your curiosity will outweigh your embarrassment, as mine did, and that you will scratch where you have no itch, and both your brain and your body, as well as those of your partners, will get the most out of this book. The adventure continues.

1

Sex in Our Cells

Sandra and Jacob are naked in an empty room. They don't really know how they got there, or what they're supposed to do. They are complete strangers. They have never seen each other before, so it's all very confusing. They also have no idea that I am observing them and taking careful notes on their behavior, nor that the researchers from Concordia University, in Montreal, have ensured that Sandra's hormonal levels will make her aroused and receptive to "mating," as they like to call it.

Despite that, Sandra keeps her distance, moving about the room as if exploring it, and seems to ignore Jacob's presence. They both avoid eye contact. They are noticeably nervous, until the action starts a few seconds later. Without saying a word, Jacob takes the initiative and walks decisively over to Sandra. She reacts by moving away. Jacob stops for a couple of seconds but soon tries to approach Sandra again. This time he manages to get fairly close when, suddenly, he smells a peculiar scent. Sandra smells like almonds. It is an intense perfume that Jacob has never smelled before. Strange, but neutral. He's indifferent to it, he doesn't particularly like it or dislike it. Jacob remains focused on Sandra's body and starts to follow her in circles around the room. Sandra keeps moving away but she's not running anymore. She seems like she's pretending to escape every time Jacob touches her or brings his face close to her. When Sandra feels his touch, she jumps and moves away. Jacob insists for a couple of minutes, but then gives up and stays in one corner.

Sandra looks at Jacob out of the corner of her eye. She is still to one side, but after a few seconds she walks slyly past him. He turns enthusiastically toward her, and Sandra leaps away again. "Solicitation!" shouts one of the scientists that I'm with, as he jots it down. It's a very strange situation. They explain that her coming closer to generate attraction and then turning away is very typical of female sexual behavior. "I know, I know…," I respond, betrayed by my unconscious. The experiment continues, as Jacob keeps chasing and Sandra's rebuffs become increasingly less convincing. In one of his approaches, Jacob traps Sandra from behind as if he wanted to copulate. She slips away, but not before the researchers notice something. "Lordosis!" they shout, writing it down. Sandra had reacted to Jacob's touch by arching her back, tilting her pelvis outward and her neck backward. It seems that this reflexive act is an evolutionary vestige that's been well preserved in mammals, and means that the female is aroused and prepared for penetration. In fact, the tension in the room is growing and, in each of his increasingly aggressive charges, Jacob manages to partially penetrate Sandra. "Intromission!" exclaims a scientist. I am totally gobsmacked. Especially because Sandra moves away again, takes a few steps, stops and allows Jacob to repeat his "intromission." And it goes on like that, several times, alternating with lapses where they seem to rest. The scientists jot down the number of penetrations, and I observe the situation as dumbfounded as you might be feeling now. It was surreal. We must have been 11 or 12 minutes into the experiment, when suddenly in one of his thrusts Jacob seems to cling tightly to Sandra's back. He is frozen like that for a brief second, and the two scientists with me shout in unison: "Ejaculation!!!" I hadn't noticed anything. "Really?" I asked, more to myself than to them. "Yes, yes… definitely," they reply. Then Jacob separates from Sandra, gradually retreats without even a caress and lies down on the floor and falls asleep. Sandra seems uneasy and keeps moving nervously around the room. Jacob no longer pays any attention to her. Half a minute passes and a giant hand enters the room through the roof, picks Sandra up by the nape of her neck and places her in another room. The experiment is over.

Sandra and Jacob are two rats in the laboratory of behavioral neurobiology run by James Pfaus at Concordia University in Canada. In fact, as frisky as he'd seemed, that was Jacob's first sexual encounter. The study consisted in making the male rats have their first copulation with female rats infused with almond oil, then exposing them several times to females in heat but without the almond scent, and after some time placing the males into cages with some rats that smelled of almonds and others that didn't to observe to what degree they preferred the almond-scented ones. If their preference was very marked, it would mean that first sexual experiences can condition part of an adult rat's sexual behavior.

I observed this curious experiment during my first visit to Jim Pfaus's lab in July 2010, when I was starting to realize that, hidden in various universities, there were a good number of researchers who firmly believed that science had much to contribute to the multidisciplinary study of human sexuality, and that it wouldn't be a bad idea to write a book describing this scientific perspective of sex. Two years later, during my second visit to Concordia University in June 2012, when I was already well into this book project, Jim explained to me that the male rats whose first sexual encounters were with the almond-scented females did have a very marked preference as adults for females imbued with the almond oil. And not only that: if they put the almond scent on a little wooden ball and placed that into their cage, they would gnaw on it and even rub their genitals against it. It was as if they had generated a sexual fetish for the smell of almond oil. And the same thing occurred when they exposed virgin males to females dressed in leather jackets: if, as adults, they were put into a cell with one female wearing leather and another naked, they would head straight for the one with the jacket.

Even more surprising: jackets and almond oil are neutral stimuli. But what would happen if the stimulus were negative? James Pfaus's team carried out similar experiments but this time with female rats with a hint of cadaverine, a substance produced by rotting flesh that unequivocally drives away every rat in the world. Cadaverine is a very strong sign of the risk of infection: any rat who smells it, immediately flees. If you put an adult male into a cage with females in heat, some with cadaverine and some without, he won't even go near the ones that smell bad. Yet, a male whose first sexual experiences were with rats that smelled of cadaverine,[1] shows no preference when exposed as an adult to females with cadaverine and without. And when one end of his cage is sprinkled with the putrid substance, unlike the other control male who runs off in fear, the conditioned male will go into the area with no problem. Sex manages to revert their aversion to a stimulus genetically programmed to be repugnant in order to prevent life-threatening infections.

It is obvious that we can't extrapolate this conclusion directly to humans. I am not suggesting that at all. When I ask Jim if there could be a similar conditioning in young women whose first satisfactory sexual experiences were with men with a lot of hair or strong-smelling armpits, or men who preferred masturbation with their partners because their first orgasms with their girlfriends didn't involve coitus, he responds: "Possibly, it's not a ridiculous hypothesis. We know that first sexual experiences leave some sort of an impression. Of course there are many other factors involved in human sexual conduct, from biological ones to cultural ones; but, of course, the reinforcements conditioned in adolescence can influence adult preferences." We will try to tackle all

---

[1] Sexual instinct is so strong that, even though he finds it repellent, a young virgin male will have sex with females that smell of cadaverine if that is his only option.

those factors in this book, but by observing people more than rats. However, let's not rush to dismiss laboratory studies using animals. Historically they have offered us very important clues, and they are just as valid for investigating some of the variables of sexual response as they are for studying diabetes, addictions or depression. Perhaps even more so. From a strictly physiological standpoint, we aren't so different from rats. Let's look at hormones for example. The human menstrual cycle is regulated by a system very similar to that of other mammals. The level of estrogens increases until the ovum is mature, the pituitary gland sets off ovulation by secreting the luteinizing hormone, little by little the progesterone increases and there are other chemical signals that regulate a cycle that is so fundamental to reproduction of the species that natural selection hasn't made any big changes to its most essential mechanisms. The basic physiology of sexual instincts and reproductive function are very evolutionarily protected.

On top of this basic endocrinology in our species, there is the whole cultural influence, experiences had during development, learning, and freedom of action. A clear example of this is that, unlike the rest of female mammals except for bonobos and dolphins, women have sex for pleasure throughout their menstrual cycle and not only around ovulation for reproductive means. Let's stop for a moment to reflect on that last point. How does a lab rat know she is in a fertile moment and should copulate? How does she know when she isn't and instinctively avoids any kind of courting behavior? In her case it isn't about trying to find a particularly attractive male, or societal pressure, or feeling more euphoric on a Friday than on a Monday. In her case the appearance of desire is an internal message conditioned exclusively by her hormonal levels. Let's also ask if this internal chemical evolutionary vestige could have a role in human behavior and whether in some cases of sexual dysfunction the loss of desire could originate in low levels of testosterone, or vaginal dryness could be due to a lack of estrogens, or the variations in progesterone levels during the menstrual cycle could make a woman, without really knowing why, prefer a more masculine man on one day and a less masculine one the next. We don't doubt that that's true. And even though in most cases it is a negligible effect compared to socialization and everyday experiences, in others it does seem to be important. It's worth studying. How? Obviously we can observe the changes in sexual response in post-menopausal women, or the effects of lessening testosterone from the use of contraceptive pills, the loss of libido from antidepressants that increase serotonin levels, or the hypersexuality generated by some medications that regulate dopamine. But we can't force humans to have their first sexual relations with partners wearing leather or smelling of almonds, and we can't inject them with different hormonal combinations to separately analyze how each substance affects sexual response. And although at first glance it seems strange, since the physical mechanisms of sexual response are something so primitive and evolutionarily conserved, rats are a good animal model for investigating very basic aspects of the endocrinology of sex.

Searching for the Hormones of Desire

My interest in the science of sex came about in a sudden and rather comical fashion one November morning in 2008, in Washington DC. I was wandering around a huge neuroscience conference, searching for something new among the more than 30,000 researchers who were presenting their work, when I saw a scientific poster with the title "Clitoral Stimulation Induces Fos Activation in the Rat Brain." In front of it stood the young researcher Mayte Parada. I couldn't help going over and asking her, as seriously as I could, how one stimulated the clitoris of a rat. "I use a paintbrush, but there are several techniques. It's easy to do because rats have very large clitorises," replied Mayte, moving her hand up and down with her index finger and thumb pressed together. "With a paintbrush... I see... but rat sex is pretty different than human sex, right?" I said, trying to tone down my skepticism. "Well, that depends. In terms of hormonal mechanisms—which is what we are studying—we aren't that different," she answered.

Mayte explained that in her laboratory they removed the ovaries of several rats, injected them by group with different combinations of estrogens, progesterone and other hormones, and then analyzed their sexual response. It was like reproducing the menstrual cycle but altering some specific hormonal level each time. Rats avoid having sex when they aren't fertile because it's risky and a useless expenditure of energy, and Mayte wanted to find out what exact hormonal combination made a rat feel internally aroused. Perhaps it could give clues into some of the cases of low libido in women. She told me that in her lab at Concordia University they were researching everything from hormonal influences to preferences conditioned by experience, both in rats and in humans, and she told me: "The scientific study of sex is fascinating because it's actually very new, and it's filled with highly interesting research subjects. You should come visit us."

I continued walking around the conference. But the virus of curiosity was already beginning to spread and reproduce in my brain. I knew that I wasn't going to be able to forget that conversation. Months later I went to Mayte Parada and Jim Pfaus's neurobiology laboratory in Montreal. There I watched as Mayte placed a rat in a cell, lifted its tail every ten seconds, and rubbed its enormous clitoris three or four times quickly. (The fact that it's enormous is not surprising; rats' testicles are almost as large as their brains.) She repeated the action and let the rat explore the cell. The key was that these cells had a particular characteristic (a scent, for example) that, after several days of the experiment, the rat would associate with sexual stimulation. After some time, Mayte would place the rat in a new cell with that same scent, but with a door that gave it the option to stay there or go into another room. Its decision would indicate whether the clitoral stimulation had been positive, negative or indifferent.

Conditioned Place Preference is a common procedure in psychology to see if a rat "likes" or "dislikes" a particular stimulus. Imagine we place a rat in a room with light and a door that leads to a dark room (or a white cell to a black cell, or one with scent and the other without). Instinctively, the rat will explore and circulate freely through both rooms. But if in our experiment we always give it a stimulus in the room with light, and we get it to associate that stimulus with that room, when after a few days we place it there again there are three possible outcomes: if the rat liked the stimulus, it will remain in the room with light, waiting for it. If the rat found it repulsive it would immediately scurry through the door and spend much more time in the dark area. If it was indifferent to the stimulus, it would alternate through both rooms with no preference. That is how Mayte was researching which hormonal cocktail made clitoral stimulation more pleasurable.

I found it very intriguing, and knowing that human sexual behavior depends on many more factors beside hormones, my interest in the biology, psychology and sociology of our sexual conduct grew exponentially.

In fact, I saw that if not for the shame that it provokes in us, this biopsychosociological perspective of sexuality would be a fabulous example to discuss how body, mind and culture interact on different levels: sex allows us to decipher the way our genes and hormones regulate our body's inner balance, explain basic anatomical principles, the functioning of the nervous system and the brain; understand the influence of learning, the effects of the mind on the organism, the cultural determinants of our behavior; discuss our evolutionary past and a long list of etceteras. It could even be a brilliant way to teach science in the schools, because in addition to perfectly displaying different scientific methodologies, ranging from experiments with lab rats to sociological surveys asking how old we were when we had our first sexual encounter, the positions in which we reach orgasm most easily and how many times we've been unfaithful. But there is much more. They can use scanners to directly observe brain activity, and physiological studies to reveal the role played by the peripheral nervous system, hormones, genes, muscles and metabolism, and of course endless sociological studies that analyze how sexuality has evolved over time and been affected by education and culture. They all offer informational clues that are incredibly useful.

In August 2010, Mayte published the results of her studies in the magazine *Hormones and Behavior* with a partially unexpected conclusion: the rats always like having their clitorises stimulated! The different combinations of estrogen and progesterone didn't affect how much time they spent in each side of the cell; in fact, they always went to the place where they thought they would receive stimulation. What was strange about it was that the rats only displayed lordosis and sexual solicitation behavior when the hormonal cocktail mimicked the moment of ovulation. Which is to say, the hormones affected their behavior by determining whether the rats gave mating signals or not, but apart from that, and even when they weren't receptive, they still enjoyed clitoral stimulation. This likely seems logical to us from a human perspective yet is quite novel in animal behavior: it implies that the hormones condition behavior but not sexual response. Making a speculative extrapolation to humans, it is consistent with other results that women enjoy sex equally in all phases of their menstrual cycle, although they feel slightly more desire and display more subconscious seduction signals around their days of ovulation.

Mayte's results also suggest a distinction between desire and physical arousal that will be important in this book when we differentiate between disorders involving low desire and arousal deficit disorders. In humans we think of desire as having more of a psychological component and arousal as the physical response to desire. In men it would be very easy to distinguish: feeling eagerness versus having an erection. The mechanisms of desire and arousal overlap, and sometimes arousal can lead to desire instead of the other way around. But everything suggests that they can both be distinguished hormonally, and that this internal chemistry can explain some dysfunctions and atypical cases as well as part of our diversity in desire, attraction and sexual behavior at various points.

I should mention that hormones are only part of the equation and that it doesn't make sense to study them without taking the sociological context into account. But we should acknowledge that you can't claim to have a complete understanding of human sexuality without a neurobiological perspective. Which is why we begin with that and from there we will fan out into all of the fascinating levels of the study of our human sexual response.

Few Differences Between the Male and the Female

Four centuries before Christ, the Greek philosopher Anaxagoras declared that boys came from the right testicle and girls from the left. It is said that, following his logic, in the Middle Ages some French aristocrats with several daughters went so far as to remove their left testicle in order to create boys. Aristotle, on the other hand, believed that the key to conceiving a male child was to put a lot of effort and emphasis on copulation, and that babies only arose from the father's semen; the mother didn't transmit any genetic information and her uterus was simply a vessel where the fetus grew until birth. Fortunately, science has made considerable advances and is now able to explain all of the genetic and molecular determinants responsible for sexual differentiation between males and females. Let's go over them.

Inside almost all of the cells in our body we have 46 chromosomes, 22 pairs of autosomes and one pair of sexual chromosomes: XX for a woman and XY for a man.[2]

The ova and sperm are exceptional in that they only have one copy of each chromosome. So when they come together with our partner's respective gamete, there are 46 chromosomes in total. In women the sexual chromosome of their ova is always X, while half of men's sperm carry an X and the other half a Y. That determines whether it will be a XX, girl, or an XY, boy. But, play close attention, because in the ultimate instance sexual differentiation is led more by hormones than by genes. If a female rat embryo is injected with testosterone during the first days of its development, it will be born with ambiguous genitals and male behavior. If the testosterone is injected at a later stage of the gestation, it will be born female but will try to mount other females. And testosterone injected into an already full-grown female rat will make it more aggressive but not affect its sexual preferences.

When a fertilized ovum begins to divide, at first all its cells are identical. Later they begin to differentiate into nerve, blood, muscle, and bone cells, and gradually start to build the different organs that make up the human body. In the case of the anogenital region, the first structure that forms at four weeks of pregnancy is an orifice called the cloaca that has a genital tubercle above, the urethra inside and a swelling around it (Fig. 1.1). At six weeks, a part in the middle of the cloaca will have closed, separating two orifices that will become the anus and genitals. Within the genital area there are a pair

---

[2] Later on we will look at cases of intersexuality and transsexuality where the sex chromosome doesn't correspond with the gender identity.

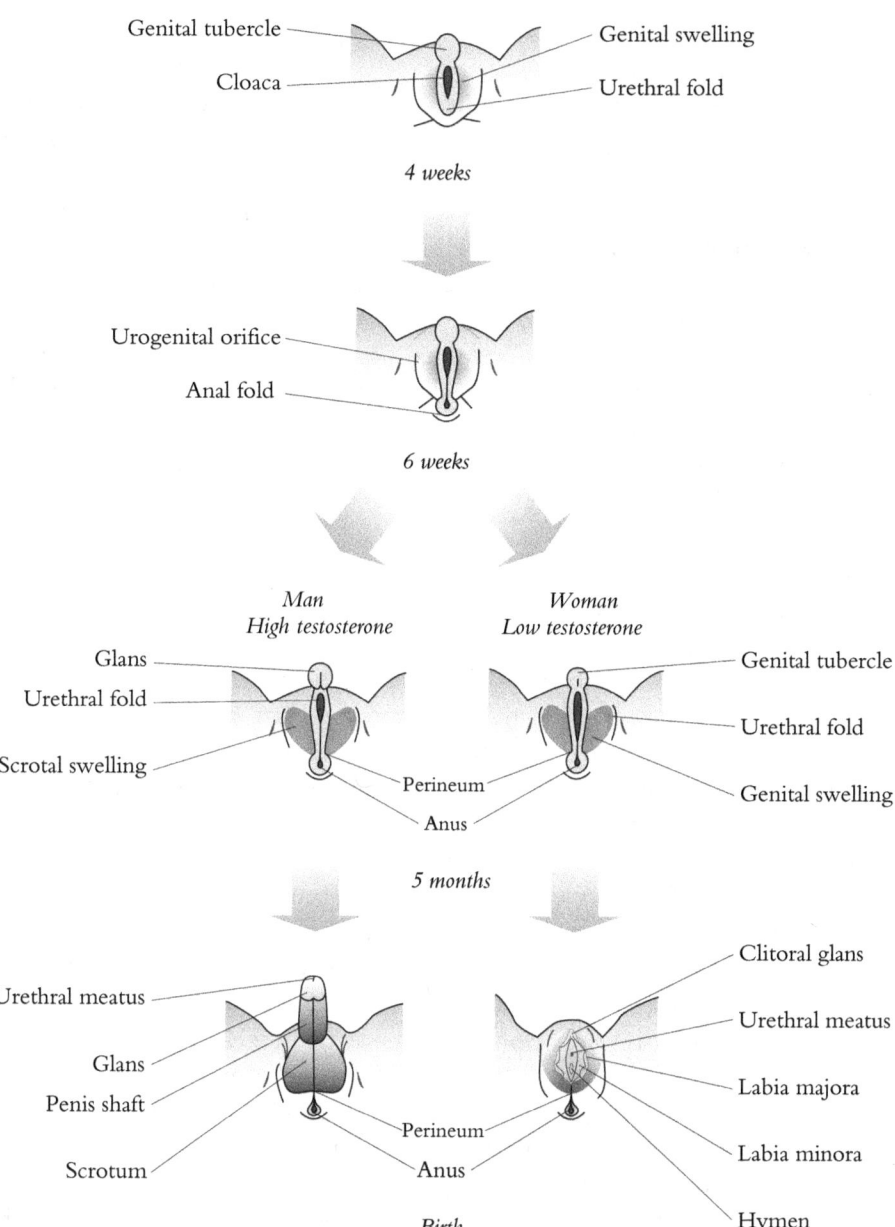

Fig. 1.1 Embryonic development of male and female genitals

of conduits connected to two gonads that will later form testicles or ovaries. But until that moment, the structure is exactly the same for future boys as for future girls. Sexual differentiation starts in week 6: if the fetus carries the SRY gene (or sex-determining region) on the Y chromosome, it releases a hormone called anti-Müllerian (AMH), forcing the gonads to become testicles. If the fetus is XX and there is no AMH released, the same gonads will develop into ovaries and fallopian tubes. This is the first stage of sexual differentiation. But the onset of the overall masculinization of the fetus begins in week 8 when the still internal testicles start to secrete testosterone. That makes them grow and begin their descent into the inflamed skin around the cloaca, which will become the scrotum. If there is no testosterone, that same skin will form the vaginal labia. Masculine and feminine genitals come from exactly the same anatomical structures, except they develop and situate differently.

Testosterone's next effect is making the genital tubercle above the cloaca begin to grow outward, closing the cloaca's orifice and taking the urethra along with it to form a penis with a glans and two cavernous bodies that swell in the event of an erection.

If the fetus is XX and there is no release of testosterone, the same exact process will occur except towards the inside and on a smaller scale. The cloaca will remain open, creating the vagina and uterus, and the tubercle will grow towards the interior of the body. Only the glans will remain outside, becoming the external clitoris, and the cavernous bodies will develop into two inner arms of the clitoris that will pass by the sides of the vagina in a v-shape, swelling with blood at the moment of arousal. Penis and clitoris have the exact same embryonic origin. In fact, the male glans is the head of the female clitoris but with the same nerve endings concentrated in a much smaller space. Really: if we observe images of the male and female genitals in their entirety, we will see that the clitoris is almost identical to the penis, that ova and testicles clearly come from the same structures, and that one of them became external first because of the action of the Y chromosome and later due to testosterone (Fig. 1.2). They are so similar that some researchers consider that vaginal orgasms are actually clitoral because penetration presses on the internal structures of the clitoris. To put it very crudely, clitoral stimulation would be like only stroking a man's glans, and vaginal stimulation would be like stimulating the shaft. Or, depending on the position during coitus, all at the same time. We will discuss this aspect further on when we talk about the G-spot. So, the message is clear: on a genital level, men and women are much more alike than we imagine.

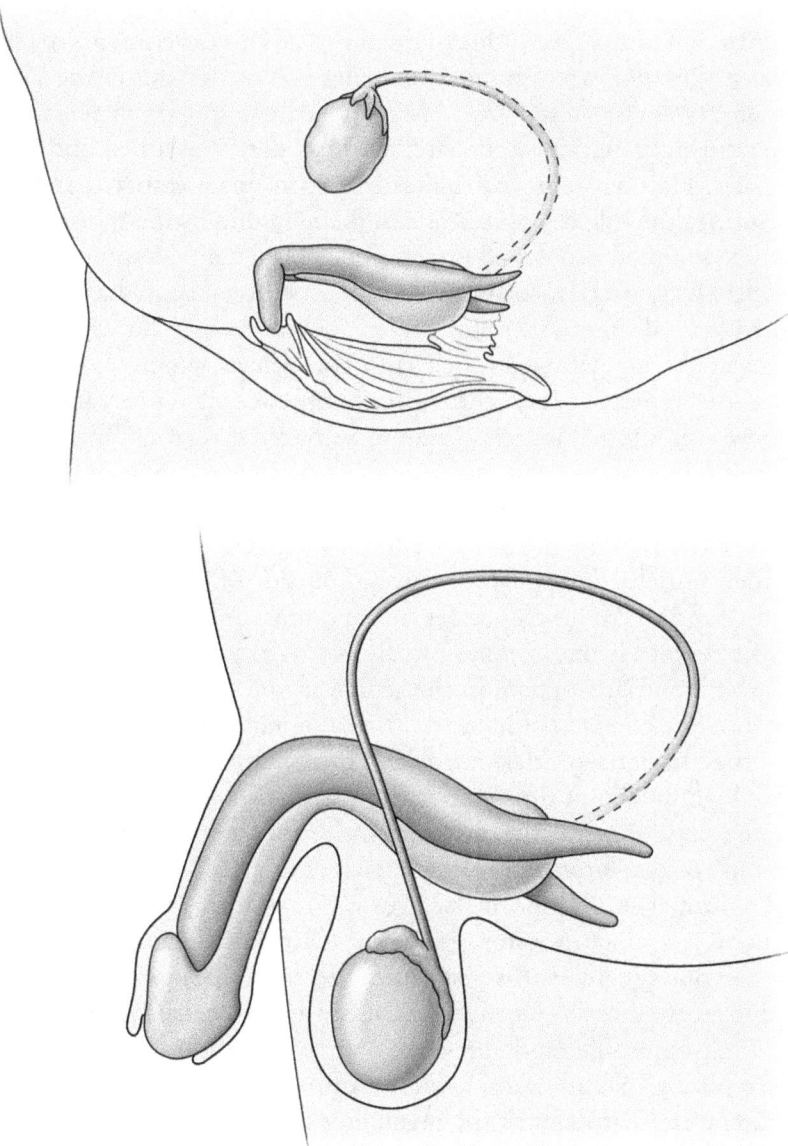

Fig. 1.2 Comparative structure of the development of the penis and the clitoris and the testicles and ovaries

It's the androgens and not the Y chromosome that ultimately lead to the masculinization of the fetus. One of the most famous experiments to confirm the greater relevance of hormones over genes was carried out in the 1940s by French embryologist Alfred Jost. Jost removed the ovaries of female rabbit embryos during gestation, and observed how they developed and behaved like females. He then removed the testicles from male embryos and, as was expected since the role of testosterone in the masculinization of the body was already known, they also ended up developing as females despite being XY.

But there is no need to resort to research on rabbits. People with androgen insensitivity syndrome (AIS) are women despite having XY chromosomes. Those affected with AIS begin their embryonic development as male fetuses in which the genes of the Y chromosome provoke the appearance of testicles. Those still internal testicles begin to secrete testosterone, but something anomalous happens: women with AIS have a genetic mutation that generates a lack of androgen receptors in every cell in their organism. Which is to say, there is testosterone in their blood, but their cells don't recognize it. As a result, their testicles don't grow or descend outside of the body, their genitals and brain continue to develop as female, and most are born indistinguishable from girls despite being chromosomally XY. Many cases are only detected later in childhood due to pain in the genital region, and sometimes not until puberty because of a lack of menstruation. Women affected by AIS are sterile because they have no ovaries, but with hormonal replacement therapy they can have high quality of life. This example brings us to another key factor in embryonic development: the masculinization of the brain by the effects of testosterone or the feminization by the lack of it.

Under normal conditions, if the fetus is XY, starting at week 12 of a pregnancy there is a spike in testosterone that masculinizes the structures of the central nervous system as it is just starting to form. The genes express themselves differently due to epigenetic marks; in men they silence various genes on the X chromosome, and some brain circuits are modified, conditioning part of typically masculine or feminine future behavior. There is solid scientific literature that defends the possibility that transsexuality originates in alterations during this phase that define an important part of sexual identity. In some cases even sexual orientation could be conditioned by different levels of hormones during pregnancy. There are so many nuances that it is worth delving into further later on, but what is undisputable is that under normal conditions this phase is essential to defining a brain predetermined towards masculine or feminine behavior. Similarly to female rat embryos injected with testosterone behaving like males and even, without a penis, trying to mount other females, if a male embryo is injected with substances that deactivate testosterone, it will display lordosis, arching its back around other males. It should be pointed out that researchers do not equate this with human homosexuality as we know it. Our sexual orientation is not defined merely by biological factors and testosterone spikes during pregnancy and the first weeks of life, but rather, as we will see in later chapters, by many other additional aspects of human development.

The true lesson here is that sexuality is a *continuum*. We have seen that defining men and women as XY or XX is sometimes simplistic, since what defines masculinity and femininity on a biological level is the hormonal environment and how various parts of the brain are configured over time. And those definitions are not set in stone; there can be many intermediary degrees between a strictly masculine brain and strictly feminine one. Add to that all the cultural and educational influences, and we have a puzzle that defies unilateral explanations of sexual behavior.

This is important in the eternal debate over whether men and women are very different or not, and in what ways. Here, after long conversations with all sorts of experts, I wanted to express some ideas. The first is that the only academic extremism I encountered was in some circles that deny any biological influence and state that sex is only a cultural construction. This position is already out of date and shows absurd dogmatism. On the contrary, no recognized scientist who studies genes, hormones or neural circuits denies that educational and socially learned factors upend all the physiological aspects of our sexual behavior, from attraction, the range of sexual activities, repression and the conscious decision to have sex or not. Scientists don't really have reductionist minds, as they are often described.

That said, we must also recognize that the biological view has also been very exaggerated. Innate differences between men and women's minds do exist, and we can see them both in the different conduct of boys and girls from a young age and in the neuroanatomical variations in each gender's hypothalamus. But we must keep three things in mind:

1. The most rigorous psychology shows us that the rift is not as great as more sensationalistic psychology has suggested.
2. Everything that we are discovering about epigenetics and brain plasticity forces us to assume that biological determinants are less stable than we believed and are definitely subject to influence from the environment.
3. The neurophysiological differences between men and women primarily affect the more primitive structures in the brain. This last point is not usually taken into consideration, and it is very relevant for understanding sexual behavior.

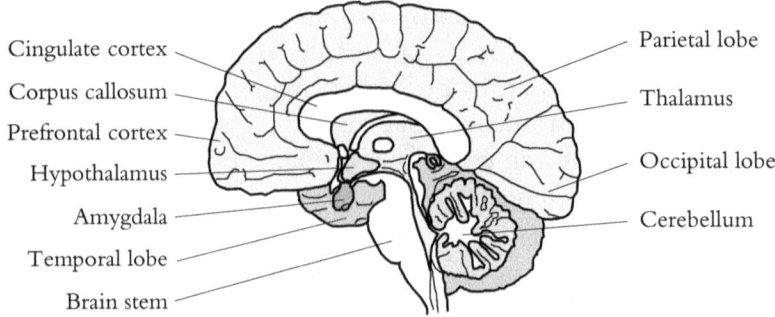

Fig. 1.3 Main anatomical structures of the brain

If we look at a diagram of the brain (Fig. 1.3) we see that it seems to be made up of a series of layers. The more internal ones, like the amygdala and the hypothalamus, regulate emotions and basic functions, and they are incredibly similar between us and other animals. However, in some aspects they do show a considerable dimorphism between males and females, particularly in the hypothalamus, where it has been shown that in the embryonic stage 5 % of the genes expressed there respond differently if the fetus has testosterone in its blood. These inner layers are key to the sexual instinct, they are the headquarters of our most "animal" behavior, and culture has little direct power over them. But on top of this primitive brain we have the more external layers of the cerebral cortex, which are responsible for all the higher functions and more sophisticated reasoning. This more evolved part of the brain is extremely plastic and is born conditioned to be modulated by learning. Perhaps there are a few predispositions based on sex, but these are much more permeable to education and cultural gender roles than to biology.

In 2005 psychologist Janet Hyde of the University of Wisconsin published a provocative study entitled "The Gender Similarities Hypothesis." It was an exhaustive review of 46 scientific meta-analyses that examined psychological differences between men and women. It is the most rigorous work published to date, and its conclusion was convincing: the phrase "men are from Mars and women are from Venus" is mostly a myth unduly promoted by the media and popular culture. In most psychological characteristics and cognitive functions, men and women are much more alike than some isolated studies indicate. Somehow, what Hyde suggested was that there is greater internal diversity in sophisticated functions between groups of men or groups of women than generic differences between men and women. Where Hyde did find significant differences was in some emotional reactions such as aggressiveness, certain motor aspects and particularly in sexual behavior.

The message to be taken, in the context of this book, is that in terms of instinct and emotions we are biologically different, and in terms of more sophisticated behavior we are culturally different. And it is important to add another fundamental factor to that: on occasion, instinct, desire and primitive behavior can emerge strongly from the more emotional depths of our brain and create a conflict if our cerebral cortex, the base of rational thought, is sending us inhibitory messages. But unless we are primitive beings with little self-mastery or with brain lesions, our species has freedom of control. That is to say: perhaps we can't decide what we like, but we can decide whether or not to do what we like.

Let's start at the beginning and learn how hormones constantly remind us who's who, conditioning our sexual and loving relationships as adults.

The Chemistry of Our Sexual Behavior

If our blood sugar levels are low, the hypothalamus will send chemical signals to spark our appetite. And when our glucose levels are restored, the insulin secreted by our pancreas and the leptin in our fatty tissue will make us feel sated. Or, if a dog attacks us, the adrenaline released by our adrenal glands will increase our heart rate and our muscular tension so we can face up to the threat. In many aspects our organism functions like a machine, just like any other animal, in which the hormones of the endocrine system are the messengers whose work is maintaining inner balance, regulating basic functions and making us respond to both interior and exterior changes. This is an important message: hormones and behavior go hand in hand. Let's retire the idea that they are opposed. Hormones are the internal language of the body. How else would the adrenal glands know to secrete cortisol when a car is about to hit us? Hormones and behavior are extremely correlated, which is why their study is so valuable. In fact, their relationship flows in both directions. Sometimes meeting an attractive man will increase dopamine levels in women and other times hormonal changes during ovulation will make them feel more flirtatious.

So let's start with the basics: the male sexual hormones or steroids are androgens and the female ones are estrogens and progesterones. In those groups, the main androgen is testosterone, while estradiol and progesterone are the main female hormones. But they aren't exclusive to the different genders. Progesterone is the only exclusively female sex steroid; men also have low levels of estrogens and women of testosterone. In further detail:

Testosterone For the purposes of this book, we can call it the hormone of sexual desire. In men it is the one that masculinizes the body and mind by spiking during embryonic development, right after birth and in puberty. Secreted primarily by the testicles and the adrenal glands, it has anabolic effects that create more muscle mass; its by-product 5-alpha dihydrotestosterone (DHT) affects the distribution of body hair and the shape of the genitals. As for behavior, in addition to increasing aggression and conditioning male conduct, it is responsible for maintaining sexual libido. In his famous experiments in the 1930s, Calvin Stone observed that castrated rabbits lost their sexual response over time, but that later injections of testosterone or testicular extract brought it back. The same loss of sexual interest was observed in eunuchs; although it was not complete in all cases because of adrenal testosterone. And if a pederast or paraphiliac is administered an antiandrogen such as cyproterone, which blocks androgen receptors in cells, we also achieve a decrease in sexual desire. While the jury is still out on their true effectiveness and their possible long-term effects, testosterone creams or injections usually return sexual vigor to men with andropause. Testosterone is produced in much smaller quantities in the ovaries and adrenal glands of women, and their relationship to female sexual desire is not as direct as it is in men. Studies have shown that there are women with much more sensitivity to the arousing effects of testosterone. With them, testosterone therapies significantly improve sexual response, but in others the effect is minimal. Scientists are confirming that there is a wide range of reactions, observing that there are people with high libido but low absolute values of testosterone in the blood, and they suspect genetic polymorphisms in the genes that codify the cellular androgen receptors. This would mean that there are women with more testosterone receptors in their cells and others with fewer, which would explain their differing responses.

Estrogens The hormones of femininity. Produced in the ovaries, they feminize the body during puberty and regulate the menstrual cycle. Their levels increase progressively between days 5 and 14, the follicular phase of the ovarian cycle. Their levels decrease after ovulation, remaining minimal with a small spike on days 18–20. Estrogen doesn't seem to directly influence sexual desire, but it contributes to vaginal lubrication, increases of blood flow to the genitals, and conditions behavior by generating more feminine, and seductive, attitudes that foster wellbeing. Low levels following menopause can generate apathy, vaginal dryness and emotional discontent. In men, its levels are very, very low but important for bone calcification and semen production. When estrogens are injected into laboratory animals they induce feminine behaviors.

Progesterone The hormone that maintains gestation, it could be called the hormone of pregnancy. Its levels increase in the luteal phase after ovulation and lessen at the end of the cycle if conception doesn't take place. It has multiple physiological effects; as for sexual behavior, it is speculated that it favors desire. In a sense it is more opposed to testosterone than estrogens are. Injected in large doses into male rats at birth (a moment when there is a strong release of testosterone), it increases feminine behaviors, including lordosis. The birth control pill that millions of women around the world take is nothing more than a combination of progestin (a synthetic form of progesterone) and estrogens. It works because the progestin inhibits the release of gonadotropins and therefore prevents ovulation. Since falling levels of estrogens and progesterone cause the shedding of the endometrium and therefore menstruation, the pill is also used by women who want to reduce cramps and heavy flow. A controversial effect is that it reduces testosterone levels, and there are scientists who suggest that the birth control pill can have a negative impact on the sexual response of some women.

Those three are the sex steroids, which all chemically come from cholesterol. In fact you may find it odd to know that the cholesterol molecule transforms into progesterone after four enzymatic reactions, and into testosterone with three more steps, and that the only difference between testosterone and estradiol is a double carbon bond created by the aromatase enzyme. But chemistry aside, there are many other hormones involved in the reproductive cycle:

Prolactin The desire-inhibiting hormone. Secreted by the pituitary gland in the middle of the brain, prolactin sets off the production of milk in the breasts and has a slight inhibiting effect on sexual desire. In fact, it is secreted in large quantities during orgasm and is believed to influence the feeling of sexual satiety and the refractory period. It is a highly interesting hormone under intense study. It has been found that during masturbation less prolactin is released than during vaginal intercourse, and that it is responsible for the lessening of desire in certain moments of gestation, and in men who have just become fathers.

The pituitary gland also produces the luteinizing hormone (LH), which marks the start of ovulation, and FSH, which contributes to the maturation of the ovarian follicles. But neither would be released if not for the signal received from the hypothalamus in the form of the gonadotropin-releasing hormone (GnRH). There are many more signals, but since our focus here is on the sexual response, we will be talking about those hormones and the neurotransmitters that are vital to the process of arousal and sexual pleasure.

Dopamine The motivation hormone. Dopamine is the hormone of euphoria, pleasure, and motivation in the search for stimuli, and has multiple effects on the functioning of the brain area it acts upon. Involved in the process of addiction, it stimulates the production of testosterone and is what, in the build-up to a potential sexual encounter, makes us feel increasingly excited, impulsive, enthusiastic, happy and determined to move toward copulation. If its levels are increased by the consumption of cocaine, methamphetamines or dopaminergic drugs, the eagerness for sex skyrockets.

Noradrenaline The hormone of physical arousal. When in moments of arousal prior to intercourse we notice a sudden energy that generates a certain type of muscular stress and seems to be preparing our body for action, that is due to the release of adrenaline in the adrenal gland above the kidneys. However, when stress is excessive and in addition to noradrenaline there is an excessive increase in cortisol or the stress hormone, it can set off the opposite situation and inhibit our sexual response.

Oxytocin The love hormone. It is called the love hormone because of the attachment feeling it produces between mother and child, sexual partners and even friends. Produced in the hypothalamus and transported to the pineal gland, there it releases into the bloodstream in large quantities during orgasm. It is said to be responsible for generating that feeling of wellbeing and attachment between lovers who had only planned on sharing casual sex.

Endorphins Pleasure molecules. Secreted during exercise and orgasm, they are the neurotransmitters most clearly related to physical pleasure and pain reduction.

Serotonin The mood molecule. This neurotransmitter modulates mood, and its low levels are associated with depressive states, lack of appetite and memory loss. High levels of serotonin can be a problem because they inhibit sexual function. Loss of libido is a secondary effect of antidepressants that act as serotonin reuptake inhibitors to increase its synaptic concentration. In fact, in some cases of premature ejaculation they use small doses of antidepressants to try to delay ejaculation.

I want to reiterate here that the relationship between hormones and behavior flows in both directions. Under normal conditions, debating which is the conditioner has an element of absurdity. It depends on the moment and the situation. But their abnormal levels in pathological or atypical situations can result in serious disorders, both medical and sexual. And knowing about them is of vital importance, even if later we try to reestablish them with behavioral therapy instead of pharmaceuticals.

To sum up, if a woman goes out on the town during the second week of her cycle with her estrogens sky high, she may subconsciously have a more flirtatious attitude than she would have if she were in the high progesterone phase. We know that raised testosterone levels increase sexual desire in both men and in women. If she feels an attraction towards her partner or a stranger, gradually dopamine and noradrenaline will increase arousal and a feeling of energy and wellbeing, causing her attention to focus on the sexual encounter. If they are both feeling very stressed, a good kiss can relax them, lowering cortisol levels. If they end up in bed together, their dopamine and adrenaline levels will progressively rise until orgasm, the moment when they will both release an enormous amount of opiates and pleasurable endorphins, along with oxytocin that will make them feel spiritually linked to the person with whom they've shared that wonderful moment. Further on we will look at how alcohol, marijuana, cocaine and methamphetamines increase or decrease these chemical sensations, but there isn't a shadow of a doubt that the attraction we feel for a person we've just seen for the first time or the wellbeing we experience when hugging our partner has a neurochemical manifestation in our organism. And that is what scientists continue to study, since there is still so much to be learned about it.

2

Sex in Our Genitals

Reaching climax is harder when we're tipsy, for the same reason we'd have trouble reacting if a car came rushing at us: our nervous system is inhibited and unable to activate the sympathetic nerves responsible for both unleashing orgasm and making our body react to sudden stress.

Jorge is really taken with Sonia. She is an intelligent, attractive, stylish woman with a lot of personality who's in very good shape for her age, 38. And, every time they've seen each other with mutual friends for an after-work drink, Sonia complains that meeting people in the big city is getting harder and harder. Even just for hookups, which was what she said she was looking for at this point in her life. She argues that there are more women than men, and that the few men left are in relationships, or gay or weirdos, and that they're scared off by someone like her, with so much character. Jorge is always left thinking that if he didn't have a girlfriend, he would be up to the challenge of her strong personality. Some day, he thinks, some day…

A couple of months later, fate sets a nasty trap for cocky Jorge. His girlfriend goes on a trip with a friend to reflect on their relationship, and he finds himself at the regular bar with several of his coworkers. Sonia happens to be there that day too.

After a few drinks, one of his colleagues announces that he's calling it a night, and when Jorge says "I'm staying, my girlfriend's out of town," he thinks he sees Sonia turn her head and shoot him a mischievous glance. A release of motivational dopamine activated the attention in Jorge's brain. Conversation and laughter flowed between the people in their group. Jorge started to flirt with Sonia, who always responded with an open smile. Their body language became increasingly revealing: bodies leaning forward, heads nodding, exaggerated gesturing, dilated pupils, restless lips. Gradually the others left, and when they were alone, Sonia asked Jorge if he wanted to have one for the road. "It'll be my fifth, but sure," replied Jorge, his forehead shiny, with a seducer's gaze that anyone would have found pathetic, except Sonia. After some more flirting in the safety of a public space, a discussion of how unnatural societally imposed monogamy is, Sonia's hand on Jorge's knee as she leaned unnecessarily close to whisper something in his ear, and increasingly bold looks, Sonia said in a candid, natural tone: "Can I invite you over to my place?" Jorge wasn't expecting her to say that, at least not so frankly. He hesitated for a moment, feeling as if the blood in his face and the energy in his eyes had suddenly retreated but, with his expression slightly out of joint, he answered, "Sure... of course..."

Once they were in the taxi Jorge's muddled mind began to flip back and forth with indecision. He is going to the apartment of a tremendously attractive woman, who is well aware of his situation, it's obviously just a fling, which should be allowed while his girlfriend is thinking things over. Well... or maybe not. Damn these doubts! Jorge feels uncomfortable. Tense. Something's not right. He's filled with a strange combination of excitement and nervousness. When they get out of the cab Sonia takes Jorge's hand and leads him to her door. They go up in the elevator and start to kiss. Jorge is aroused, but he can't seem to shake the tension. His mind is racing. It's as if he's thinking too much and can't just go with the flow. They enter her apartment, still kissing passionately, fall onto the sofa and Sonia starts to take off her clothes. Jorge is captivated by her body; it's much more sensual than he'd been imagining. Still caressing each other, Jorge starts to undress somewhat clumsily and with unnecessary urgency. Soon he's only in his underwear. But he notices that something's off. Something between his legs isn't as happy as it should be. Sonia's caresses slowly go further and further, finally dipping below Jorge's waist. When she finds him flaccid, she stops for a second and starts to kiss him again as if the foreplay was just beginning. She knows exactly what she's doing. Jorge is fascinated by Sonia but he is starting to be anxious about his lack of an erection. He can't understand it. How can this be? It's never happened to him before. After a few minutes it seems that Jorge has gained a bit of volume, but he is still tense and stressed. Sonia lowers his boxers with the intention of giving him some more direct stimulation, but his member retreats even further. Sonia is unfazed and begins to stroke Jorge's genitals. There is no response and sweat starts to appear on Jorge's forehead as he contracts his muscles as if trying to send blood towards his penis. It doesn't work. Sonia asks Jorge to relax, begins kissing his neck, then chest, then abdomen, but she stops when she realizes that Jorge is completely unnerved. "Are you okay? You're all sweaty," she says. Not only sweaty; Jorge's heart rate is high, he has hyperventilating dilated bronchioles, adrenaline in his blood, muscular tension, a tight sphincter and high blood pressure. All the signs that his body has activated its sympathetic nervous system. And under such conditions, no matter how excited Jorge's mind is, no matter how much it wants to take control over his organism, an erection is almost impossible.

The sympathetic nervous system is a mechanism designed by evolution to make us react in the face of sudden stress. It is the same system that allows us to run like the dickens in fractions of a second when we're strolling leisurely through the forest and a bear jumps out of nowhere, or leap suddenly if a car drives up onto the sidewalk. The heart accelerates in milliseconds, the lungs take in as much oxygen as possible, the adrenal glands secrete adrenaline at their own discretion, glucose metabolism dials up to the maximum, testosterone and cortisol are through the roof, the pupils dilate for better vision, the skin perspires and all the blood moves into the muscles. It is not a moment for digestion, excretion, or reproduction. Jorge's internal organs are blocked, his sphincter is clamped shut and his penis is as white as his face. Blood enters it but slips right back out. When the threat passes, a few minutes later, the parasympathetic nervous system will again take control of the body and it will be able to regain normal physiological functions. That is if—as could happen to Jorge—the experience isn't traumatizing and the stress reappears every time when faced with a similar situation.

Jorge assures Sonia that he doesn't understand what's happening, and he tries to explain it as a problem of having drunk too much. But no, it was his own sympathetic nervous system that did him wrong. Sonia hugs Jorge, and lies to him, saying "it's no big deal" and she wonders if she should give him a second chance. She does, and we'll see what new surprises his sympathetic system has in store for him then.

Sympathetic and Parasympathetic Nerves in Sexual Arousal

There are two systems that govern our body's internal functioning and communication: the endocrine, which is regulated by hormones, and the nervous, regulated by electrical signals and neurotransmitters.

The nervous system can be classified in two different ways. Structurally, we can distinguish between the *central nervous system*, made up of the spinal cord and the encephalon (brain, medulla oblongata and cerebellum), and the *peripheral nervous system*, made up of all the nerve fibers that connect the internal organs, muscles and skin with the spinal cord. This communication can move in both directions: the peripheral nervous system has *sensory nerves* that send information from the skin and internal organs to the spinal cord and brain, and *motor nerves* that transmit the instructions of the central nervous system to the muscles. The pudendal and the pelvic nerves, for example, form part of this peripheral nervous system.

The nervous system can also be classified into: *(a) somatic nervous system*, which we control and operate consciously (moving an arm), and *(b) autonomic nervous system*, which automatically regulates the involuntary reactions and processes of our organism. It is the latter that interests us now, because it has two clearly differentiated states: the *parasympathetic autonomic nervous system* and the *sympathetic autonomic nervous system* (Fig. 2.1).

Let's think of how different our body's needs are when we are digesting our food and when we are being attacked by a lion in the savannah. Under normal conditions, the body is relaxed and functioning according to the orders of the parasympathetic nervous system: activated digestive function, relaxed sphincters, blood flow to internal organs, predisposition towards sexual activity… But when a stressful situation suddenly arises and our entire organism must immediately react, the sympathetic system takes over: bronchioles dilated to allow in more oxygen, increased heart rate, secretion of adrenaline in the adrenal glands, pupils dilated to increase vision, inhibition of the digestive and reproductive functions, blood and glucose to the muscles, etcetera.

As for the sexual response, the pudendal and pelvic nerves are regulated by the parasympathetic ganglia. Which is to say, in order for them to function correctly we must be relaxed under the control of the parasympathetic nervous system. But if we are stressed by fear or any other type of pressure, the sympathetic system activates and everything changes. From thoracic vertebrae in the middle of the spine comes the signal to restrict the muscular cells around the arteries that allow the entrance of blood into the corpora cavernosa of the penis and the clitoris, so physical arousal is hindered.

The Erection of the Penis and the Clitoris

The mechanism by which an erection is produced and maintained is very simple. I could show you a diagram like those in textbooks. But I sincerely believe that it is better that you imagine it in your own body or in the body of your partner.

Let's put to one side for a moment the part of arousal and inhibition that stems from the mind. It is undoubtedly fundamental, and we will later see how it interacts with the physical part. For the time being let's focus on the muscles, nerves, chemical signals and blood flows.

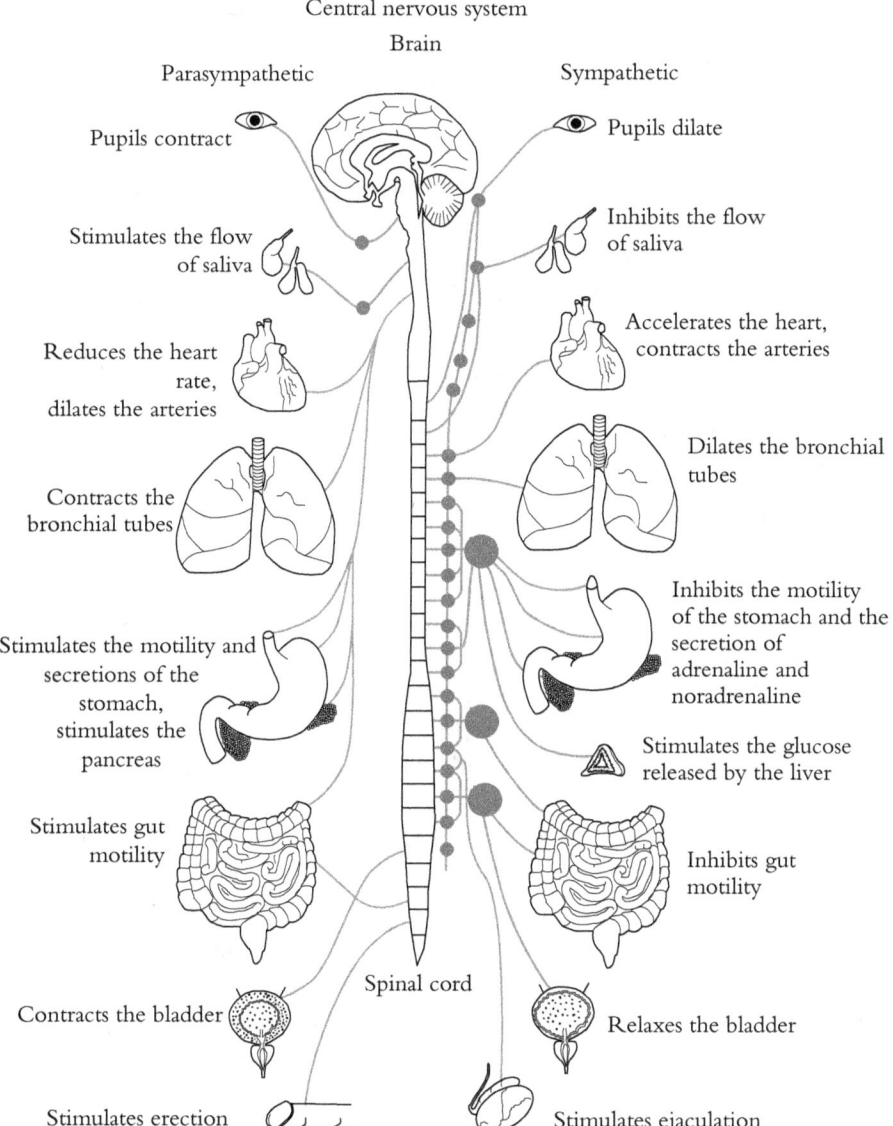

Fig. 2.1 The sympathetic and parasympathetic autonomic nervous system

After all, an anesthetized man can ejaculate on reflex,[1] and someone with a spinal injury can have an erection and an orgasm without any information coming or going between his genitals and his brain.

---
[1] In order to obtain semen samples from animals, electroejaculation is used. An electrical charge is administered inside the anus, near the prostate, activating the PC (pubococcygeus) muscle, thus bringing about ejaculation as a reflexive act.

Right below the skin of the penis—and it is the same in the clitoris, which we have already seen is not as different as it seems—are some very special nerve endings. They end with a sort of corpuscle or knot made up of a dense coiling of nerve fibers that offer an enormous sensitivity to touch on the surface of the penis and the clitoris. They are called the *genital end bulbs* or Krause corpuscles, are primarily concentrated on the glans and the frenulum, and are what make both the penis and the clitoris so sensitive.

Let's imagine that the action gets started. When the penis receives continued friction of sufficient intensity, the pudendal nerve sends a signal to the lowest part of the spinal cord, in the sacral segment. There some interneurons receive the sensory information and form synaptic connections with the pelvic nerve, which sends motor information back to the erectile tissue of the penis. Obviously, if the arousal occurs as a result of a fantasy or erotic thoughts, visual stimulation or insinuations from a lover, it is the brain that sends the information to the interneurons, which communicate with the pelvic nerve with no need for tactile stimulation.[2] And if we combine physical and mental stimulation, even more signals are sent to the interneurons of the sacral region. The physiological response begins with the activation of the pelvic nerve and the signals it sends to the erectile tissue (Fig. 2.2).

Erectile tissue is made up of two corpora cavernosa located on the left and right sides inside the penis. The corpora cavernosa are like sponges; they are composed of cavities that start to fill with blood when they receive information from the pelvic nerve. But how does the pelvic nerve transform the electrical signal from the spine into an increase in blood flow to the penis? By the release of various neurotransmitters. The most relevant being a gas called nitric oxide, which in turn induces the secretion of some nucleotides called cGMP and cAMP (keep them in mind when we discuss Viagra). The function of cGMP (cyclic guanosine monophosphate) and cAMP (cyclic adenosine monophosphate) is nothing more than relaxing the muscle cells that surround the arteries to allow a greater flow of blood into the corpora cavernosa. In fact, when the blood enters the arteries faster than it can come out of the veins, the penis starts to increase in size. And if stimulation continues, it grows so much that at a certain point it constricts the veins through which, when flaccid, blood circulates out of the penis (Fig. 2.3). When the erection is complete, those veins remain closed and the blood cannot escape. It is a true hydraulic

---

[2] Scientists still are not entirely clear on all the neurophysiological factors implicated in mental stimulation. They do know that, with sensory stimulation or fantasies, the hypothalamus can secrete oxytocin and send a signal through the spinal cord that activates the pelvic nerve and the hypogastric nerve, and that they can cause an immediate erection in a mouse by electrically stimulating a particular part of its brain. But, as we will see later in this book, our bodies have reactions that leave even the most experienced physiologists stumped.

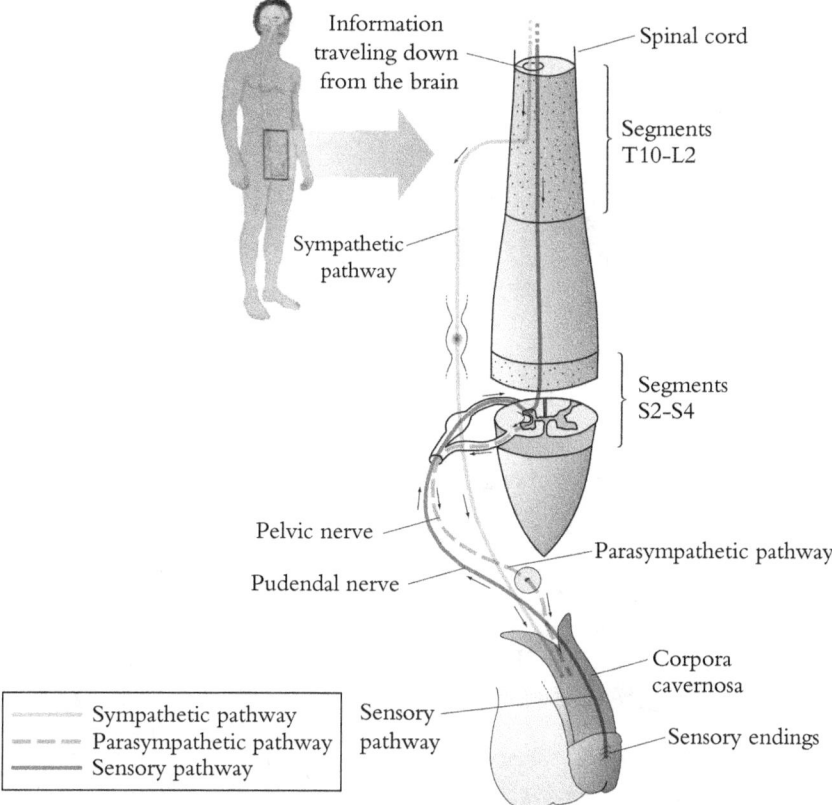

Fig. 2.2 Nerves involved in erection

system: entrance pipelines expand when they receive a large flow, blocking the exit pipelines. Until the next chemical command.

When we see the penis getting darker after a few minutes, it is because the blood flow has stopped and the blood is losing oxygen. It is the same effect that makes a drowning person look purplish when oxygen isn't reaching their skin.

Perhaps you are wondering why the blood is unable to leave but the semen can be ejaculated. You should imagine the erect penis from above, visualize the two hard corpora cavernosa on the left and right, and beneath a lower, central part that is slightly softer: the corpus spongiosum, a different cavity above which the seminal conduits pass without being compressed.

In this erect state the stimulation of the pudendal nerve can continue until orgasm and ejaculation (which usually coincide in men, but not always) are reached. Then another series of nerves are activated, relaxing the erectile tissue

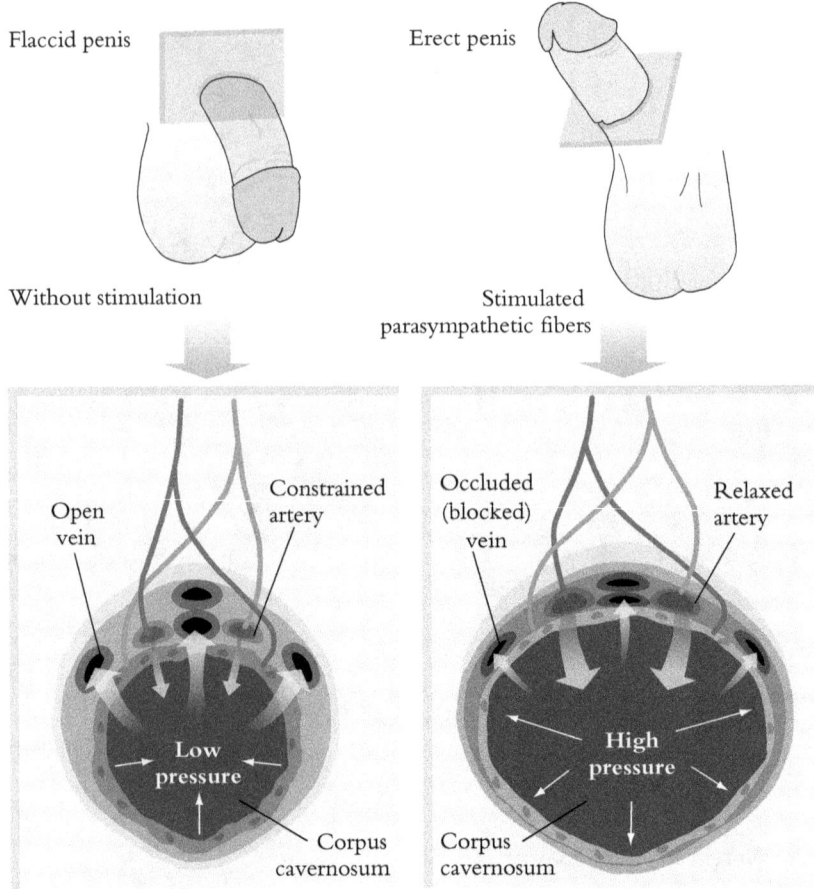

Fig. 2.3 Schematic diagram of the cross section of the penis in flaccid and erect states (with blocked veins)

and allowing the release of blood, despite attempts to retain it. Curiously, they are the same sympathetic nerves that were frustrating Jorge. Let's have a look at how that works.

During the entire sexual act, the nervous system that governs our body is the parasympathetic. We may sweat and have an increased heart rate if the exercise is intense, but it still falls under normal physical activity. It's not like running stressed as fast as our legs can carry us because we are trying to escape a fire, but rather like running for pleasure, in a relaxed state. In the case of men, when we are enjoying sex, our blood flows well through our internal organs and we move at will. But as our excitement increases, a very peculiar point approaches in which suddenly the fibers of the sympathetic nervous system are activated: orgasm. During orgasm, blood pressure rises suddenly to more than 200 mmHg, our muscles remain in tension for a few seconds, our pupils suddenly dilate, our faces reflect that strange mix of pain and pleasure, and the PC (pubococcygeus) muscle is activated, causing ejaculation. For a few moments, the nerves of the sympathetic and parasympathetic systems are activated at the same time, but after a few seconds the sympathetic system will take control. The pudendal and pelvic nerves are parasympathetic and no longer respond. On the other hand, from the highest thoracic segments of the spinal column, the sympathetic nerves give the order to constrain the muscular cells around the arteries that allowed the entrance of blood into the corpora cavernosa, and the blood begins to leave the penis unimpeded.[3]

That is what happened to Jorge, long before he became aroused. His sympathetic system was activated by the stress of the situation and there was no way that his pudendal nerve could transmit messages to the spine, or the pelvic nerve could send the signal to allow the massive entrance of blood to his penis. The physical response is very similar to a post-orgasmic situation in which a man wants to immediately obtain an erection and his body doesn't cooperate.

Erection Problems and Premature Ejaculation Due to Stress

Jorge left Sonia's apartment depressed, feeling his manhood threatened. A few days later, when he recovered his confidence somewhat after convincing himself that it was the alcohol that had betrayed him, he felt the absurd necessity to have a new date with Sonia to show his virility and erase that embarrassing incident from his memory. From his memory and from Sonia's, of course.

Sonia liked Jorge. He was fun and not bad looking at all. She accepted his invitation for dinner. The evening was going along really well: the setting was pleasant, the meal appetizing, the conversation good. Sonia drank wine and Jorge drank water, saying that he had gone out the night before and overdone it. Sonia smiled to herself at how obvious Jorge was. His choosing a restaurant in her neighborhood "so she wouldn't have to take a cab" didn't fly. And ordering water… Sonia knew exactly what was going through Jorge's male mind.

---

[3] The opposite occurs with nocturnal erections. They usually take place during the REM phase, and coincide with a decrease in activity in a specific area of the brainstem that radically deactivates the sympathetic system, generating a very favorable balance toward the parasympathetic, leading to an erection.

And she decided to play along. When after dinner Jorge suggested they go have a drink, Sonia very elegantly offered a nightcap at her place. That way the situation would flow more gradually. They both strolled calmly back to Sonia's apartment, but soon Jorge started to feel a certain pressure in his abdomen again. He was a little nervous.

Once at her house and sitting on the couch, Sonia served two glasses of wine. Jorge only took a tiny sip and asked for a glass of water because he was feeling hot. He was tense. Not as much as the first time, but he did feel ill at ease and insecure. He couldn't forget about what had happened or stop worrying about whether his equipment was going to function properly. Soon they started kissing. But Jorge's mind was still too active. He wasn't relaxed and his movements were awkward, as if he were thinking about where to place his leg instead of just enjoying the moment. Nevertheless, things went differently that time. Sonia took a much more sensual approach, with more gentle caresses, slower motions, subtle moans and indirect contact. She hugged Jorge close while pressing his genitals against her thigh, and she helped him undress at his own pace. The truth is that both of them were highly aroused. And Jorge, despite being hesitant and having an elevated heart rate, managed to get an erection. That increased both of their confidence. Jorge was thrilled and Sonia just wanted to be penetrated. Their stroking intensified and suddenly Jorge leapt upon her. Sonia was wet and warm, and when she felt Jorge entering her powerfully, she immediately countered with vigorous movements of her hips. Jorge met her momentum with his own, but after a few seconds he felt incredible arousal in the glans of his penis that… "Oh, oh no… No! wait… not yet…" Jorge came. He didn't even dare to express his pleasure, and for a few seconds he tried to continue as if nothing had happened, so he could keep stimulating Sonia. But it was impossible. His penis languished and he had to keep his head down to avoid looking into Sonia's face. She was frozen stiff, her eyes wide as saucers.

What happened? Jorge didn't understand it. He didn't know that his sympathetic nervous system had betrayed him once more, but this time in the opposite way. What happened is that, because of his stress prior to coitus, his sympathetic nervous system was constantly on the verge of activation, and that led to a very quick orgasm after the intense stimulation of penetration.

If Jorge had known about the effects of alcohol on the sexual response, maybe he would have had a few drinks. In large quantities alcohol can impede an erection, but when it doesn't it delays orgasm. In fact, alcohol acts in a very strange way. According to the Dual Control Model, the sexual response involves the balance of excitatory and inhibitory systems. In this balance, alcohol is a powerful disinhibitor that makes us feel much more relaxed and mentally excited, but in excess it's a depressant to the central nervous system that reduces physical arousal and hinders the activation of the sympathetic system. Which is to say, the mind feels more desire but the body doesn't respond in the same way. That is why it is hard for us to reach climax when we've drunk too much.

The activation of the sympathetic system is the key to triggering orgasms, in both genders, but much more so in females. Many women who have trouble reaching orgasm use different techniques to prepare the body and generate more stress in the moment of pleasure. Some people need extra stimulation and shout obscenities, seek out risky situations or resort to aggression. They need to increase tension in order to be able to activate the sympathetic system and reach orgasm.[4] The paradox is clear: nerves and stress prior to sexual arousal can block it, but after arousal they can favor orgasm. Surely you've experienced some situations in which this can be applied.

How Does Viagra Work?

Jorge doesn't need Viagra. Nor do most of the young patients with erectile problems who visit urologists like Doctor Michael Werner, the director of a male sex clinic in midtown Manhattan. At his office, Doctor Werner explains to me that Viagra is a solution for the physical problems that impede erections in older men, when it is not a case of some surgery or illness making sexual function impossible. But in most cases involving patients younger than 40, the problem has a clear psychosomatic origin. On many occasions they've been traumatized by a negative experience, and the usual course of action is to reassure them and recommend that they use Viagra a few times until they regain their confidence. Nevertheless, when there are any doubts whether the causes of the problem are entirely mental, the patients are sent home with a curious contraption.

The Rigi-Scan's only function is to detect nocturnal erections. It is comprised of a sort of elastic ring that is placed around the penis at bedtime, and is linked with a cord to a small device that is tied to the leg. The Rigi-Scan registers the changes in penile width throughout the night. If the patient is in good health, each time he enters into the REM phase he will have an erection for several minutes, usually three a night. The patient carries out the test for four or five consecutive nights to gather sufficient information and then returns

---

[4] In fact, the relationship between the sympathetic system and orgasm is so clear that if you are having sex with a woman and she moans and squirms, but her pupils don't dilate, her heart rate doesn't shoot up and the sudden increase in blood pressure doesn't break a few capillaries leaving her cheeks and the upper part of her chest red, you should probably be suspicious.

the Rigi-Scan to the clinic for analysis of the recorded data. If there were no erections registered, it is a physical problem that needs immediate attention, since it could be a first sign of cardiovascular disease. If, on the other hand, he did have the usual nocturnal erections, the problem is psychological and not physical. The most common is the performance anxiety that shows up when a failed erection like Jorge's provokes an anxiousness that hampers new sexual relations. In those cases behavioral therapy is necessary, but Dr. Werner admits that if the patient is healthy he usually recommends Viagra to help him recoup confidence in his sexuality. But how does Viagra work?

First of all, it should be noted that Viagra is the commercial brand name and that the active ingredient is called sildenafil. Since it was commercialized by Pfizer and approved by the FDA in 1998, other comparable brands have appeared on the market like Cialis (tadalafil), Levitra (vardenafil) or generics that can be bought online.[5] The most general name that you may have heard of is PDE5i or phosphodiesterase inhibitors. This term is derived from the way it works, by ultimately increasing blood flow by releasing nitric oxide.

Phosphodiesterase inhibitors act on the nucleotide cGMP that we mentioned earlier. Remember how the process works: there is stimulation and the pelvic nerve releases nitric oxide in the penis, which in turn induces the secretion of cAMP and cGMP, which provoke arterial relaxation and allow more blood into the corpora cavernosa. But this process must be regulated in order to avoid an erection that lasts too long. An enzyme called phosphodiesterase-5 takes care of that, by constantly breaking down the cAMP and cGMP. While there is physical or mental stimulation, more cAMP and cGMP will be released and the penis will remain erect. But if the stimulation disappears, the phosphodiesterase will degrade the nucleotides, stopping the entrance of blood, and the erection will start to subside. Here is where oral medications containing phosphodiesterase inhibitors come in: they block the action of the phosphodiesterase and allow the levels of nucleotides to remain high and, therefore, the arteries constantly relaxed. It is important to stress that if there is no sexual desire, no matter how much Viagra someone takes, the pelvic nerve will not activate the whole chain of signals and he won't get hard. But if desire and high libido are there, the blood flow will indeed be much greater and the erection more intense and long lasting. Because, desire is one thing and arousal quite another.

---

[5] There are also natural products and dietary supplements that work very well for erectile problems. Interestingly, according to a study realized by Pfizer, in many of those products sold as alternatives the manufacturers have added small doses of the active ingredient in Viagra. It is also interesting to note that sildenafil was a pharmaceutical that was being studied to counteract arterial hypertension, and when they began the clinical studies with patients, the scientists noticed a curious side effect: it provoked erections. They changed the target of their studies and it became a multimillion-dollar business for Pfizer.

Desire and Arousal Are Not the Same

The matter of disorders of desire and arousal is so complex—including the pharmaceutical interests involved—that we will gradually deal with them in more detail as the book progresses. But for now I would like to establish some basic concepts. After speaking with therapists, urologists, gynecologists, sexologists, psychologists and researchers, and visiting clinics specialized in male and female sexuality, we can sum it up this way (although there are diverging opinions): desire is the motivation and interest in sex; and arousal is the physical response that accompanies sexual activity. Obviously they are both deeply interrelated, but distinguishing them is very helpful when diagnosing a problem.

In men, it is simple because it is easy to see when mental arousal is accompanied by genital arousal, or not. But in women it is more complicated to figure out whether arousal problems stem from a lack of desire or mask problems in physical response related to lubrication, sensitivity or blood flow to the genitals.

Starting with men, male lack of desire is usually brought on by psychic aspects such as relationship issues, insecurities, depression or stress. But there can also be physical causes such as the lessening of androgen levels, which are the hormones involved in the sexual response. These fluctuations don't affect everyone in the same way, and for many men the progressive loss of desire associated with less secretion of testosterone by the testicles is a normal part of the aging process. But for those who refuse to accept nature's rhythms, testosterone supplements can help them to regain the sexual vigor and motivation they are looking for. During a conference of the International Society for Sexual Medicine (ISSM) that I attended in Chicago in August 2012, Irwin Goldstein was very clear: "You are alive, you will age, your androgen levels will drop and you will lose sexual desire. If you want to, you can intervene in that process." In the United States, testosterone supplements are very common, in the form of patches, creams or injections, and their increase in male desire is empirically proven, but they should always be taken on the advice of a doctor because of the side effects and risks they entail.

It is also possible for desire to exist and erection (arousal) to fail. We will talk more about that and its importance as a possible first symptom of cardiovascular disorders further on in this book, but when erection problems are related to diabetes, reactions to medications, injuries or are simply caused by age, a solution should be sought (Fig. 2.4), of course always under medical supervision. The most common is first trying oral medications such as Viagra, Levitra or Cialis, and if they don't work, trying injections of papaverine and

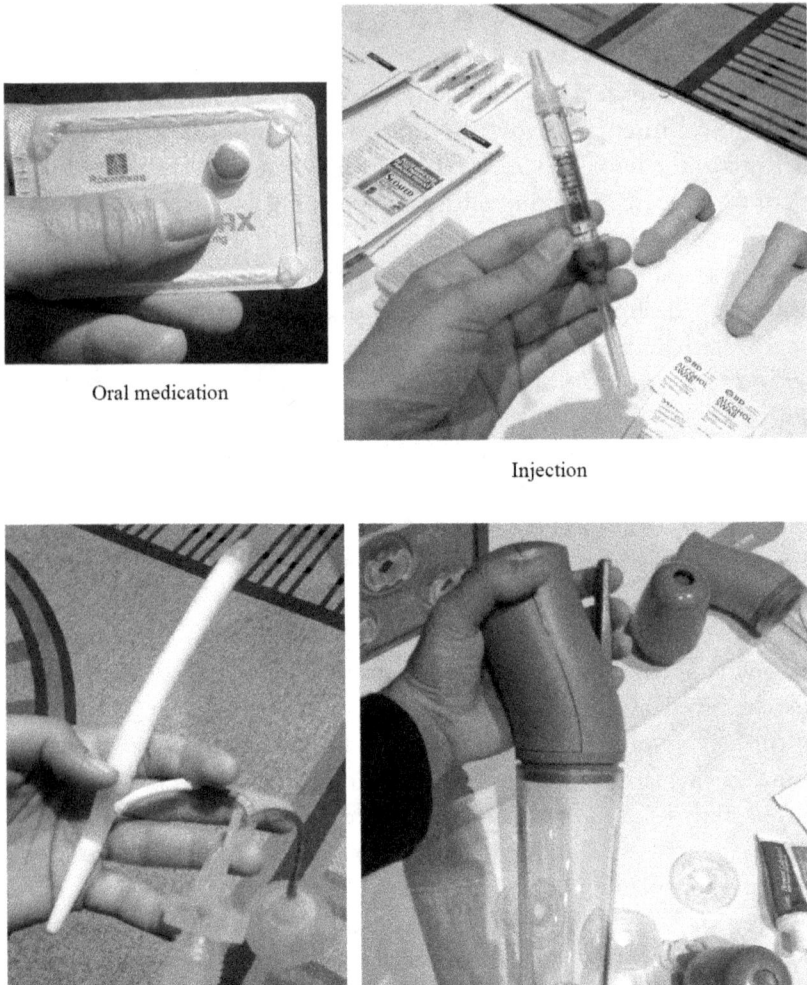

Oral medication

Injection

Implant

Vacuum pump

Fig. 2.4 Treatments

vasodilating substances, which are very effective. When the erectile dysfunction is permanent, one can use an implant: a sort of long plastic balloon is inserted into the penis and connected to a reservoir that holds salt water and a button in the scrotum. When the button is pushed, the water enters the balloon and voilà! There is an immediate erection that lasts until the button is pushed again and it deflates. Another option are devices that create a vacuum and send blood to the penis: the member is introduced into some sort of tube that is closed at the top, pressed against the inner thigh, pumped to remove the air inside, and the blood flows in to create an erection. Before the vacuum device is removed, a ring is placed around the base of the penis that keeps the blood from flowing out. It's a practical and inexpensive technique if you are in a relationship, but for casual sex the whole process can kill the mood.

The situation for women is more controversial for several reasons, and in Chap. 6 we will go in depth about the differences between lack of desire (*hypoactive sexual desire disorder*, HSDD) and *female sexual arousal disorder* (FSAD), and the possible merging of both in the next edition of the manual of mental illnesses, the DSM-V. But it is worth mentioning that a lack of sexual desire should only be considered a problem if the woman perceives it as such, and that there is a debate because many groups consider this dysfunction to be a medical construct promoted by pharmaceutical companies to create a potentially highly lucrative market. Accordingly, in 2010 the FDA did not approve a Boehringer drug called Flibanserin for use combatting lack of desire in women. Boehringer sold the patent to a new company named Sprout that did further studies and presented the drug to the FDA again in 2013. It was again rejected. But on the third try, in 2015 after a campaign in its favor, it was approved, becoming the first medication ever to treat female lack of desire.

Every one of the experts I consulted agreed that in most cases a lack of sexual desire is rooted in relationship or work issues, lack of sexual education, blockages, family concerns, stress or all sorts of problems that detract from one's ability to focus their energies on sex, and that following the instructions and exercises of an experienced sexologist can be very useful for reactivating the libido and one's sexual life, both individually and in a couple. But they insist that you shouldn't rule out possible physiological and hormonal causes; I particularly remember the words of Doctor Bat S. Marcus, the head of the Medical Center for Female Sexuality in New York, who told me during my visit, "I've had patients who come to my clinic after years of therapy and when I analyze their blood, I find they have the hormone levels of a ten-year-old girl." Once again we must insist on a biopsychosociological approach.

But lack of desire is only one aspect of sexual dysfunctions. Later on we will discuss vaginism, vulvodynia and other disorders that cause pain during sex.

While problems of physical arousal can be clearly related to a lack of desire, there are many women who feel sexual desire but do not experience enough arousal and physical pleasure. This can be due to lubrication problems, failure to climax, or a lack of sensitivity or blood flow to the vulva and clitoris. It is as if the physical response fails. Here individual and couple exercises such as sensate focus developed by Masters and Johnson help women to get to know their own bodies better; lubricants and vibrators are very helpful, but hormonal and physiological factors must also be explored. The decrease in estrogens after menopause diminishes lubrication, and various diseases can also reduce blood flow to the genitals, as happens in men. The diversity of female sexual response depends on many more factors than male response does. In fact, while science may someday solve the riddle, at this point there isn't a single sexologist, no matter how experienced, who can explain what lies behind this diversity.

Orgasm Based on the Distance Between the Clitoris and the Vagina

A friend of mine almost stopped speaking to me after I made an innocent comment about the position of her clitoris. I was in no way trying to make fun of her. While I was researching this book I began to see all things sexual with aseptic and curious scientific eyes, often forgetting all modesty and sometimes saying inappropriate things. Believe me, a point came where, for me, sex became a fascinating world to discover and experience from an academic perspective, without any further connotations. It was interesting to detach myself briefly from the exaggerated importance our society gives to sex and put aside taboos, restrictions and preconceived ideas in order to analyze it with the distance and objectivity of a scientific observer. It allowed me to reflect more freely on the subject and I feel I've learned quite a bit. But back to what I was saying. My friend was explaining to me how she rarely had orgasms during penetration. She remembered having a few with an ex-boyfriend she was with in her 20s, and others recently if she stimulated herself manually while her lover penetrated her. She usually reached climax when masturbating, but never during coitus. It was as if she was missing an extra degree of arousal to reach sexual ecstasy. "Oh, that's very common," I told her. "There are little vibrators designed to stimulate you while making love. It could be that your clitoris is very far away from your vagina and there isn't enough contact." Man oh man did she get mad! She almost stormed off. "My clitoris is very far away from my vagina? Who do you think you are?" My friend thought that I was making fun of something that for her was a very serious concern. Thank goodness she gave me time to explain that, even though it may have seemed inappropriate or insensitive, I had mentioned it because of a scientific study I'd just read.

There is a curious story about a grandniece of Emperor Napoleon Bonaparte, Marie Bonaparte, who had the same issue as my friend. Marie was married to the Prince of Greece, and never had orgasms while making love but always when masturbating. This was in the 1920s and there were no doctors she could go to for a consultation. She was practically obsessed with what she took to be highly abnormal, and she imagined that perhaps her clitoris was too distant from her vagina to allow her to reproduce with her husband the same friction she enjoyed while masturbating. If we think about it, her hypothesis is not so far-fetched. The clitoris is an organ designed exclusively to generate pleasure, but it isn't ideally situated for reaching orgasm through penetration. Evolutionary biologists interpret this fact by saying that while pleasure is fundamental as a motivation for wanting to have sex and reproduce, unlike male ejaculation, the female orgasm isn't evolutionarily necessary and could even be counterproductive if it leaves a woman sexually sated. This would explain that natural selection had placed it close enough to be stimulated during intercourse, but not so close as to readily facilitate orgasm.[6]

Princess Marie Bonaparte sought out the help of a doctor and made one of the most peculiar scientific studies ever. It would have won an Ig Nobel Prize if they had been awarded in the early twentieth century. Bonaparte measured the distance between the clitoris and vagina of 243 women, and asked them each how frequently they had orgasms during the sexual act. Based on this study she established a distinction between three groups of women: the *mesoclitoral* represented about 10% of the sample and were those whose clitoris was situated some 2.5 cm from their urethra; in the *teleclitoral* the clitoris to urethra distance was significantly greater than 2.5 cm; and the *paraclitoral* were those whose clitoris was closer to the urethra. In 1924 Marie Bonaparte published her results under the pseudonym A.E. Narjani in the scientific journal *Bruxelles-Médical*, concluding that there was indeed an inverse relationship between the frequency of orgasms during coitus and the distance of the clitoris. In fact, in 1940, American psychologist Carney Landis tried to replicate the study with a sample of his patients and found the same correlation: the further the clitoris was

---

[6] Evolutionary psychologists love to propose hypotheses consistent with natural selection that, later, are rarely put to test experimentally. One that was very popular for interpreting the logic of the female orgasm was the up-suck theory, according to which the muscular contractions during orgasm would carry the semen to the cervix and increase the possibility of pregnancy. While at first the up-suck theory was widely circulated because it was logical and generated the typical "oh, of course" response, later experiments refuted it. Having an orgasm doesn't significantly improve the possibilities of fertilization, and many evolutionists still think that the female orgasm has no evolutionary function, that it simply exists because the physiology of female genitals is the same as male ones.

from the vagina, the fewer orgasms there were from penetration. Both studies are old and employed very limited methodology, but in 2011 researchers Kim Wallen, from Emory University, and Elisabeth Lloyd, from the Kinsey Institute[7] reviewed the information from Bonaparte and Landis with contemporary statistical methods and published an article in the magazine *Hormones and Behavior* concluding that the distance between the clitoris and the vagina could indeed be one of the many factors involved in the frequency of orgasms during intercourse. I had the chance to speak with Kim Wallen during the International Academy of Sex Research conference held in Estoril (Portugal) in the summer of 2012. Wallen did not wish to speculate on whether this could be due to the proximity of the clitoris allowing more contact with the penis or to aspects of internal physiology, and she admitted that it could be a mistaken correlation. But she did seem confident that the effect could have some importance among the many other more relevant ones and she lamented that this possibility had not been better studied by medical science. This is one of the many examples that show how sex is a taboo subject even for scientific investigation.

Once again, obviously many other factors influence in the ability to reach orgasm, including a lover's skill, the woman's mood, stress level, degree of arousal, tiredness, worries, the rapport between the partners, and the situation. We know that sometimes orgasms can be produced through stimulation of other parts of the body and in this book we will meet women who practice Tantra and reach ecstasy with the help of breathing and meditation. But it is also true that the explanation that *it's all in your head* has been overused. No, it's not all in your head. The genitals of a person with a high spinal cord injury can become aroused in a reflexive response if they are stimulated, and that person can feel an orgasm without any information being transmitted to their brain. Our genitals can also surprise us by acting on their own steam, and becoming aroused in the most unexpected circumstances. The physical factor is important. We can conclude that there *are* women who are more physically predisposed to have one type of orgasm or another, a controversy which increases when we discuss the dichotomy between the vaginal and the clitoral orgasm. But in order to do that we have to quickly review female physiology.

---

[7] Founded in 1947 by researcher Alfred Kinsey (whom we will discuss in detail in Chap. 4), the Kinsey Institute at Indiana University is perhaps the most emblematic research center worldwide devoted exclusively to the study of human sexuality from a multidisciplinary standpoint. Currently headed by Julia Heiman, its staff contains biologists, psychologists, anthropologists, sociologists, historians and experts in sexual behavior. Its library has one of the largest collections in the world of printed and audiovisual material about sex, and it is a point of reference for any and every sex researcher.

Vaginal Orgasm, With and Without Clitoral Stimulation

First of all, we must remember that the clitoris is a much larger organ than the little external glans that we know, and that internally it extends like two arms that run along the sides of the vagina (Fig. 2.5).

Knowing this, the big question is not whether there is such a thing as orgasms from penetration alone, without direct contact with the clitoris during coitus. Of course, exclusive stimulation of the interior of the vagina can provoke orgasms. The question is whether such stimulation stems from the nerves that surround the vagina itself or if what is really happening is that penetration is indirectly stimulating the internal parts of the clitoris. It is a debate that seems never-ending. In fact, in the 1960s, the researcher couple William Masters and Virginia Johnson defended, with their experiments on the physiology of sex, that the clitoris was the only organ responsible for pleasure, and that the vagina was relatively insensitive. They claimed that had to be the case in order to give birth. The feminist movement embraced the idea because it meant that penetration wasn't necessary for more intense stimulation, and because it put paid to Freud's absurd idea that clitoral orgasms were childish and vaginal ones were mature. But not everyone agreed, and the debate heated up with the discovery of the G-spot, an extremely sensitive area in the upper inside of the vagina that can provoke orgasms in some women that are different from those brought about through clitoral stimulation. We will talk more about the G-spot later. But the controversy is due to some studies, such as the sonographs done by the Frenchwoman Odile Buisson, that defend the idea that penetration actually stimulates the internal part of the clitoris (its shaft) instead of the glans, and that exclusively vaginal orgasms do not exist.

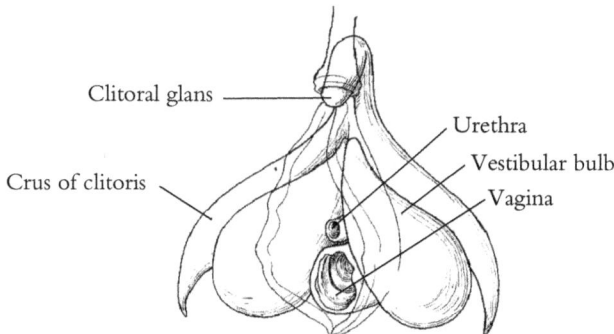

Fig. 2.5 Complete structure of the clitoris

Other scientists don't deny that this indirect stimulation could be happening, but they support the idea that vagina does have sufficient sensitivity to produce vaginal orgasms, which many women describe as less localized or as "full body." In fact, the clitoris is only innervated by the pudendal nerve, while the vagina receives information from the pelvic and the hypogastric nerves as well. And studies using brain images of women stimulating the inside of their vaginas show that the areas of the sensory cortex that are activated are different than those activated by clitoral stimulation. If to that we add the fact that orgasms can be produced by stroking the breasts, the neck, the anus or the ears, plus all the opinions of so many women who describe vaginal and clitoral orgasms as different, the current consensus is that Masters and Johnson were wrong, and that vaginal orgasms could indeed exist as such. The brain offers us more evidence in the next chapter.

3

Sex in Our Brains

Scientific knowledge progresses through rigorous investigation and new technologies that allow us to observe the previously invisible. But also through hunches, unexpected discoveries and a willingness to think differently from those who've come before us. It is a marvelous adventure filled with intellectual feats, both personal and collective, like those of neuroscientist Barry Komisaruk, which I find enthralling. His work is an example of deep rigor, the application of new technologies, intuition, unexpected discoveries and provocative ideas, all within a unique and fascinating scientific trajectory.

Science Is More Interesting than Sex

One of the first studies in Barry Komisaruk's career dealt with pseudopregnancies: when a woman believes herself to be pregnant, and despite no ovum being fertilized or any hormonal signal traveling from the uterus to the brain, her mental activity is capable of modifying the state of her organism and beginning to set off signs of pregnancy. Among those signs is an uninterrupted secretion of progesterone, the hormone responsible for maintaining gestation. It was in the 1960s, and a young Barry Komisaruk was doing research at the University of California, in Los Angeles, on how a sensory stimulus modified brain activity, and how that translated into a release of hormones to the blood stream that set in motion a specific behavior pattern. Psychological pregnancies undoubtedly offer an extremely interesting example to unraveling the relationship between sensory activity and neuronal and hormonal activity. Komisaruk knew that occasionally the lab rats could also have the initial phases of pseudopregnancy after copulation, and he thought they could function as a model.

Barry began to vaginally stimulate the rats in order to analyze what hormones they secreted just by being penetrated, even when there had been no mating. He also measured neuronal activity in areas like the hippocampus and the hypothalamus, always with the objective of distinguishing the specific effects the vaginal contact had on the brain. But during the experiments he observed something that puzzled him: all of the rats without exception exhibited lordosis when they were stimulated. Lordosis is the reflexive act of arching your back to favor copulation, which in theory rats should only display when they are ovulating (or at least that was what was described in scientific literature). But all of Barry's rats exhibited lordosis when they were penetrated, whether they were ovulating or not. Intrigued, Barry removed the ovaries of some of the rats to keep them from being able to secrete the estrogens believed to be responsible for this behavior. To his surprise they continued to display lordosis. And not only that, Barry also noticed that during penetration the rats remained rigid, immobile, as if blocked. He could push them or pester them, but they didn't respond. He even tried lightly pricking one of their legs and he saw that they didn't react. That gave rise to a deeper dilemma in Barry's scientific mind: did they not react to the prick because the vaginal stimulation produced paralysis or because it lessened the pain? The vaginal contact clearly provoked an effect in the rats' behavior, but what type of effect exactly: muscular blockage or analgesic? Barry continued measuring the rats' neuronal activity and saw that the neurons of the sensory cortex responded identically to the prick with vaginal stimulation or without it, but activity in the brain areas involved with the pain seemed to diminish. That could imply that the vaginal penetration had an elevated analgesic power, and if one could discover the mechanism, perhaps one could even identify new therapeutic means for treating pain. In that period, now the late 70's, Barry's wife was being treated for cancer, and seeing her suffering forced him to redirect all his investigations toward this specific study about the possible analgesic power of vaginal stimulation.

But before he could completely discard the possibility that the rats didn't remain stock-still just due to a paralyzing effect, and since there was no way of asking them, he had to put aside rodents for the moment and begin to do research on women.

Now installed in his laboratory at Rutgers University in New Jersey, Barry asked a few women to stimulate themselves vaginally while being pinched on the finger to varying degrees of intensity. The results were revealing: the threshold at which they said they felt pain was significantly lower when they were being aroused. In fact, they tried different degrees of stimulation, from the most neutral to close to orgasm, and the more excited they were the more pain resistance they experienced. And with other tests they proved that their sensitivity to the touch didn't lessen in the slightest, their skin distinguished every contact or change of temperature as ever, it was only the intensity of the pain that changed. That was very important because it implied that the vaginal stimulation really acted as a specific analgesic against pain, not like an anesthesia that decreases all sensitivity in general. But how did it work? What was the trigger? This was the question whose answer could have profound implications. In order to explore differences in neurotransmitters and the effects of different nerves that transmit information from the vagina to the spinal cord and the brain, Barry went back to experiments with rats. He began to cut specific nerves such as the pelvic and the pudendal in order to trace all the changes that resulted in terms of arousal and pain. He published several hypotheses on the specific role of the pelvic nerve in this pain relief but, once again, the rats weren't a valid enough model. Their interaction between sexual pleasure and pain is very different from ours.

Komisaruk returned to studies with women, focusing on a group with a very particular characteristic: spinal cord injuries. Barry and his research team recruited women with injuries at different heights on their spines, which affected specific nerves. These women offered two big advantages for the study: on one hand, since they had no conscious sensitivity in their vaginas, there was no possible pain mitigation from pleasure or psychosomatic effects; on the other hand, they allowed the researchers to discern the concrete implication of each nerve on the reflexive reduction of pain produced by vaginal contact. Within the group of women there were some with injuries very high up on their spines, they formed the control group since no nerve fibers reached from their brains to their external or internal genitals. The results from this control group were a new, unexpected surprise in Barry's scientific career. Some of these women began to notice that when the penetration was very deep they did feel a slight response; in fact, when they applied the pain test, their resistance was also significantly higher. That didn't make any sense. It wasn't possible that women with such a high spinal injury had sensitivity in their vaginas. He consulted with doctors and experts in spinal injuries and no one was able to give him an explanation; the height of their injuries should block any sensation coming from the pelvis. The only possibility was that, unlike what was stipulated in medical manuals, the vagus nerve was involved. The vagus nerve doesn't run along the spine but rather through the inside of the body, transmitting information to the brain from various internal organs such as the lungs and the intestines, but it isn't believed to reach the uterus or the cervix. This could be very groundbreaking. The first thing that Barry had to do was rule out possible psychosomatic effects, or that in the analyzed cases the injuries were incomplete and left some medullary fibers connected. In order to confirm or disprove this hypothesis, Barry injured some lab rats by removing whole sections of their spines, and observed that deep penetration induced a dilation of their pupils, a reaction associated with sexual stimulus. And not only that, when he surgically cut the vagus nerve, the dilation disappeared. The fact that the vagus nerve could reach the cervix was so groundbreaking that Barry decided to continue his experiments on women with complete spinal cord injuries. And the least invasive way to do so was to use functional magnetic resonance imaging (fMRI) brain scans that analyzed activity in different areas of the brain's sensory cortex during the stimulation. Barry remembers those studies as the most outstanding of his career. Not only because they did indeed confirm that the vagus nerve transmitted information from the cervix to the brain, but also because several women with complete spinal injuries felt sexual arousal for the first time since their accidents, and some of them even reached orgasmic feelings again. Obviously he found the situation moving on a personal level, but it also made Barry realize that, as surprising as it seemed, no one had ever before analyzed brain activity during sexual response and orgasm. And we were already in the twenty-first century! The scientific void was enormous, and Barry began his pioneering studies sexually stimulating women—this time without spinal injuries—while using fMRI brain scanners. One of his studies focused on determining whether stimulation of the clitoris, vagina and cervix sent signals to the same areas of the sensory cortex, or to different ones. The results indicated that there is an overlapping, but that the clitoris, vagina and cervix each activate specific areas of the brain, which also correspond to the signals from the pudendal, pelvic, hypogastric and vagus nerves. This allowed Barry to determine that the clitoris only sends information through the pudendal nerve, that the labia and the entrance of the vagina only transmit through the pelvic nerve, and the deepest part of the vagina and the cervix through the pelvic, hypogastric and vagus. This was irrefutable proof that the stimulation of the clitoris and the vagina travel along different routes, and could offer much information about the diversity of the female sexual response. In fact, Barry is investigating with women who reach arousal but never orgasm in order to see what differences there are in their brain activity as compared to other women who do climax. He began by registering all female brain activity from the initial resting phase up to orgasm, observing that from the start of the stimulation there is progressive activity first in some areas and then in others, leading up to the generalized explosion of cerebral activity produced during orgasm.

Barry acknowledges that simply observing brain activity on its own does not tell us anything about the function that area of the brain is carrying out, but that establishing this sequence and comparing between various women and types of stimulations, it can be intuited, for example, what difference there is between ordinary friction and friction of the same intensity that is pleasurable, or at which moment an action is placed in an erotic context, and above all to see what part of the sequence of cerebral activity fails in women who do not become lubricated or do not reach climax.

Another of the groundbreaking projects being worked on is neurofeedback. Biofeedback, which has been well known for some time, is our ability to modulate some of our vital signs if their rates are shown to us on a screen. Seeing those rates acts as a reference point that helps to focus our minds on our body, and allows us to try different techniques, such as holding our breath or concentrating on certain thoughts in order to see how they affect the heart rate, blood pressure or the physiological response we are analyzing. It is like training to control bodily functions by using a constant point of reference. It has limits, but it works, and there is evidence that suggests that something similar could be done with brain activity by watching our own fMRI in real time. Barry believes that a field like sex, in which the mind is so important, can be an ideal model for testing neurofeedback. And who knows? Maybe anorgasmic women or those with arousal problems could try different types of fantasies or actions while they observe their brain activity, seeing what they react to and thus train themselves to improve their sexual response.

It's still speculation but, in his sequences of brain images, Barry observed how the genital sensory cortex activates first, the thalamus seems to awaken the signal of physical arousal, the activation of the amygdala increases desire and physiological reactions of tachycardia and hypertension, the hippocampus could be involved in fantasies, the nucleus accumbens (responsible for pleasure) skyrockets, at the moment of orgasm the hypothalamus suddenly activates and releases oxytocin, the cerebellum responds to the muscular tension, and even specific areas of the prefrontal cortex light up.

These relationships are still not completely established, and other investigators, like Janniko Georgiadis, at the Dutch university of Groningen, have observed a specific decrease in orbitofrontal cortex activity right before orgasm. The orbitofrontal cortex is an area related to the awareness of one's own body and self-control, and over a few beers, during the European sexology conference held in Madrid in September 2012, Janniko told me that "it's as if right before orgasm a region of the brain involved in body awareness shuts off." I asked him if a neuroanatomical signal could be responsible for that momentary feeling of losing control, of getting "carried away," and the later haziness, and whether the feeling many women describe of sometimes being on the edge of orgasm without achieving it might be due to this signal not being sent. Reluctantly, with the caution typical of the most rigorous scientists, Janniko replied, "Perhaps it could be involved in the loss of control you mention. But we need more experiments. And we always have to be careful to not reduce sexual experience to just brain activity." No one is suggesting that. We know that a change in cerebral activity can be a consequence and not a cause of a change in behavior. But it is fascinating how bringing together all these pieces of information from different experiences and academic disciplines perfectly illustrates how scientific knowledge advances.

In science, each researcher knows that he or she is carrying a dim flashlight that allows him to illuminate only one part of an enormous dark room. One researcher focuses in one direction and another on another. Separately they can illuminate different corners and reach totally disparate conclusions. There are also those who try to conquer the unknown without a flashlight, using their imagination. But gradually, when the scientific lights increase and band together, the room starts to reveal its contents. And only those who don't want to see will refuse to shift their preconceived ideas. Often, when illuminating a part of the room, one finds a window onto another, even larger dark room, which paradoxically increases our ignorance of reality. There is no choice but to open up that new window and start from zero to reveal the mysterious room that it leads to. This is the slow process that we are following with nature, the universe and the human brain, waiting for the day when scientific light replaces the darkness and theories.

My Orgasm as Revealed by fMRI

When I finally agreed to take part in Barry Komisaruk's study, my main concern was whether I would be able to manually bring myself to orgasm under a scanner that measured the activity of the different parts of my brain. Performance anxiety and obvious dread of erectile difficulties. The situation got worse when on the Thursday prior to the experiment I got a call from Nan Wise, a researcher in Barry's lab at Rutgers University, explaining the procedure and insisting that the most important thing was that I keep my head as still as possible. "It would be good if you could practice," she told me. It was 11 in the morning when I got Nan's call. I was working in my apartment in New York, not sexually aroused in the slightest, and I decided to give it a try. I grabbed a bottle of vitamins, went to my bed, placed the bottle on my forehead, and tried to see if I could give myself an erection without the bottle falling off. After several minutes of trying to balance I gave up, laughing at myself and the kooky situation I'd gotten myself into. "There is no way I'm going to put this in the book!" I said to a friend. But here goes.

The Monday after that I took the train to Rutgers University in New Jersey, using the travel time to think about what images I would conjure up under the scanner. Obviously, I'm not putting that in this book. It was my third visit to the psychology department headed up by Barry Komisaruk, but that time I was feeling shy. "I don't know if I'm going to be able to do it, Barry," I told him. "Don't worry, if it doesn't work that's okay," he replied. He introduced me to the technicians and department staff, and I imagined them all thinking "this is the guy who's going to…" To top it all off, a Korean television crew was there working on a documentary, and they asked for my permission to record "just before and after, and the computer images." Ugh… Even more pressure: it would be even worse to fail on camera! But I'd gotten that far so… I gave them the go-ahead too.

A magnetic resonance brain scanner is an enormous machine with a hole in the middle where a supine person inserts the upper half of his or her body. It is in an isolated room that faces another one where the scientists are monitoring with their computers and giving information via microphones. Barry and Nan had covered the windows with cardboard to make it more private and ensure that no one else could see me. "That would be the final straw," I thought.

Before we started, they went over the experiments with me—I was actually there to do two tests—and then they accompanied me to a room where I could change my clothes. I use that term loosely, because I came out in a tee shirt with a towel wrapped around my waist, feeling completely ridiculous, and headed straight for the scanner. I lay down and placed my head in some sort of cushioned holder. They placed styrofoam blocks around it to keep it steady, asked me to extend my hands, then came the rhetorical question about whether I was comfortable, and the "bed" went into the scanner. The test was starting.

There was a screen before my eyes where I read the instructions, and a small speaker conveyed Barry's comments to me. While they prepared the program, a muted episode of *The Simpsons* played on the screen. It was all very surreal: the Koreans filming, me watching *The Simpsons* inside a scanner, and charged with the mission to manually stimulate myself to… but how did I get myself into this mess? "Ready?" I heard Barry ask. "Ready," I answered. "Remember that if you want to stop all you have to do is press the button in your left hand," said Barry. "OK, full steam ahead!" The mission was beginning.

The first study was very easy: every 20 seconds some instructions would appear on the screen. When I read "Air" I had to bring together my left index, middle and ring fingers to my thumb five times. That way I would see the area of the sensory cortex that represented those fingers. When the words "Glans-easy" appeared on the screen, I had to gently touch the upper part of my glans with those same fingers. The sequence of first "Air" and then "Easy" was repeated for five complete cycles, during which the scanner registered which area of my brain the pudendal nerve was sending information to. Then we repeated the same operation but with "Air" and "Midshaft-easy": five times just touching fingers, and then gently touching the central part of my penis five times, repeated over five cycles. The scanner again registered where the pudendal nerve was sending the information. The third series of cycles was identical for the "Scrotum-easy": this time I had to very softly pinch the skin of my scrotum, and the fourth for "Testicle-easy" I carefully stroked my testicles.

Then we repeated the entire process, but with "deep" instead of "easy." I had to press my glans with force but not to the point that it was painful, the shaft noting that I was not only touching the skin but also the internal corpora cavernosa, pinch my scrotum harder and squeeze my testicles until I felt slightly uncomfortable. This was the essence of the experiment: the glans is only connected to the pudendal nerve, both inside and out. And even though the rest of the surface of the penis is also connected only to the pudendal, deep pressure could active the pelvic nerve, and it is unclear what degree of information it can transmit to the brain. And something similar happens with the testicles, which are connected to both the pelvic nerve and the fibers of the sympathetic nervous system. In fact, I was going to be a healthy control subject and they would compare my scans with those of patients who had been through prostate operations and suffered nerve damage, especially to the pelvic nerve, with different degrees of erection capability loss. The experiment would serve to shed light on the mechanisms of erection, add what injuries are produced by prostate operations, perhaps helping to minimize them.

Once the first experiment was over, Barry asked me if I wanted to rest or begin the second one. Honestly, after all that gentle and harder touching I felt physically—if not mentally—prepared. It would be a question of seizing the moment. "Let's continue, let's continue…," I said, convinced. "Perfect! Then wait a few seconds, and when you see the green light appear on the screen you can begin the stimulation. Take your time, there is no rush or time limit. Just relax and when you are finished press the button in your left hand. And try not to move your head…" The truth is that with my head perfectly fitted between styrofoam blocks it was much easier to concentrate on the task at hand than with a bottle of vitamins slipping down my forehead. And to my surprise, either because of the prior stimulation or because of the situation I was recreating in my mind, I soon had an erection. Moving my hand was more complicated, especially in moments when I felt I needed to go faster than I could without moving my body. But 9 minutes later, I completed my mission. Gruesome details aside, after I finished they asked me to stay in the scanner for a little longer. I think I almost fell asleep. Between that and the magnetism that my brain had received I left the room in a bit of a daze. I went to clean myself up and get dressed, collected the 200-dollar check that was my compensation for taking part in the experiment, said goodbye to the Korean cameramen, Barry and his team, and I went back to Manhattan with a feeling that's hard to describe. I was the first man in the world to have an orgasm beneath an fMRI scanner (Fig. 3.1). I didn't know if I should feel proud or embarrassed, but I was definitely very curious to see the results. That would take a couple of months. Meanwhile, I told the anecdote on more than one occasion, and got all sorts of reactions. Very, very interesting…

The study was completely anatomical. Barry had already realized studies with women having orgasms from stimulation of only the clitoris, vagina or cervix. This had helped him prove that the stimulation activated different areas of the sensory cortex and that the pelvic, pudendal, hypogastric and even vagus nerves were involved. He had also carried out an experiment in which he saw how first some areas of the brain activated and then others before reaching massive activation during orgasm, but a similar study with a man had never been conducted. Not by him and not by anyone. On one hand, Barry wanted to see which areas of the sensory cortex activated in connection with which nerves and, on the other, he wanted to establish a sequence of brain areas from the start of arousal to orgasm, similar to the study he'd conducted with women. Again, I would be a healthy control that he would compare to men with different dysfunctions, such as premature ejaculation, anorgasmia and difficulties maintaining an erection. Seeing what happens differently in the brain of a premature ejaculator as compared to mine could provide a lot of information. Or maybe not, but science had to give it a shot.

Several weeks later I sent an email to Barry asking him if he had the results. He replied that Nan had already processed them and they were going to analyze them the following week. Barry said he would call me. And when he did, a few days later, he was in quite a state. He was looking at the results right then, and he said they were of excellent quality. It seems I had managed to keep my head still, and that you could observe very clearly how various areas of the brain lit up gradually until reaching orgasm, and how they soon switched off afterwards. Barry was thrilled. He said that my cerebellum had been very active throughout, that the insula showed a very interesting pattern,

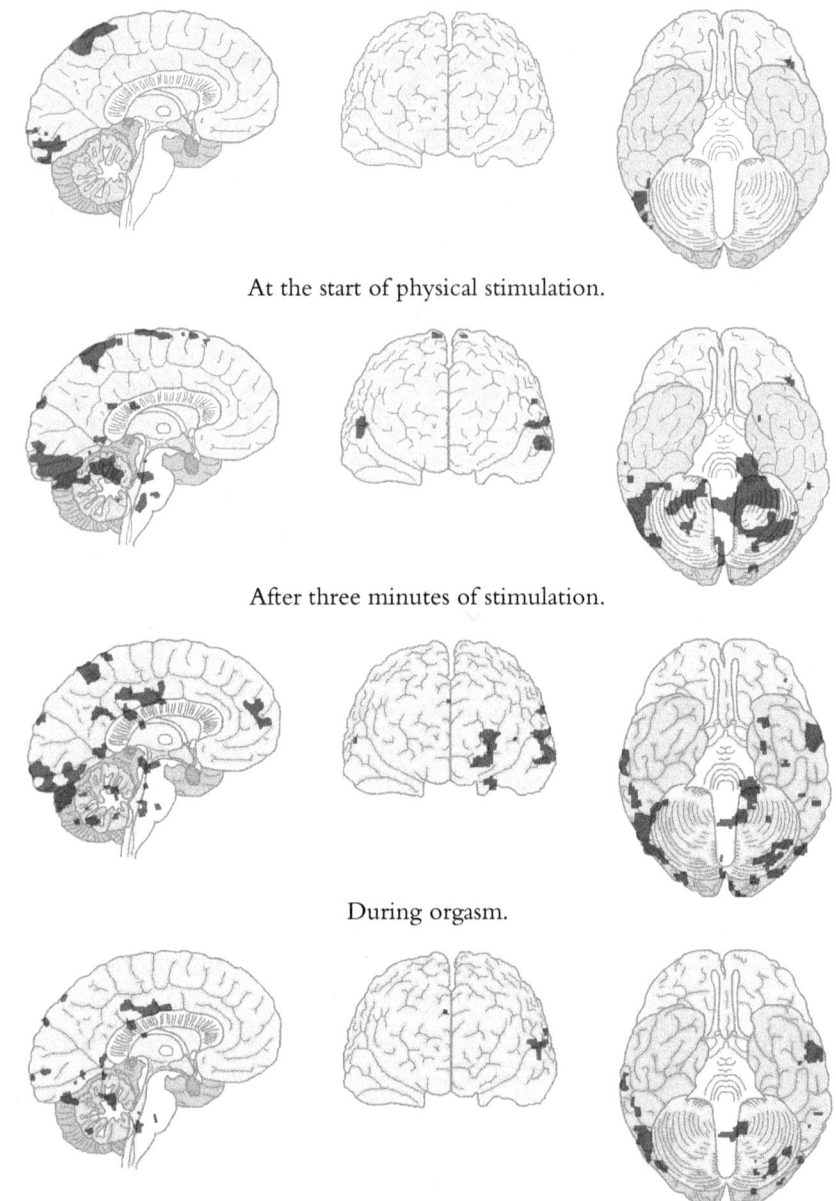

At the start of physical stimulation.

After three minutes of stimulation.

During orgasm.

One minute after orgasm.

Fig. 3.1 fMRI scans of the brain

as did the amygdala; he talked about the changes in the prefrontal cortex and the posterior one, in the hippocampus, the nucleus accumbens and the areas related to itching, and while he still needed to analyze the material further, it seemed like he had excellent findings to publish. Or even make a film about, like the one on the results with women that, a few months earlier, had been picked up by many media outlets and seen by millions of people around the world. That was in May of 2012.

Science advances slowly, and when weeks before I finished this book I visited Barry Komisaruk's laboratory again, the results were still not completely processed. Nevertheless, he told me that they had found interesting patterns. It was a Friday in late October, and we spent all afternoon looking at and commenting on the activity in specific areas of my brain at different times during stimulation, orgasm and later relaxation. Barry's first conclusion was that "as I imagined would be the case, they are very similar to the images of arousal and orgasm in women." Keep in mind that this was the first time that anyone had registered the complete and continuous sequence of cerebral activity from initial stimulation to orgasm in a male brain. One of Barry's objectives was comparing it to his prior studies with women and, from what he said, on a brain activity level, "the similarities were much, much greater than the differences. The only major difference is that, and we'll have to confirm this with more volunteers, your hypothalamus didn't activate during orgasm, the way it did with all of the women."

The hypothalamus is the area that secretes oxytocin, the so-called love hormone. We already knew that it is secreted in greater quantity by women than by men and also in greater quantity during coitus than during masturbation. Barry didn't dare to draw conclusions based on the scans of a single male brain, but he did say that "if it were to be confirmed that the male hypothalamus does not activate during self-induced orgasm (as they call it), that would be a very important difference." Various areas reacted predictably: the ventral tegmental area that produces dopamine in the midbrain; the nucleus accumbens that is directly involved in pleasure; the frontal cortex; a secondary area of the sensory cortex called the *operculum*; the amygdala that is home to the emotions; and an area called the reticular formation. They all progressively increased their activity as the experiment advanced. The cerebellum, which is related to movements, and the genital sensory cortex in the paracentral lobule, were active from the very beginning; the anterior cingulate cortex showed little but constant activity, and activity in the hippocampus shot up only at the moment of orgasm. I asked Barry if that could suggest that, since the hippocampus is involved in memory, it activates to make us remember all the circumstances surrounding an orgasm, and if that's what makes us have such vivid sexual memories, and he answered that "that's a valid interpretation, but it cannot be inferred from the results at this point."

It was truly fascinating to see how all of my brain started to light up with colors, even though it was following the expected process. The only surprises for Barry were the inactivity of the hypothalamus (producer of oxytocin), and a highly interesting pattern in one specific part of the posterior cingulate gyrus, which was associated in other studies with itching. "It is a concrete region that very clearly activates when you have an itch somewhere on your body, and in your images we see that activity there constantly increases from the third minute on. This is really interesting. We will have to confirm it with other volunteers," Barry told me enthusiastically.

Lastly, he mentioned another unexpected result, which he didn't deem very important: "It is very strange that your visual cortex was active from the beginning. In fact, in the images from minute one it is far and beyond the most active zone. It's hard to understand, because you weren't looking at anything…" I think I blushed. Suddenly I remembered other studies that demonstrated that, even with your eyes closed, when you imagine something vividly your vision areas activate, and I asked Barry "that… what if… just for example… from the beginning, I'd been remembering something that I'd experienced, you know, like the previous night… could that activate my visual cortex as I reconstructed those images?" "Oh, of course!" said Barry with a slight smile. Oops… turns out you can't hide anything from a fMRI scanner…

Want-Like-Learn, and the Empire of the Senses

We have actually already discussed what the field of neuroscience knows about human sexual behavior, which isn't very much. Psychology and sociology, which we will start to talk more about in the following chapters, are far ahead of the physiological brain studies. But this is precisely why neuroscience is so fascinating: it is the field that can give us the most information and answers in the next few years, and this work has already begun.

The chemical aspects of the role of neurotransmitters and hormones are pretty well defined, in large part thanks to studies with laboratory animals. In earlier chapters we've talked about the chemistry of sex, and we will apply that to concrete cases in further chapters.

One of the hottest topics in neuroscience is the study of brain connectivity, which is to say: not just focusing on individual neurons or activity in parts of the brain but understanding how the various neuronal networks interrelate and communicate with each other. It is a very new field that is proving to be highly important for topics on learning and developmental problems such as autism or schizophrenia but, of course, sexual behavior has not yet reached that status.

Perhaps the methodology that is contributing the most interesting information, of which there is much more to be mined, is the study of brain activity using fMRI scanners. I remember reading an article by Roy Levin on the refractory period following orgasm and immediately picking up the phone to call Barry Komisaruk. I asked him why after my orgasm under the scanner we didn't continue the experiment, trying to stimulate my penis again to see what changed in my brain as compared to the movements of the first stimulation. Barry replied, "Of course! Why didn't I think of that? We could see how identical contact before and after orgasm is interpreted differently in the brain of the same person, and compare it with people with erectile dysfunction. It might offer very interesting information. No one has done that yet!" I was surprised, but over time I saw that it was no exception. For example, no one had yet compared the brain activity of hypersexuals and asexuals, even knowing that there are no significant differences in their hormonal levels. Honestly, it is very difficult to find such simple aspects of human physiology that no scientist has investigated before, especially when they can offer extremely interesting information not just about sex, but also about the neurobiological bases of our behavior. And I'm not just saying that last part because I like hearing the sound of my own voice.

Various Types of Pedophile Brains

Let us look, for example, at the promising studies with pedophiles carried out by a German scientist of Spanish descent, Jorge Ponseti. Jorge was one of the first neuroscientists to identify concrete differences in the brain activity of pederasts when showing them images of children, as compared to men who felt no attraction towards minors. When we met in Portugal, Jorge explained that in some cases these techniques could be used to identify risks, and perhaps to analyze the progress of some therapies, but mostly to better understand the origin of different types of pedophilia.

When we spoke again in October 2012, Ponseti was starting to work on the largest multidisciplinary study project on pedophilia to date, which—financed with two million euros by six German research groups—had already recruited 250 pedophiles who would have their genes, hormones, neurotransmitters, psychological analyses and brain activity analyzed. Jorge explained to me that, as strange as it seems, "There is very little prior scientific information."

Let's imagine the simple fact of comparing the brain activity of pedophiles who, although they feel sexual attraction, have never abused a child because they know it is despicable (Ponseti tells me that in Germany 0.7% of the population are pedophiles, people who feel sexual attraction towards minors), to that of pederasts who, while they also know that, are unable to control themselves and end up committing these crimes. Exploring these differences, distinguishing between impulsiveness and ability to control, can give very valuable information on the neurobiology of human conduct. As can comparing the two main types of pederasts: those that abuse teenagers already in puberty and those who are only aroused by prepubescent children. Some hypotheses maintain that the areas of desire in the former wouldn't be very different from the healthy control group's and yet that the latter would have some specific mark. Then there is another big group: the men who are fascinated by children and want to be around them all the time, but would never abuse them or harm them in any way. While under the scanner, Ponseti will show them images of infants and adults, both human and animal. One hypothesis suggests that paternal love in humans is something very new in evolutionary terms, and as such is subject to more variability and possible problems, and that perhaps in some cases the obsession with children stems from a fascination and compulsion to protect childhood. If these subjects react both to human children and baby animals, the hypothesis would be closer to confirmation.

And lastly there is the analysis of whether the pederast is predetermined from birth or not. There are three times more left-handed pederasts than right-handed ones, they are generally less intelligent, 20% of them are homopedophilic, and there are indications that the hormonal effects of pregnancy can alter the circuits responsible for attraction and sexual orientation. Ponseti will analyze the brain activity at rest in pederasts to see if it is different from that of conventional people. In patients with a predisposition to depression or other disorders they have already revealed some differences, and it is suspected that the same will happen with pedophiles. Finally, Ponseti insists that the most important thing is understanding and classifying them to see which will benefit most from one psychotherapy or another, and who might need a pharmacological treatment to reduce their desire. The subject is incredibly interesting and it's surprising—given how common such abuse is according to recent polls, and the problems they generate in those affected, especially the older and more aware they are—that medical science has taken so long to devote important resources to its study. The American National Institute of Mental Health still doesn't, as its director Thomas Insel admitted to me in person. More evidence of the fact that sex is also taboo in science.

Dual Control Model: Excitation Versus Inhibition

Scientists need theoretical models onto which they can fit pieces and test their hypotheses, and one of the most commonly used in sexology is the *Dual Control Model of Sexual Response*, initially put forth by Helen Singer Kaplan and further developed by John Bancroft and Erick Janssen of the Kinsey Institute (we'll meet them later). According to the dual control model, our sexual behavior responds to a balance between two systems of excitation and inhibition, which are in turn conditioned by physiological and psychological factors. Researchers use this paradigm to examine all the hormonal, behavioral, relationship and atmospheric influences, until they can distinguish, for example, whether a loss of sexual activity is due to more inhibition or less desire, or whether the fact that a pedophile commits an abuse is due to overwhelming excitation or failures of inhibition. The model is a bit too simplistic but, as we will see, it is very useful for proposing and analyzing concrete cases.

Another model is the one that separates sexual response into three phases that we could call "want, like and learn." The "want" phase deals with attraction and the origins of desire, the "like" with pleasure and the physical response during the sex act until orgasm, and the "learn" with all the cognitive ences, analyzing how the stimuli that accompany an experience (depending on how satisfying it was) are reinforced or inhibited. It is a bit theoretical, but researchers have established that different cerebral mechanisms influence each phase.

Some of the more interesting work here is that of Janniko Georgiadis, who compares the mechanisms of desire and pleasure between sex and food. Janniko is one of the world's experts on the neurophysiological study of sexuality. When we began talking about the interest or "want" phase, he described a compelling study, published in 2010 in the *Journal of Neuroscience*, in which French investigators wanted to compare the cerebral reaction to visual stimuli related to sex and money. Since monetary value is something of such recent appearance in our species, they thought they would find important differences in how the brain processed the information. And they found some. For example, the amygdala, site of the most visceral impulses, lit up much more intensely only with the images that suggested an erotic encounter. There were also differences in the activation of the orbitofrontal cortex (OFC) related with the awareness of internal states and the control of actions: the posterolateral area of the OFC was activated by erotic stimuli and the anterior area by monetary ones. But what was surprising was that all the rest was almost identical. The striatum, the midbrain, the insula, the anterior cingulate cortex and the rest of brain areas related to motivation and circuits of reward reacted the same way to both stimuli. This led Janniko to think that actually sex was similar to other stimuli, such as food.

In a review of scientific literature published in 2012, Janniko concludes that except for obvious particular characteristics, such as the fact that food is necessary to survival and sex isn't, and that hunger exists throughout life while sexual instinct only begins in puberty, the functional neuroanatomy of sexual behavior and its phases of "want, like and learn" are comparable to other circuits of pleasure-reward such as that of food.

Although this might seem irrelevant, the final reflection it leads to is definitely not: there are no physiological mechanisms or neuronal networks specific to sex or other rewards. They share the same circuits of motivation and satisfaction. Which is to say, beyond context, brain cells do not know if they are activating due to an erotic, nutritional or other type of trigger, and this could imply relationships between various disorders that are apparently very disconnected. This is still in the hypothesis phase, but the study of sexual response will undoubtedly offer very important information.

Pheromones, Caresses and Mirror Neurons Activated by Pornography

When studying the nervous system we cannot forget about the senses and perception. It is a vast subject, but we can cite some interesting and groundbreaking results. With regard to touch, a very basic question could be why do we experience a soft caress so differently from rubbing of a different speed or intensity. Obviously context and interpretation play a fundamental role, and we will see how in certain circumstances we even eroticize pain. But in 2009 *Nature Neuroscience* published a hugely interesting work by Swedish and English researchers, documenting how specific nerve fibers are implicated in that distinction. Our nerve endings have various types of mechanoreceptors, and it seems that gentle caresses only activate the nonmyelinated ones, while more intense or rapid contact activates those as well as the myelinated fibers. The findings were also perfectly aligned: when the study participants' definition of the contact as pleasurable, the devices registered that only the nonmyelinated nerve receptors had been activated. In their article, the researchers write "these results are, to the best of our knowledge, the first demonstration of a relationship between positive hedonic sensation and coding at the level of the peripheral afferent nerve, suggesting that C-tactile fibers contribute critically to pleasant touch." In other words, when we softly and slowly stroke our partner's skin, we are only activating certain fibers involved in pleasure.

A big discussion in the study of the senses is about the possible role of pheromones in sexual desire. Anatomical studies ruled out that we as humans, unlike many other mammals, have a vomeronasal organ responsible for codifying olfactory signals related exclusively with the cortex and sexual information. That led some scientists to state that pheromones played no role in humans. Others, however, considered the vomeronasal organ unnecessary, and that pheromones and other chemical signals involved in social conduct could be detected by the olfactory epithelium, as is the case in rodents, for example. In recent years, the role of pheromones has been the subject of interest after discovering that smelling them can affect activity in the hypothalamus and the amygdala, and that the responses vary based on gender and sexual orientation. The conclusion is that, beyond perfumes, body odor itself can have an influence on the start of sexual response.

In any case, in our species the sense most involved in sexual attraction is, far and away, sight. Birds emit sophisticated calls, insects secrete pheromones to attract their future mates, and some primates in heat exhibit swollen genitals and move them in such a way as to incite procreation. We are visual beings in a constant state of alert toward sexual stimuli. Perhaps the best evidence of this is recent studies showing that we even react to erotic information only captured by our unconscious. When we are chatting and we suddenly turn to the right without consciously knowing why, and we find ourselves looking at someone attractive, we really didn't do it on purpose. It is our unconscious that picked up on it before we did. We have this excuse thanks to a study published in the scientific journal *PNAS*.

All the information captured by our senses reaches our brains, but we are only conscious of a small part of it. We have selective attention, which allows us to discriminate between different visual stimuli and ignore what we deem irrelevant. But the unconscious does receive them, and on occasions forces us to act. Researchers carried out a simple experiment in which they would run images of naked men and women past heterosexual men, gays and women, but beneath their detection threshold. Imagine looking at the center of a screen in which identical images appear on the left and right, but every once in a while there is a hidden subliminal nudie pic. We don't even realize, but by analyzing the head movements and shifts in attention the researchers observed that the heterosexual men responded positively to the naked women and ignored the naked men, while the opposite occurred with the gay men and the women. Some heterosexual men even looked away from the unconscious perception of the nude men. When someone says to us, "your eye is wandering!" and curiously it only seems to wander toward someone attractive, we can say with confidence, "Of course! I can't help it. It's my unconscious. It says so in an article in the journal *PNAS*."

A provocative, and much more speculative, idea is that the great appeal of pornography, especially to men, comes from a supposed activation of circuits responsible for imitation, which may involve mirror neurons. These neurons implicated in learning and imitation exist in many mammals and are neuronal circuits that activate when the animal sees another carrying out a determined action. Their true influence on humans is not yet known, but some authors suggest that they could be related to empathy, or the fact that seeing someone smile brings a smile to our faces, along with the corresponding sense of wellbeing. In part we feel others' experiences as our own. If this logic were correct, looking at pornography could be not only visually stimulating but also a mechanism for activating the pleasure circuits, making us feel—to a certain extent—part of the action. Again, this is speculation, but a study published in 2006 by Jorge Ponseti in the journal *Neuroimage* observed that pornographic images activated the premotor frontal cortex, the area where—if they exist—the mirror neurons would be located. If this were the case, porn would not only excite us because of what we see but also because it activates parts of the brain related to imitation, making us believe that we are involved in the sexual act. What surprises me is that, given the powerful response it elicits, neuroscientists don't use porn to analyze the existence and role of mirror neurons and empathy circuits.

What is clear is that while the contributions of neuroscience to the study of sex continue to be impressive, they are still in the initial stages and are nowhere close to offering an overall explanation of human sexual conduct. So we must make the leap to experimental psychology, which analyzes our mind and behavior from the outside instead of the inside.

4

Sex in Our Minds

Sex is not a rational act. Let's imagine we are single: would we go to bed with someone we found attractive but who we know has had unprotected sex with ten different people in the last 2 months? If we ask the question hypothetically and not in the heat of the moment, the answer might be no. Why take the risk. But what if we're asked the same question while being shown the smiling face of someone we are very interested in? Not even then, right?

Sex Is an Irrational Act

Kathryn Macapagal, a researcher at the Kinsey Institute at Indiana University, was explaining to me the experiment I was about to participate in: she would show me the faces of various women, along with the number of men they'd had unprotected sex with in the last 2 months. I had to respond by choosing a number between 1 and 4 according to my predisposition for having intercourse with them. Kathryn asked me to imagine that I was out partying on a Friday night and in the mood for a casual hookup. The test seemed easy: no matter how attractive she was, no way was I going to sleep with some stranger knowing she'd had sex without a condom with seven or eight different guys in less than 2 months! There was certainly no need; I wasn't that hard up.

The experiment begins and a woman with a serious expression and a flashing number 6 appear on the screen. Without giving it much thought, I press the 4: "I would not have sexual relations." Then a pretty attractive girl appears with a 2, and I press 2 as well, telling myself that I would use protection. And it goes on like that, until all of a sudden I get stuck on the image of a gorgeous woman, with a great smile, clear gaze and a 10 flashing on the lower left side of the screen. Soon the message "Please, respond more quickly" also appears. My mind is still searching for some rational justification for my desire to press a 1: "I would definitely have relations with her." I react by marking a highly unexpected 2.

Obviously I don't know how I would react if the situation were real, but the truth is that I had less of an idea minutes earlier when I was making the decision with a clear head, and I'd bet that the same thing would happen to most of you. This is one of the messages that we'll be repeating most throughout the book when we talk about sexual behavior: sex is an irrational act, and it is usually accompanied by very intense emotional states. And our brain is terrible at trying to predict how we would react in future situations that are emotionally new. Every time someone describes a situation to us, we imagine how we would feel and how we would act, but often reality surprises us. Sometimes positively and sometimes negatively.

That is what we are learning from all the psychologists and neuroscientists who work in behavioral economics, or in the newer areas of neuroeconomics and neuromarketing. The objective of all these disciplines is understanding how we make decisions in different areas of our lives and what factors are actually the most influential. Two of their main messages are:

1. Our internal emotional state enormously conditions our rational thought;
2. Initial perceptions are usually more important than all the later judgments.

To simplify, we can say that the decision-making process goes through the following phases: first there is the perception of the situation we are going to face (an attractive woman I know has had a lot of casual relationships is hitting on me); secondly there is an analysis of the different possibilities (trying to go to bed with her, not doing it, looking for somebody else…); thirdly there is a reasoned evaluation of the pros and cons of each option (the possibility of infection, the protection offered by a condom, how much fun it'll be…) and, finally, the decision itself. In this sequence, we believe that the key step is the third, in which we put all of our intelligence and brain power to work for present and future benefits to the rational beings that we are. Well, no, maybe in monetary or work terms we are a little bit that way, but in the more basic aspects that activate our most primitive and emotional brain, like food, sex and fear, the basic instincts send powerful early messages that often defeat the calculations of our prefrontal cortex. Especially if we've been drinking or our rational "brakes" are otherwise inhibited. Undoubtedly the key phase is the first one: perception and initial desire. That is why it is so hard for us to change our minds during an argument if we are already convinced of something, and no matter how good the reasons we're given are, we twist them to fit in with our original idea. How we perceive a situation conditions our reasoning in the later phase. I saw that young woman who I found incredibly attractive, and I started to think that those ten sexual partners over the last 2 months were surely trusted friends, that being so good-looking she could pick and choose and wouldn't just sleep with anyone, that if I used protection from the start the risk was practically zero, and I searched for any other excuse to rationally justify my impulse and my decision to press a 2. Perhaps seeing the same girl but with a different expression on her face my reasoning would be the opposite. Who knows, we are really a bit lost when making these decisions. We believe that first we calculate rationally and then we decide, but actually we decide based on our emotions and then we justify them rationally. To take a common example, think about infidelity and the difference between having committed it or having suffered it.

Let's imagine two couples whose marriages are going through moments of crisis. One of the husbands (let's call him Joe) meets the wife of the other (María) and they start to take a liking to each other. Joe suggests to María that they have sexual relations, he is quite insistent, but she resists despite being tempted. In parallel, María's husband (let's call him Pablo) feels attracted to a coworker and ends up having a little affair with her. Some weeks pass and Pablo confesses his infidelity to María. She gets furious, deems him a swine, and leaves him. A few weeks later Joe separates from his wife and starts to date María. All apparently normal. But… wait a minute… let's be fair… Joe is as much of a swine as Pablo, right? In fact, if María had let him, he would have slept with her while he was married to his wife. Isn't that the same thing Pablo did? Why does María now think that Pablo is a selfish liar and Joe the man of her dreams? María would answer that nothing really happened between her and Joe, that Joe was in love, and a thousand other excuses to hide a simpler and more likely reality: we evaluate situations subjectively based on whether we like them or not, and how much they affect us directly. Perception and emotion versus reason.

The test where they showed me photos of relatively attractive women and gave me information about their sexual past was just one part of the experiments led by Julia Heiman, director of the Kinsey Institute. The one I did was repeated on men and women with different personality traits and conditions: homosexuals and heterosexuals, young people and teenagers, drunk and sober, ones who'd been shown erotic images before the test to arouse them and ones who'd been given different messages of prevention and sexual education designed to reduce risky behavior. In addition to the responses given by the participants, they also measured their reaction time. Different variables can be studied, and in this particular case they were trying to identify which interventions are more effective in curbing sexual risk-taking, especially in the context of HIV contagion.

In the United States many researchers find public funds to investigate sexual behavior thanks to HIV. Later on we will discuss the occasions in which the US Congress has forced cuts in financing to studies on homosexual attraction or sex among the elderly, because of pressure from conservatives. Due to this, just like many environmental projects are contextualized around climate change to facilitate funding, something similar happens with sexual behavior and AIDS. In fact, when the AIDS epidemic broke out in the early 1980s, a large amount of resources were devoted to the preventive study of risky sexual conduct. The team led by veteran sexuality researcher Anke Ehrhardt, at New York's Columbia University, is the one that has received the biggest grant to date for the scientific study of behavior and AIDS contagion. In her office at the HIV Center for Clinical and Behavioral Studies, Doctor Ehrhardt tells me: "When AIDS broke out in the early eighties, the department of health started to ask us about rates of homosexuality, statistics, habits among young people, the degree of condom usage… and they saw that there were very few rigorous studies with reliable data." Like the vast majority of researchers that I've met, Dr. Ehrhardt points out that psychology was very late to investigate sexual behavior, and that, in addition, it offers tremendously interesting challenges: "It is incredibly complicated to guess how we will react in sexual situations. We can't simply look at logic and our suppositions, it is too complex and irrational."

Julia Heiman also was very insistent on that point: it's one thing to take surveys on past experiences, but quite another to ask how we would react in certain future situations. Deductions are not reliable, people respond without knowing how arousal affects our decisions, and in many occasions they lie or deceive themselves by responding what they would like to feel instead of what they feel. Julia Heiman is adamant that "as scientists we need to have a more experimental and objective approach to the study of sex, and to be able to measure degrees of arousal without them being subject to personal interpretation." And there are various methods of doing that.

Measuring Sexual Arousal at the Kinsey Institute

Kathryn Macapagal showed me another of the Kinsey Institute's laboratory devices: a vaginal photoplethysmograph (Fig. 4.1). This is merely a little tube the size of a tampon that is able to measure the changes in blood flow to the

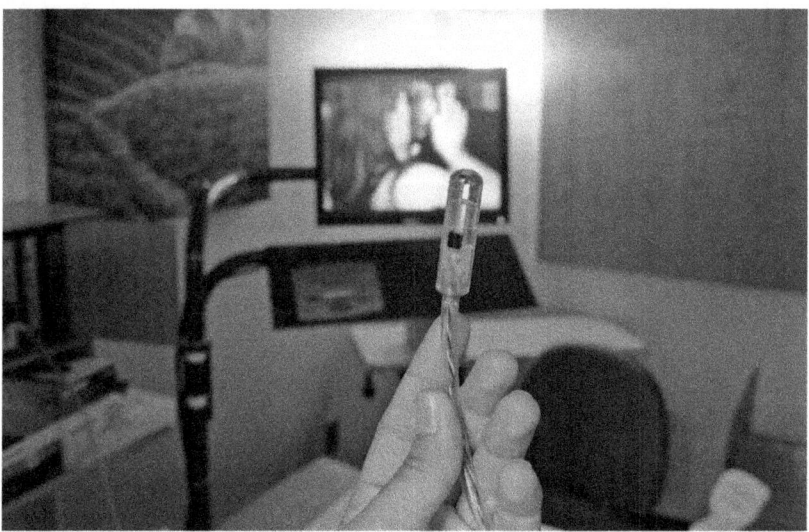

Fig. 4.1 Vaginal plethysmograph

vagina. It does that by transmitting small points of light and gathering them up again with a built-in detector. The more blood circulating through the vaginal tissue, the more light will be reflected, marking changes in the vagina's size and blood flow, and which the photoplethysmograph exports through a small cord. With this device the sexual physiology laboratories are able to measure the degree of genital excitation provoked by stimuli. It sounds like a joke and is the source of much disbelief, but all the researchers assure me that for the time being it is the most standardized way to measure the not always conscious reactions of the female genitalia. Obviously it is a method with many limitations, but at the Kinsey Institute they are testing a similar device to directly measure erections of the clitoris, and there are others who are searching for very different alternatives. I myself have spread my legs in a recliner in Irving Binik's lab at the Canadian McGill University with an infrared camera pointing into my crotch, wearing glasses that showed different erotic images (Fig. 4.2). Irving told me that this equipment, which measures temperature changes in the genital area, is more expensive but more precise than vaginal photoplethysmography for women, or the penile plethysmograph for men (Fig. 4.3). Because they can also precisely measure the subtle changes in our penis size while they show us a conventional erotic scene, then one of homosexual sex, or with transsexuals, a violent scene, or one of sex with minors. They do so by placing a flexible cloth ring some five centimeters in diameter around the penis and attached by a cord to a detector that registers

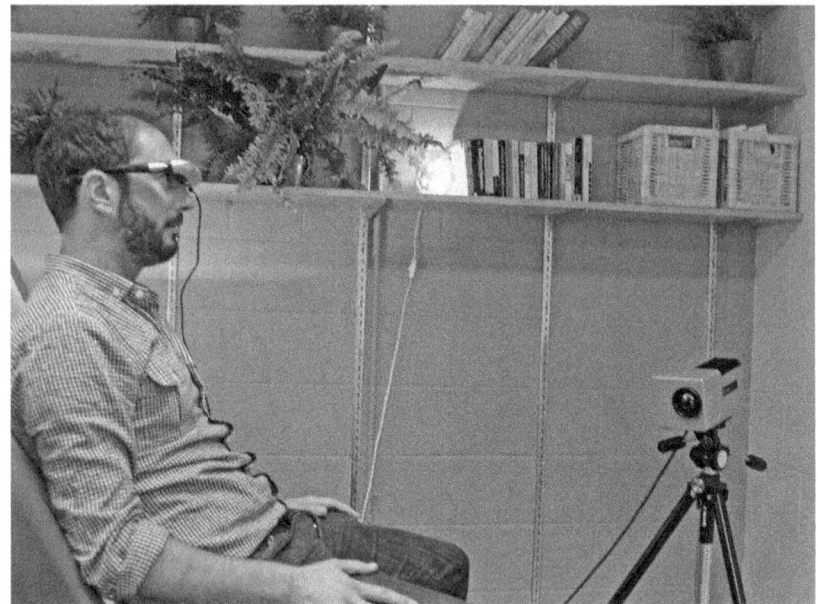

Fig. 4.2 Thermograph

Fig. 4.3 Penile plethysmograph

small changes in thickness. All of this takes place in the male sexuality lab at the Kinsey Institute, in a gloomy room with an old brown armchair in the middle, placed in front of a television, and a shelf to the left filled with porn VHS tapes and DVDs. You have to pull down your pants, place the little ring around your member, relax first with a nature documentary and then see what happens when another type of images appear. It's a bit rudimentary, honestly, but the head of the laboratory, Erick Janssen, assures me that the penile plethysmograph and vaginal photoplethysmography have been used for a long time and offer reliable measurements of male and female excitation.

It may seem easier and more reliable to just directly ask us if what we see excites us more or less. But it isn't, and this is the important part of the question: first of all our mouth and our brain can lie more easily than our penis, and secondly our genital reaction does not always coincide with our subjective perception of being excited. In men there is an easier reference point because we can tell when there is the start of an erection, but in women this point of reference doesn't exist, and it isn't so easy to know when the vagina begins to lubricate or stiffen from greater blood flow. The mind can decide that a violent sex scene is horribly unpleasant while the genitals can react without our realizing. Careful! That doesn't mean that the woman condones the scene or is interesting in imitating it, but her mind could be saying one thing and her vagina another. In fact, that is one of the main things empirically proven by recent sexual research: in more than a few women the agreement between genital and subjective excitation is very low, and that could explain many things.

The Lack of Agreement Between the Female Mind and the Female Genitals

Think of me what you will, but I admit I tried it: I asked a friend if she would come over to my house to do an "experiment" without telling her what it was about beforehand. Once she was relaxed on my couch, I asked her if she liked porn. She told me that she had watched it occasionally but that in general it didn't do much for her. Perfect, that was what I had figured from some of our previous conversations. Then I asked her, "Do you think you would be excited by scenes of lesbian sex?" She wrinkled her nose, looked at me with surprise and said flat-out, "no."

Then I turned on my computer and asked her to watch 4 min of a video on pornotube.com that showed two naked women caressing each other in a bed, kissing and explicitly masturbating. It wasn't very hard-core, but it did have close-ups, moaning and lustful attitudes.

When the video ended I asked my friend if it had excited her. "No…" was her response. "Not even a little?" I insisted. "No, really. As it went on I was curious to see what they were doing, but there was no point where I felt sexually excited." Then I told her that I was going to go into the next room and I asked her to calmly explore her genitals to see if she noticed any reaction.

When I came back, my friend was blushing. "I don't understand…," she said shyly. Her vagina had lubricated, her labia were slightly swollen and when she touched her clitoris she found it "very sensitive," soon generating a certain physical arousal. The rest is off the record.

Obviously, it wasn't a very rigorous experiment, but it was relatively similar to tests that Meredith Chivers, who I was set to interview the following morning, had carried out in her laboratory of sexuality and gender at Queen's University in Canada.

Chivers is one of the world's main experts on the study of "sexual agreement." She measures the degree of correlation between our genitals' physiological response and our subjective experience of feeling excited or not. Let's put aside for the moment my use of lesbian sex images. The primary message is that we can be unaware that our genitals are reacting to stimuli that our minds do not interpret or experience as exciting. Believe me when I say that idea is already widely accepted by the community of sex researchers.

What Meredith Chivers does is show different types of erotic and neutral stimuli to men and women, measure the response of their genitals with a penile plethysmograph and a vaginal photoplethysmograph, and after the experiment ask them how excited they felt. In men the responses and the change in the penis thickness were usually very correlated, but between women there was a very wide range. Some showed the same sexual agreement as any man, but many others displayed much less.

Chivers acknowledges that the methodology is not infallible and she is testing other devices that measure changes in blood flow directly in the clitoris or infrared sensors that analyze the temperature changes in the genital area. But she assures me that "the lack of sexual agreement in women was first observed in the late sixties, and has been confirmed on many occasions independent of the type of erotic stimuli employed." And she has data that backs that up: in 2010 Meredith Chivers published a very thorough study in which she analyzed a hundred and thirty-two peer-reviewed scientific articles that contained data on physical excitation and subjective perception in a total of 2,505 women and 1,928 men. That is an enormous sample, and the meta-analysis concluded that, indeed, sexual agreement between the female mind and the female genitals is much more limited than we imagine.

In addition, it didn't seem to be caused only by the fact that in men it is easier to identify the start of an erection and associate it with a feeling of arousal while some women don't know they are beginning to lubricate until they touch themselves. While this is undoubtedly a factor, it is not the only explanation. In a study published in 2012, Canadian researchers showed 90-s erotic film clips to 20 men and 20 women while measuring both their genital response and the changes in their heart rate and breathing. They asked them to estimate their psychological excitation, and then compared it, and they concluded that the pattern of differences in sexual agreement in genitals remained but also extended to other physical changes involved in the sexual response. This lower correlation between female sexual excitation and subjective perception was something specific and independent of other physical processes and, according to the authors of the study, could respond to some selective pressures in the evolutionary past.

In any case, Chivers insists that "the diversity of the response is vast. In the laboratory we find women who are completely aware of the state of their genitals, and others who really don't perceive any change. Now we are investigating what factors could condition this variability." Chivers tells me that age and marital status don't seem to have an influence, but that still unpublished results indicate that sexual agreement is linked to a higher level of education and frequency of masturbation. She also observes that women who practice meditation or relaxation techniques can develop greater agreement. This is important because it would imply that it is something that can be trained. What for? In science, correlation doesn't imply causality, but "we also observe that women with higher agreement say they are more satisfied with their sexuality." It's speculative, but it could be argued that being attuned to our genital response can give useful information about some fantasies or practices that could appeal to us without us being entirely aware of them.

Chivers's team is doing another very interesting study that aims to detect what specific aspects of sadomasochism are the ones that provoke the erotic response. The experiment is based on showing men and women (both sadomasochistic and conventional) two types of images: some are scenes of pain, dominance and submission without any erotic charge; the others, in addition to their sadomasochistic component, contain nudes or people in suggestive clothing, or are being practiced by attractive people. What they are observing is that most of the sadomasochists—though not all of them—usually get genitally excited in both cases. The interesting thing is that some of the conventional women who do react to the erotic images of sadomasochism claim that they don't like what they are seeing nor find it exciting. Chivers assures me that they aren't lying, that subjectively they do not feel any sort of arousal despite the fact that their genitals are reacting. And they observe that some women are even more excited than when looking at similar erotic scenes without a component of dominance and submission. In those cases, if a person were predisposed to exploring some games of dominance and submission with their partner, they might discover some new interesting facet. But, at this point there is an essential nuance to point out.

Women's Genitals Can Be Bisexual, Even When They Aren't

It is the mind that dictates our sexual preferences, not our confused genitalia. My shocked friend asked me two very valid questions: "Does the fact that my vagina is lubricated mean that I was excited?" and "Does that mean that I actually like women a little bit?" The second question is easier to answer: no. Many times our genitals react on their own, and it is our convictions that have the final word.

When I finally met Meredith Chivers in person, during the International Academy for Sex Research conference in Portugal, she told me about another of her observations, outside of the field of agreement: women react genitally to a much, much wider range of sexual stimuli than men do. And among them are same sex erotic images. When she showed images of lesbian sex to firmly heterosexual women, many of them displayed increased blood flow to their vaginas independently of what their minds were saying.

Meredith assures me that the typical calendar photo of a hunk with a muscular torso generates much less genital response than seeing two naked women caressing each other. She says that perhaps it is empathy at seeing another woman who could be them excited, the greater erotic charge or that maybe they do actually have some attraction to women, but that her results are conclusive: "In heterosexual women what is important is the sensuality, not the gender of the protagonists of what they are watching." And she immediately clarifies that this doesn't imply a closeted bisexuality: "Just because something excites you physically, that doesn't mean you like it. The two things can go together, but not always," she explains, citing unconscious triggers, repeating that the key is the sensuality and the empathy, insisting that if someone is convinced of their sexual orientation, a genital reaction shouldn't make them doubt that, and that sometimes our genitals really can do their own thing and disregard our will, which is what counts.

That reminded me of the case of a friend who, in her first visit to a swingers' club, after several minutes in a corner watching all types of sex acts around her, she told the man with her that she felt uncomfortable, that she wasn't aroused at all and that she wanted to leave. She was shocked when she touched herself and found that her vagina was wet and both her labia and her clitoris were incredibly sensitive. "Textbook example of lack of sexual agreement," said Chivers when I told her about it.

Surveys and Statistics on Sexuality

Let's review: we've talked about hormones, the nervous system, muscles, brain activity, rat clitorises, dysfunctions and measuring the blood flow to the genitals. We still haven't discussed the evolutionary perspective on our sexual behavior and applied all this knowledge to concrete situations from the limited but always curious and informative scientific perspective. But none of this would have any meaning without a more sociological analysis of sexuality, and a multidisciplinary approach that tries to bring together all these perspectives and document the enormous diversity of sexual expressions in our private lives and in society.

Even though we began this chapter with the more physiological studies by Heiman and Janssen at the Kinsey Institute, in the offices of that internationally renowned center we find many other specialists who are studying sexuality from other disciplines. The evolutionary biologist Justin Garcia researches online dating, the explosion of hookup culture among youth, and patterns of unfaithful behavior. Stephanie Sanders investigates why people don't always use protection in risky situations and tries to find out which techniques she can recommend to men who lose their erections when putting on a condom or what messages are most convincing to each social sector. The Croatian sociologist Aleksander (Sasha) Stulhofer analyzes the prostitution situation in his country, teenage relationships, the increase in anal sex among heterosexuals, the impact of the Internet and the rise in risky behaviors. Bryant Paul objectively studies whether porn has negative effects on its consumers, and Liana Zhou is in charge of the impressive collection of printed material, recordings and videos of all kinds that Alfred Kinsey started gathering in the 1940s, and that has become one of the largest archives of sexual material worldwide. When strolling through the more hidden rooms of its library, paging through books and publications from years gone by, you feel like you are in a real academic museum of sex.

Crossing a couple of streams in the middle of the Indiana University campus, we reach the Center for Sexual Health Promotion where Debby Hebennick directs the research most akin to Alfred Kinsey's original work:

surveys focused on finding out how we think and act on a sexual level behind closed doors in our deepest intimacy. This is the type of study that most frequently reaches the mass media: statistics that reveal that oral sex activity is increasing among teenagers; what percentage of men or women practice sadomasochism; differences between cultures and social sectors; or what percentage of people over 60 have an active sex life. It is surprising to note two things: first of all, that until Kinsey's studies were published in the mid-twentieth century no one had made any systematic analyses documenting the diversity of human sexual behavior. Secondly, that Kinsey's work is still, after 60 years, one of the most comprehensive ever published. It is easier to get funding to study the behavior of apes than for studying human sexuality.

When I entered the office of the young, smiling Debby Hebennick at the Kinsey Institute my gaze was drawn to a plastic penis on her desk. I looked at it more carefully and found a metal plate on what would be the upper part of the shaft, about a third of the way up from the base. I was intrigued and asked her about it, feeling comfortable because of her friendly appearance and proximity in age. "Oh, yeah, some sex toy company sent that to us to see if we would try it out. That metal part gives off small electrical shocks and they want to see if that increases stimulation of the G-spot," responded Debby. "Wow, and what do you think?" "I don't know, we haven't tried it out yet. I have my doubts, but they say that in moments of extreme arousal it might help to reach orgasm. We'll see."

That's not the kind of study that Debby normally does, but it perfectly reflects the range of research projects and requests that the Kinsey Institute receives. In fact, Debby Hebennick's main job is as co-director of the National Survey of Sexual Health and Behavior (NSSHB) that, financed by the Trojan condom company, in 2010 published data about the sexual behavior of almost 6,000 Americans between 14 and 90 years old. It isn't one of the surveys with the largest sample size, but the rigorous selection criteria, follow-up and breadth of questions covered made it one of the most important done to date. The survey was carried out online and even though that could seem to make it less rigorous, Debby explained that they decided on that methodology because people lie more often on the telephone or in front of an interviewer, even when they are completely sure that their answers will be anonymous. The key is in selecting the participants well. Just posting a questionnaire online and waiting for volunteers to fill it out is not the same as beginning with well-defined groups of participants.

Having said that, statistics do not always offer relevant information, especially in a field such as the sexual, in which normality is such a vague concept. Saying, for example, that American men between 30 and 44 years of age have had an average of seven sexual partners over the course of their lives may feed our curiosity but it doesn't mean that it's uncommon to have had just 1 or 50. And it gives the feeling that we're missing something when the same study establishes that, in the same age range, women have only had four different sexual partners.

Statistics are particularly useful when they focus on a specific population that can be compared with another of a different culture, age range or social class, and especially if they show us aspects that were not previously well documented and might surprise us. For example, the NSSHB survey co-directed by Debby informing us that 71.5% of American women between 25 and 29 have masturbated over the past year doesn't give us groundbreaking information, but that 46.5% of women between 60 and 69, and 32.8% of those over 70 have does indicate that sexual activity in the elderly is much more frequent than is assumed in some circles, and that medicine should pay much more attention to their sexual health than it currently does.

Again, there are curious and not particularly relevant facts about the frequency of infidelity, that the higher one's level of education the more they masturbate, or that 1 in 4 men and 1 in 25 women have looked at pornography online in the last month, but knowing that 28.1% of men report having lost an erection when trying to put on a condom in one of their last three attempts is useful. This less obvious piece of data clearly shows that as an important aspect to keep in mind. It is also relevant that 10% of women have felt deep, inexplicable sadness after intercourse in the last month. Learning from the NSSHB that more women under 25 have practiced oral sex at least once than women over 60 indicates that oral sex has become a much more frequent practice in recent decades. The fact that the percentages of sexual practices between young men and women between the ages of 25 and 29 are not very different indicates that gender roles are less important than they once were. And that 85% of men say their partners reached orgasm during their last sexual encounter, but only 64% of women say they did, should leave us very, very worried…

We will offer more data on fantasies, anal sex, the use of vibrators, paraphilia and genital size in specific chapters, but it is worth reviewing here some of the main surveys on human sexual behavior that have been made. Alfred Kinsey's work in the 1940s with nearly 18,000 personal interviews has some failings, such as the lack of representation of African-Americans and various social classes, and some biases that mean the data isn't reliable enough to establish comparisons with contemporary data. This lack of knowledge about the past is one of the limitations of the study of sex. But, without a doubt, Kinsey's data was revolutionary and his work revealed a new image of the sexual intimacy of Americans. The strangest thing is that his work wasn't updated and expanded on for decades. It wasn't until the 1980s that the couple Samuel and Cynthia Janus made a detailed analysis of almost 3,000 questionnaires on sexuality to publish their Janus Report. In it, they indicated that, since the sexual revolution, the age at first coitus had decreased significantly, that sexuality had also extended further into adulthood, and that extramarital sex and practices such as fetishism and sadomasochism had become more tolerated. The Janus Report has its detractors, but it is still considered the most exhaustive study since Kinsey's.

Obviously, the United States is not the only place where these surveys have been carried out. In the late seventies Swedish researchers made the first broad survey on sexuality on the European continent, repeating it in 1996 to compare results. Finland also began to make periodic surveys in 1971, eventually becoming one of the best points of reference for analyzing the changes in sexual patterns during recent decades. It showed, for example, that in 1971 very few women had more than ten lovers over the course of their entire lives, but in the early 1990s the percentage was 20%. Another essential point of reference is the British survey made in 1990 to almost 20,000 people, which was repeated with a smaller sample in 2000 and 2010. Among many other things, that survey showed a progressive increase in oral and anal sex among heterosexuals. In 2003, in Australia, researchers Richter and Rissel published an extensive work based on 19,307 telephone interviews with Australians on sexual function and behavior that generated 18 scientific articles on specific subjects, and in many other regions they have done smaller, more specific studies on concrete themes. For example, it has been seen that in Latin America the rates of sexual dysfunction are not very different from those in the United States and Europe, and that anorgasmia in women in Southeast Asia (Indonesia, Malaysia, Philippines, Singapore and Thailand) is at 41%, while in Southern Europe (France, Israel, Italy and Spain) it is at 24.

This last piece of data comes from the Global Study of Sexual Attitudes and Behavior headed by sociologist Edward O. Laumann at the University of Chicago, one of the few efforts to bring together various surveys and offer a global perspective on sexuality. Laumann gathered data from 13,882 women and 13,618 men between 40 and 80 years old in 29 countries and was able to make countless comparisons, many of which surpass the intentions of this book. Perhaps most interesting was that he did not find big differences in sexual dysfunctions among the cultures analyzed. While he observed a slightly higher proportion of premature ejaculation in men from Mexico and Brazil (the only Latin American countries represented), very high levels of sexual dissatisfaction and erectile dysfunction in Southeast Asia, age and health continued to be a much more determining factor than nationality in terms of male physical dysfunctions. In women, the sociocultural aspects seemed to play a much more important role, since careful examination of the tables showed that lack of sexual desire is practically double in Southeast Asian and Middle Eastern women (43.4 and 43.3%) than in women from northern Europe (25.6%).

There are countless surveys that we will cite throughout the book in their appropriate context. Some of the more global ones compare tendencies and other more specific ones analyze concrete sectors of the population. But if there is one study devoted to documenting human sexual conduct that is essential to our discussion, it is Alfred Kinsey's, one of the pillars of modern sexology that serves as a brief review of the history of scientific research on sex.

A Brief History of Scientific Research on Sex

From the first cave paintings to the art of ancient civilizations and works like the *Kama Sutra*, eroticism has had a constant presence in all human intellectual activity. And from Plato to Foucault, the not strictly scientific contributions to the understanding of human sexual conduct are endless. Ancient Greek philosophers were already speculating on the ethics of sexual behavior, the causes of dysfunctions and the mechanisms of reproduction. Aristotle was correct in observing that some animals reproduced sexually and others asexually, but he was wrong in his belief that all the information needed to father a child came from semen and the mother was merely a receptacle that contributed nutrients. Perhaps due to that Aristotelian influence, when in the seventeenth century the first microscopes allowed us to observe spermatozoa, they drew them as tiny men with miniature people inside, which would be deposited into a mother's uterus where they just grew until birth.

Aristotle represents a tremendous revolution in the history of human knowledge, for having attempted to understand the world through observation and the rational interpretation of nature without resorting to gods or supernatural forces. But Aristotle is a starting point for science, not a destination. The scientific method didn't appear until the incorporation of doubt and experimentation that put hypotheses to the test.

In the Middle Ages, and during the birth of modern anatomy, Leonardo da Vinci dissected corpses and drew sexual organs in works such as his famous couple drawn in a cutaway profile during intercourse. The fallopian tubes and the Cowper glands are named for anatomists Falloppio and Cowper. And curiously it wasn't until the sixteenth century that the existence of the clitoris as an organ of pleasure was anatomically described. In the more physiological aspect (anatomy is structure and physiology, function) there were pioneers such as the Italian doctor and anthropologist Paolo Mantegazza, who in the midst of the repressive Victorian era carried out experimental studies in which he measured blood flow and temperature during erection, transplanted frog gonads, and wrote medical essays on masturbation, infertility and both male and female sexual dysfunction. It is interesting to note that, while he was one of the fathers of modern sexual medicine, he never used the term sexuality to refer to sexual relations, always *amore*.

In fact, the most psychological study of sex entered a new era in mid nineteenth-century Germany, partly in response to that chaste Victorian period in which gynecologists examined their patients without looking at them. An indispensable work of that era is *Psychopathia Sexualis*, by Dr. Richard Freiherr von Krafft-Ebing. Published in 1886, the book documents 237 cases of pedophilia, sadism, exhibitionism, transvestism, necrophilia, coprophilia and other sexual deviations described in impressive detail: if we flip through the book, on page 269 we find the testimony of a 37-year-old fetishist who acknowledges: "From my earliest youth I have always had a deep-rooted partiality for furs and velvet, in that these materials cause me sexual excitement, and the sight and touch of them give me lustful pleasure. I can recall no event that caused this peculiarity [...]; in fact, I cannot remember when this enthusiasm began." On page 517 there is a description of an exhibitionist who kidnapped girls as they were leaving school, tied them up and forced them to look at his genitals. He was in and out of prison several times and continued to commit these crimes. Case 229, on page 562, is that of a man who was caught penetrating a hen; his defense before the judge was that his genitals were so small that sex with women was impossible. Case 230 is that of a 16-year-old boy who committed bestiality with a neighbor's goose, arguing that it didn't do any harm to the goose, and curiously, on page 199, a 35-year-old woman is classified as hypersexual just for frequent masturbation and her desire to be dominated by another woman after suffering a heartbreak. The vast majority of paraphiliacs are men, including murderers and rapists, and it is said that the character of Jack the Ripper was inspired by one of these cases.

A controversial part of Krafft-Ebing's work was that he initially included homosexuality as a sexual psychopathy, but later came out against the belief of the time that homosexuals were mentally ill. He did consider rapists sick, and he was one of the first to talk about clitoral orgasms and female pleasure.

Sexology as an independent discipline was only born little more than a century ago, in Germany. One of the great pioneers was the Berliner Magnus

Hirschfeld (1869–1935), who, in addition to founding the first gay rights organization (illegal at the time), in 1897 he created a committee for the scientific study of homosexuality, and in 1908 he established *Zeitschrift für Sexualwissenschaft*, the first scientific magazine devoted exclusively to the study of sex. In fact, it was another German, Iwan Bloch, who coined the term "sexology" (*Sexualwissenschaft*) a year before, in 1907, when he published the book *Das Sexualleben unserer Zeit* ("The Sexual Life of Our Time") in which he proposed an interdisciplinary study of sexuality. In 1914, Bloch would create, along with Albert Eulenburg, the Medical Society for Sexual Science and Eugenics in Berlin. Those were the first steps toward the establishment of sexual research as an academic discipline, which would gain importance in 1921 when the German capital celebrated the First International Sexology Conference.

Germany had a large community of the first sexologists, but then a dark era arrived. Hirshfeld and most of his colleagues were Jews and all their work was destroyed with the rise to power of Hitler and National Socialism in 1933. Many researchers went into exile and that meant the end of the trend toward studying sexuality initially led by the Germans.

We cannot continue without mentioning the important contribution of the Austrian Sigmund Freud (1856–1939). Despite the many criticisms of his unscientific approach and his wild exaggerations, Sigmund Freud got some things right and some things not so right in his subjective interpretation of the functioning of the human psyche. Actually, the problem isn't so much his errors and lack of methodology as the syllogism that later led many to defend Freudian theories based only on the principle of his authority. Freud was resoundingly wrong about the idea of women's castration complex, his interpretation of frigidness and his consideration of vaginal orgasms as more mature than clitoral orgasms. He undoubtedly exaggerated the exorbitant role he gave to the libido and Eros in our conduct. But he also contributed greatly by revealing the importance of the mind's unconscious processes and the development of a psychosexual theory that includes childhood sexual experiences. He was correct about the significance of childhood traumas, but he gave them too much importance and thought that everything could be solved by searching out repressed memories of the past. While psychoanalytic therapy is openly criticized by current medical science, it has evolved and is still often used in the treatment of sexually related disorders.

Another very important figure of the era was the Englishman Henry Havelock Ellis (1859–1939). In 1886 Havelock Ellis published *Sexual Inversion*, where he argued that homosexuality was not abnormal and he linked it to artistic and intellectual success. After a life dedicated to the study of sexology, his six volumes (which were initially banned) of *Studies in the Psychology of Sex* were published. In them he spoke of sex as an act of natural, healthy love. Let's not forget that in that period the medical community still considered masturbation dangerous, even going so far as to perform vasectomies on "patients" who suffered from excessive onanism or ablation of the clitoris of women with too much sexual desire.[1]

At the same time, more biological investigations into the role of sex hormones was moving forward,[2] and by the mid-twentieth century Harry Benjamin introduced the term transsexuality and carried out the first hormone treatments on people with male bodies who felt and identified as women. Ernst Gräfenberg described more intense pleasurable zones on the vaginal wall, female ejaculation, and developed the first IUD. The urologist James Semans studied the phenomenon of premature ejaculation and the Italian Giuseppe Conti examined the structure of the erection and the role of the arteries in the corpus cavernosa. John Money would make the distinction between sex and gender and delved into aspects of identity and sexual orientation. The number of sexologists grew around the world but the undisputed Einstein of sexuality, at least in terms of media repercussion and social impact, was the controversial researcher Alfred Charles Kinsey.

The Kinsey Revolution

Imagine living in the 1940s, tormented by a sexual fixation on women's shoes, thinking that you are the only being on the planet with that perversion, and then the book *Sexual Behavior in the Human Male* is published in 1948 explaining that you aren't an isolated case, but rather that there are many others with this same obsession. For many people with "inappropriate" behaviors or fantasies this was liberating. But imagine the commotion caused

---

[1] The origin of this view of masturbation as a practice with negative health consequences dates back to 1758, when Swiss doctor Simon Auguste David Tissot (1728–1797) published *L'Onanisme, dissertation sur les maladies produites par la masturbation* [Onanism, a dissertation on the diseases produced by masturbation]. Many teenagers have been raised with pressures, usually of religious origin, based on the supposed and never scientifically confirmed negative effects of masturbation.

[2] Indians, almost 2,000 years ago, already attributed impotence-curing properties to the ingestion of testicles. In the Middle Ages the simple observation of eunuchs reflected the obvious fact that the testicles had a role in sexual desire and male characteristics. In the mid-nineteenth century, the German Adolph Berthold transplanted testicles between animals until suggesting that some substance was contributing to male development, and at the end of the same century, the Frenchman Charles-Édouard Brown Séquard injected himself an extract made of animal testicles and documented how they brought back his sexual desire. In the 1930s testosterone was identified as the primary androgen, leading to all the current research into its influences on embryonic development, its role in sexual dysfunctions, its decrease during andropause, its role in female sexual desire and how the birth control pill interferes with its production.

when a researcher exposed to the conservative society of the period that there are many married men with homosexual desires, that there are multi-orgasmic women and that infidelity is extremely common. That is the revolutionary contribution of Alfred Kinsey to the science of sexuality: just as astrophysicists reveal that the universe contains many more heavenly bodies than the stars we can see with the naked eye, Kinsey revealed that the reality of common people's sexual behavior is much, much more diverse than was previously believed.

Alfred Kinsey was born in New Jersey in 1894. He earned a doctorate from Harvard in 1919, and during his first years as an entomologist at Indiana University he traveled more than 50,000 km, collecting hundreds of thousands of insects. He was a compulsive researcher, and he focused all his studies on the phylogenetics and behavior of one species of wasp. But in 1938 his scientific career took a radical turn when the president of Indiana University, Herman Wells, assigned him to teach a course on sexuality and marital conduct.

While preparing for the class, Kinsey was surprised by the enormous lack of scientific information on sex. He and countless zoologists were making all sorts of studies on animal ethology, yet the ignorance of human sexual conduct was absolute. It was inconceivable. Kinsey began to distribute anonymous surveys to his students, and was surprised to see that the rates of homosexual desire, female masturbation and traumatic pasts were much, much higher than what the academic texts stated. There was a huge void of information, and someone had to be the first to try to scientifically document human sexual behavior.

Financed by the Rockefeller Foundation, Alfred Kinsey recruited a team of researchers who began to interview couples and individuals about different aspects of their sexual practices, experiences, beliefs and fantasies. The diversity was overwhelming. Kinsey and his team traveled all over the United States until they had completed the more than 18,000 interviews that became the basis of the revolutionary and polemical *Sexual Behavior in the Human Male* (1948) and *Sexual Behavior in the Human Female* (1953).

Kinsey established that homosexual practices were frequent even among married men, he was the first to give statistics about women who never reach orgasm and he documented an enormous diversity of paraphilias, physical problems, extramarital sex, techniques of female masturbation and many other aspects that are not surprising to us today but that 60 years ago no one had scientifically documented.

Perhaps the best-known facet of Kinsey's work was the "Kinsey scale" of homosexuality. He himself was married but felt some attraction to men. He was heterosexual, but not 100%. That led him to create a scale in which sexual orientation was measured as a continuum and not as something static with two perfectly discrete categories. A 0 on the Kinsey scale was a man who was only attracted to women and a 6 is one who defines himself as absolutely gay. But a man can be a 2 or a 3 if he feels heterosexual with some desire toward men, and a 5 or a 6 if it's the other way around. While today the Kinsey scale is considered too basic and fails to reflect all the different identities, on a conceptual level it was an incredibly important milestone. In fact, in many sociological surveys in which people are asked to classify themselves as heterosexual, homosexual or bisexual, if the categories include "almost always homosexual" or "almost always heterosexual," many responses fall into one of those definitions.

After publishing his controversial works, Kinsey set up a sort of clandestine laboratory in the attic of his home. There, he and his collaborators observed couples having sex and masturbating, and they measured physiological aspects such as the change in the color of vaginal lips during excitation, the dilation of the pupils in orgasm or the ejaculation distance of semen. Kinsey's biographers explain that at that point he was almost on the edge of obsession, and that was when his work was interrupted by pressures that forced the Rockefeller Foundation to withdraw all their financing. Kinsey's work describing, for the first time in such an exhaustive and precise way, the sexual conduct of regular humans was too provocative for the time. But his legacy had already reached society.

Masters and Johnson

Kinsey's dream was finally materialized in the 1960s and 1970s by the gynecologist William Masters and his wife Virginia Johnson. Masters and Johnson built a real sexual laboratory through which, over the course of 11 years, 328 women and 312 men volunteers—both homo- and heterosexuals—visited to take part in almost 11,000 sexual encounters. On many occasions the volunteers were assigned to copulate with another volunteer who was a stranger. Located in the city of Saint Louis and using everything from transparent vibrators to all sorts of medical instruments, Masters and Johnson measured countless anatomical and physiological aspects of sexual response and then published their *Human Sexual Response* and *Human Sexual Inadequacy*. While Kinsey's work was revolutionary for its discoveries on human sexual behavior, Masters and Johnson's was for its data on the physiology and pathology of sexual function. Among many other things, they documented that vaginal lubrication stems from the vagina and not the cervix; that the sexual response consisted of four phases: excitement, plateau, orgasm and resolution; that women were multi-orgasmic because they had no refractory period; that the first muscular contractions during orgasm happened every second and then slowed; that there was no age at which sexual ability necessarily disappeared, and that the only organ purely of sexual pleasure was the clitoris. In this sense, Masters and Johnson argued erroneously that the vagina was not very sensitive and they caused controversy when they designed a program to revert homosexuality. While their techniques were limited and at this point some of their information has been corrected, their work undoubtedly was a tipping point for conceptualizing sexual problems from a medical perspective. There are mistakes in their work, but modern sexual therapy owes a lot to Masters and Johnson's findings.

Other sexologists have made great contributions. Helen Singer Kaplan distinguished between physical and psychological arousal and included the initial phase of desire in Masters and Johnson's diagram of the four phases, and she is cited as one of those responsible for integrating psychiatry and medicine in the treatment of sex. There are many more names, but Alfred Kinsey, William Masters and Virginia Johnson, along with the first modern sexologists in late nineteenth-century Germany, are clearly the pioneers of the inconceivably recent science of human sexuality.

5

Sex in Our Beds

I remember the last evening of the International Society for Sexual Medicine (ISSM) conference, held in August 2012 in Chicago. I was having a conversation with an experienced psychiatrist who knew absolutely everything about the history of sexology, who had worked directly with Helen Singer Kaplan, had many scientific publications and extensive experience as a therapist in the clinic he himself ran in New York. He was one of the conference's big names. I confessed to him that I had mixed feelings about the event, because I had heard many discussions of sexual dysfunctions, hormones, psychological problems and medical treatments, but few on pleasure, satisfaction, the range of sexual expression or ways of improving intimacy within couples. He burst out laughing and told me, "Oh, of course! If you ask my wife she'll tell you I know nothing about sex!"

Honestly, I did have the feeling that the urologists, doctors and scientists at the conference had a very limited vision of sexuality, and that at least in the United States it was pretty focused on purely physiological aspects. When I asked a urologist about techniques for obtaining pleasure in the case of irreversible sexual dysfunction, he had no bones about replying: "Oh, I don't know, that's a matter for sexologists." These phallocentric urologists are devoted only to making sure the machine works, which is no small feat actually. I remember the well-known Argentine urologist Edgardo Becher, outgoing president of the ISSM, explaining to me how years earlier, when Viagra appeared, a patient told him, "But doctor! How is it possible that after 6 years on your couch you solved it all with a pill?" Edgardo himself insisted that, at his clinic in Buenos Aires, his patients always saw a psychologist in addition to a medical doctor, and a Venezuelan sexologist who was also taking part in the conference told me: "Here in the United States they know a lot about sexuality, but nothing about sensuality."

Perhaps the most irrefutable proof of that was seeing that most of the commercial stands at the conference were hawking testosterone supplements against andropause, surgical tools, injections and implants against erectile dysfunction, but curiously the stand with the most visitors milling around it was the one for Lily brand female vibrators. Even though they were only offering the same models as ever, just nicer looking, the doctors seemed to find them the most groundbreaking thing at the conference, based on how much attention they paid to them. As such in this chapter, without ever putting science on the back burner, before we begin talking about sexual dysfunctions and problems, we will review the less orthodox research designed to comprehending and improving sexual pleasure and experience. And what better way to start than by seeing whether science can finally shine some light on the mysterious G-spot, which everybody talks about but few ever get to really know.

Even Scientists Have Trouble Finding the G-Spot

Pilar doesn't know if she has a G-spot or not. What she does know is that the most sensitive part of her vagina is towards the front about 3 or 4 cm from the entrance. In fact, she loves it when at peak excitation during intercourse her boyfriend pulls out his penis partway and rubs the glans right in that area. Occasionally that is how she reaches her best orgasms. In addition, unlike when she has an orgasm through solely clitoral stimulation, with that localized pressure she sometimes releases a little bit of liquid. She has talked about it with some of her girlfriends and it doesn't happen to all of them. The female sexual response is incredibly diverse.

Pilar does have what—beyond academic and terminological discourse—we all understand as the G-spot: an area on the vaginal wall that is notably more erogenous than the rest. There are those who still say it is a myth, but according to an American survey, 84% of women say that they believe they have an area more or less localized in the inside of their vagina that generates more pleasure when touched. Scientists are disconcerted because they have been carrying out studies for years and have yet to find any trace of any differentiated anatomical unit (for example a greater concentration of nerve endings, a physical feature or a small organ in the vaginal wall) to which they can attribute this greater sensitivity. That has led more scrupulous researchers to declare that the G-spot is an invention spread virally by the media.

Beverly Whipple,[1] the sexologist who coined the name G-spot in the early 1980s, categorically disagrees: "Of course it exists!" says Beverly when I mention this skepticism. "Maybe not in all women, but in most." Retired but still very active in campaigns to promote women's sexual health and satisfaction, renowned Beverly Whipple tells me: "We have never claimed that the G-spot is some sort of organ or different anatomical unit, just that it is a point of maximum sensitivity that, when stimulated correctly, can bring about orgasms very easily. And there is no doubt that it exists."

In fact, it was in her first physiological studies in the late 1960s that Whipple and her collaborator John Perry observed that many women said they had this more erogenous zone in the front part of their vaginas. Searching for medical bibliography on the subject they discovered that in 1950, the German doctor Ernst Grafenberg had already suggested the existence of an area 1 or 2 cm thick that, according to his texts, "induced orgasms through direct mechanical stimulation." Whipple named the G-spot for Doctor Grafenberg, and in 1982 she published the best seller *The G Spot and Other Discoveries About Human Sexuality*.

The *Kama Sutra* and other Indian texts from the eleventh century already mentioned more sensitive spots inside the vagina, but it was through Whipple's book that the G-spot began to spread in popularity in the West, with initial reactions that ran the gamut, including total bafflement, media hype, and the creation of new vibrator designs.

Initial skepticism among sexologists was due in part to the fact that the existence of this G-spot seemed to contradict the findings of Masters and Johnson, who are held in such high regard in the field. William Masters and Virginia Johnson, with their detailed physiological research, established that all female orgasms required stimulation of the clitoris, and that the vagina was not a particularly sensitive organ. This concept was embraced by the feminist movement, partly because it debunked the wrong-headed idea that had been around since Freud that not having orgasms through penetration was a sign of sexual immaturity. With Masters and Johnson the clitoris had become the almost exclusive center of female pleasure, but the arrival of the G-spot brought the debate between vaginal and clitoral orgasms back into the forefront, and as incredible as it may seem to all the women who clearly

---

[1] Beverly Whipple is another example of a scientist who began researching sexuality when she realized the gaping hole in medical knowledge on the subject. She explains that in the sixties one of her students asked her what to recommend in terms of sex for a patient with cardiovascular disease. Whipple didn't know what to answer; she asked her medical colleagues and they were stumped as well; she searched out bibliography and decided to start her career as a sexologist when she found that, as outrageous as it may seem, no one in the medical community had taken the time to do rigorous research on whether the sexual act was beneficial, dangerous or neutral for those with heart disease.

distinguish between them, three decades of investigation later there are still discrepancies among the scientific community.

In 2012 the *Journal of Sexual Medicine* published a review of all the scientific literature to date on female ejaculation, vaginal orgasm, the G-spot and the physiology of female genitalia, and after analyzing 96 scientific articles the authors concluded that:

> the surveys found that a majority of women believe a G-spot actually exists (…) but objective measures have failed to provide strong and consistent evidence for the existence of an anatomical site that could be related to the famed G-spot.

And they add:

> However, reliable reports and anecdotal testimonials of the existence of a highly sensitive area in the distal anterior vaginal wall raise the question of whether enough investigative modalities have been implemented in the search of the G-spot.

In other words: scientists might have been looking for the G-spot in the wrong place. And believe me, they've been trying…

Masters and Johnson Were Wrong. Vaginal Orgasms Exist

The whole G-spot controversy is almost comical. Perhaps the confusion stems from the word *spot*, that makes us think of some sort of magic button and makes physiologists think of an area in the vagina that has "something" different from the rest. They've been searching for that "something" for years.

First they carried out anatomical analyses to see if there was some sort of organ, gland or independent structure that could be identified as the G-spot. But no, there is no specific organ responsible for that pleasure in the way that there is a prostate, taste buds, and an eardrum. Next they performed histological studies to determine if that area on the anterior wall of the vagina has more nerve endings than the rest. But they found nothing; there doesn't appear to be a point in the vagina that has a higher concentration of sensory nerves, like in the glans or the clitoris. Then some scientists suggested that the Skene glands, structures which are common to the male prostate and involved in female ejaculation, could perhaps be stimulated from inside the vagina and be responsible for that greater sensitivity. But physiological studies also ruled that out because of a lack of sufficient sensory receptors. Finally in 2008 a French scientist named Odile Buisson came up with the most plausible hypothesis to date: she published several studies with ultrasound images that documented differences in thickness of the urethrovaginal space between women who had orgasms from contact with the G-spot and those who didn't. Buisson's interpretation is that, when exerting intense pressure on the anterior wall of the vaginal, if the urethrovaginal space is thinner, you can reach the internal structures of the clitoris, and that indirect contact with the internal clitoris would be what produces the intense pleasure.

This hypothesis is extremely interesting. On one hand it would explain why there is a more sensitive area in the vagina, but on the other it would imply that the G-spot, strictly speaking, doesn't exist as such. There wouldn't be women who "have one" and others who didn't, it would simply be an area from which the internal part of the clitoris could be reached and, doing so would depend on your urethrovaginal space, your lover's skill and other diverse factors. In fact, one interpretation of Odile Buisson's results is that the exclusively vaginal orgasm does not exist and that actually all orgasms are clitoral, whether through external or internal stimulation.

Taking into account that the clitoris is a much larger organ than what can be seen from the outside, as a hypothesis it isn't farfetched at all. The penis and the clitoris come from the exact same embryonic tissue and development; it's just that the penis extends outside of the body while the clitoris develops internally. It's as if the external part of the clitoris were equivalent to the glans of the penis and the internal part were the corpora cavernosa that, in the case of the clitoris, are separated internally and border the vagina, swelling with blood in the same way when a woman is aroused. This swelling, accompanied by adequate stimulation and facilitated by a thin urethrovaginal space, would be what allowed you to reach the G-spot, which would actually be the internal body of the clitoris.

Obviously direct stimulation of the clitoral glans and penetration feel different, but as men we also notice a difference between stroking a part of the glans and putting strong pressure further down. For many critics of the G-spot and defenders of the clitoris as the sole pleasure organ, that would mean an end to the myth of the vaginal orgasm and would give the clitoris back its primordial role established by Masters and Johnson in their physiological studies. It makes sense, but the problem is that there is much other proof that does distinguish between vaginal and clitoral orgasms.

In addition, all the surveys establish that a very high percentage of women say that they do have a G-spot and feel vaginal orgasms very differently from clitoral ones. This doesn't invalidate Buisson's hypothesis, but in studies in which thermic sensors were introduced into the vagina they observed greater sensitivity on the front wall than on the back without needing to exert any pressure that would reach the clitoris. Another theory has been put forth by the Italian Emmanuele Jannini, who seems to have identified a thicker area with a higher density of nerve endings in the anterior vaginal wall, which instead of the G-spot he prefers to call the clitourethrovaginal (CUV) complex, but which according to him could constitute an independent unit and become a new physiological candidate for the G-spot. In any case the most conclusive data in favor of the exclusively vaginal orgasm is that of Barry Komisaruk's group, with functional magnetic resonance scanners that show that two different areas in the sensory cortex of a woman's brain activate if her external clitoris is exclusively stimulated, versus the inside of her vagina. As we mentioned in earlier chapters, this reflects the fact that there are different nerves—pudendal, pelvic, hypogastric and vagus—that send different signals to the brain from the clitoris than from the vagina, and would explain why some women can reach orgasm from stimulation of only the vaginal nerves. The clearest evidence of this are women with spinal injuries whose pudendal nerve (of the clitoris) is damaged and still experience the orgasmic reflex.

I vividly remember Barry explaining to me how shocked he was the day a woman with a spinal injury began to cry tears of joy after the experiment when she discovered, 3 years after her accident, that she could still feel orgasms if she stimulated herself in the deepest part of her vagina, near her cervix. If we add to that the fact that there are women who have orgasms just from caressing their nipples, neck or even just with their imaginations, there is nothing that suggests the clitoris is the only organ responsible for orgasm. In fact, ultimately, orgasms are produced in the mind and not in the genitals.

The Two Types of Female Ejaculation

I met Micky at the Bagdad, a porn club in Barcelona. He told me that he's an expert in squirting: in making women ejaculate a huge stream of liquid in the moment of orgasm. He assures me that it's very easy, and that most every woman can do it. He has taught the technique to a lot of porn actors. It entails pressing very hard with two fingers on the G-spot, maintaining an intensity that is difficult to reach with your penis during intercourse. He told me that the most important thing, above all, is to not let up the pressure when the woman reaches orgasm, that pressing on top of the pubis can help, and that that's when the explosion of liquid and pleasure takes place. According to him, when women masturbate they don't expel so much liquid because right when they orgasm they usually relax the stimulation. I told him that seeing is believing, and he suggests we try it with a friend of mine if I want to. But strangely enough, none of the ones I asked were willing to try. Something tells me that if it were reversed and I was looking for male volunteers to experience a new sexual technique with an expert in adult films, I would have had much more success…

Anyway, the day after I saw Micky I met up with a friend of mine and halfway into the conversation I casually brought up the subject of female ejaculation. People are much more open to talking about their sexuality than we imagine, and she told me straight out that it had happened to her twice in her life. Once was with an ex-boyfriend she had dated for more than a year, but with him she only ejaculated once and she hadn't known how it happened. She thought that it was urine and she was embarrassed, but it was definitely a more intense orgasm. The second time was again totally unexpected and with someone she was having casual sex with. My sampling is insufficient, but it coincides with what the more elaborate surveys suggest: most women have never squirted, many have done it on a few isolated occasions and a few do it frequently. I have never found reliable published percentages.

Micky and other experts in female ejaculation that I've spoken to, like Tristan Taormino, a director of porn for women, and José Luis, also known as Ninja Squirt, assure me that in large part it's a question of technique.

But let's make a very, very important digression: independent of vaginal lubrication, there are two different types of female ejaculation during orgasm. Both ejaculations are expelled through the urethra, but the first is a sort of whitish, thick liquid that begins to gush out moments before the climax, and the second is what is called squirting, a much thinner liquid that is expelled in large quantity and force at the moment of orgasm itself.

The first type of ejaculation is better observed during masturbation, since during coitus it can pass unnoticed as part of the lubrication. In fact, there are those who claim that during orgasm there is always at least a slight expulsion of liquids but that, depending on the contractions of the orgasm, it very often ascends towards the bladder or the inner part of the vagina. Squirting is very different. Much less frequent, and popularized in porn films (sometimes real and sometimes faked), it is the sudden, abundant expulsion of liquid that resembles urine. Porn actresses say that it isn't, but physiologists cannot find any other receptacle for storing so much liquid other than the bladder.

The controversy over whether it is piss or not is so common that even some scientists have suggested analyzing whether both ejaculations are really separate phenomena, and if one responds to temporary urinary incontinence or if it is just a question of volume and dilution. After all, that's easier to measure than the G-spot.

In the late 1990s, sexologist Gary Schubach carried out the following experiment: he recruited several women who said they expelled a large quantity of liquid during orgasm, and introduced a thin catheter in their urethras before asking them to masturbate. If the liquid was urine it would come out through the catheter; if not, it would be outside of it. Schubach observed that in the cases of copious amounts the liquid did come through the catheter, but that another fluid of a different texture also appeared outside of the catheter. When he analyzed the second fluid he saw that it contained many substances similar to male semen, including enzymes secreted by the male prostate such as prostatic acid phosphatase. The liquid seemed to come from some glands located beside the urethra called "paraurethral" or "Skene's" glands, whose origin is homologous to the male prostate and which Schubach gave the name "female prostate."

In another research study published in 2011, the Italian E. Jannini and the Mexican A. Rubio analyzed in detail the expulsions of a 43-year-old woman who was able to produce both liquids. Comparing it with her own urine collected in the mornings, the scientists confirmed that the abundant transparent liquid that was squirted out contained urea, uric acid and creatinine, and as such was mostly diluted urine from the bladder, and that the thicker, more translucent substance was chemically different.

Since then, the few studies that have analyzed the nature of female ejaculation coincide with this distinction found between the ejaculation produced by the Skene's glands and squirting, which is a release of highly diluted urine as a result of muscular relaxation.

I met up with Emmanuele Jannini both in the International Consultation on Sexual Medicine in Chicago and in the European Sexology Conference in Madrid in September of 2012. I asked him the big question: how come, if we already know that squirting is a release of urine, so many women (many of them porn actresses) say that it doesn't smell or taste or have the color of urine? Occasionally during the sex act there can be inconsistencies in the levels of a hormone called vasopressin, which among its other functions is antidiuretic and responsible for concentrating urine in the kidneys. In fact, it is the reason why after intercourse we are very thirsty and have a strong desire to use the bathroom. According to Jannini, perhaps in function of the intensity of pleasure or the duration of the act, if the levels of vasopressin fluctuate a lot it could generate a temporary situation analogous to diabetes insipidus, where urine production in the kidneys increases greatly. On these occasions, the bladder would accumulate an enormous amount of urine, which would be highly diluted and could be expelled uncontrollably during orgasm due to muscular tension. It would be so dilute that we wouldn't identify it as urine, but it really is liquid from the bladder that comes out through the urethra. This shouldn't make women uncomfortable, since many men aren't repulsed by that but rather fascinated, reminded of some of the erotic images they've seen on the Internet. There's a reason why many actresses drink large quantities of water before their scenes…

The Genetic Component to Female Multi-orgasmic Capability

I am having a drink with three female friends who have agreed to talk to me about different aspects of their sexuality. Patricia tells me that she is multi-orgasmic, that if stimulation continues after her first orgasm she can soon have another, and only rarely does she end with one so intense that she feels she has to stop. Eva says that she rarely has more than one orgasm, that when she masturbates she feels satisfied and doesn't usually feel the need to continue, and that during intercourse with her partner she can continue if he hasn't climaxed yet but she is less excited and far from feeling about to repeat quickly. For her it's more like starting over from the beginning.

Megan, however, never has orgasms only from penetration. She says that sometimes she reaches orgasm in positions where it is easier for her to rub her clitoris against her partner's pelvis, or if during intercourse she stimulates her own clitoris with a small vibrator, but that it takes her a really long time and she's obviously never had a "vaginal" orgasm.

What is the origin of the differences between Patricia, Eva and Megan? Megan gets upset when they tell her "it's all in your head," or that she just has to relax and go with the flow. She feels that she's as liberated as Patricia, she's had several partners, her attitude towards sex is completely open, and she is convinced that it is a question of anatomy. When I explain to her that there are meditation therapies and exercises designed to focus attention on the genitals and increase sensitivity she says that that's interesting and that she will look for more information, but when I tell her about the studies Marie Bonaparte and Kim Wallen did, indicating that the distance between the clitoris and the entrance to the vagina could influence the ease with which you have orgasms through penetration, she replies with conviction: "I think it is something like that."

We are not going to reduce sexual sensations to hormonal levels or the distance between the clitoris and the vagina. Sexual pleasure is undoubtedly affected by trust, relaxation, satisfaction with your own body, degree of desire, mutual understanding with your lover and many other factors.

The new biopsychosociological paradigm of sex understands human sexuality as an interaction of behavioral, cultural, educational, psychological and developmental factors; but also physiological, neurobiological, anatomical and endocrine ones. And, though this may surprise you, even genetic ones.

Identical Twin Sisters Look More Alike than Fraternal Ones. Their Orgasms Are More Similar Too

Studies with twins have long been a wonderful tool for identifying whether a certain feature has a greater or lesser genetic component. The basic idea is that if attention deficit and hyperactivity—for example—have more of a correlation between identical twins, whose genes are 100 % the same, than with fraternal twins, who only share 50 % of their genes, then the disorder will have a certain genetic component. If, on the other hand, the correlation between identical and fraternal twins', say, musical talent is the same, then it would indicate that it is something primarily cultural or educational.

Researchers have huge databases of twins, such as TwinsUK, that they have analyzed for things ranging from a predisposition for cataracts and high blood pressure to phobias or dietary preferences. So British scientists Kate Dunn and Tim Spector were surprised when they realized that, given the notable differences in female orgasmic frequency, no one had ever explored genetic or family history aspects related to orgasm.

The methodology was simple. Dunn and Spector sent a questionnaire to 3,654 pairs of twin sisters between 19 and 83 years of age, asking them how often they reached orgasm by penetration alone, how often with a partner but with additional physical stimulation and how often during masturbation. They didn't all respond, but they managed to collect data from 683 pairs of monozygotic (identical) and 713 dizygotic (fraternal) sisters. Their results confirmed that the ability to reach orgasm both during coitus and through masturbation did have a clear genetic component. They suggested that genes could account for between 34 and 45 % of the differences in the population. How do those genes have an affect? It has yet to be investigated in detail, but it seems that they could be related to the predisposition for depression or anxiety, anatomical differences and even levels of prolactin (the hormone that influences sexual satiety and is released after orgasm).

In science, isolated studies are always viewed with caution, but an independent investigation carried out in Australia with 3,080 twin sisters offered very similar results: the frequency of orgasms was significantly more correlated in identical twins than in fraternal, even when controlling factors such as the number of parents, divorces, and socioeconomic and cultural aspects. The authors also speculated that this genetic characteristic could be associated with other character traits such as extroversion or lack of inhibition, and there are studies underway that analyze the genes involved in the metabolism of serotonin, vasopressin, estrogens and other hormones. No one is suggesting that genetics play a determining role, but the conclusion is obvious: a woman may not be able to reach orgasm during intercourse due to anxiety, stress, cultural inhibition or problems with her partner, but it may also be because she has a different physiology from other women who do reach orgasm easily. After all, many doctors are seeing patients who have wonderful relationships and no apparent psychological problem but are still worried about their lack of libido or their difficulties climaxing.

Personality, Socioeconomic Status and Sexual History Factors in Orgasm Frequency

Setting aside biological predisposition, countless studies have related sociocultural factors and life experiences with the development of female sexuality. One, for example, associated the age at first sexual encounter with the frequency of orgasms in adulthood, establishing that the younger a women was when first experiencing sex with penetration, the higher the proportion of orgasms during coitus she had as an adult. The sample was large, the study methodologically sound and the differences significant but, in science, the fact that two variables are statistically linked does not imply that one necessarily conditions the other, nor that there isn't a third influencing them both. This relationship between age at starting to have sex and orgasms in adulthood could be interpreted in various ways: one could think that full sexual activity during adolescence and teen years favors better sexual development, but also that having less orgasms in adulthood could reflect less desire in earlier stages, and there could even be a religious or educational factor conditioning them both. Establishing associations is interesting in order to come up with new hypotheses, especially in medicine, but to confirm them there must always be highly controlled epidemiological studies with a very large sample group that, obviously, in sexual subjects are complicated to carry out.

Nevertheless, in 2011, a multidisciplinary team of researchers published a large study in which 2,914 Australian women answered questions on their frequency of orgasming during sex with penetration and during masturbation, and they contrasted the data with socioeconomic characteristics such as education level and social class, personality characteristics such as impulsiveness, extroversion, political ideas or degree of neuroticism, and sexual development characteristics such as age at first coitus, libido, number of past sexual partners, predisposition to casual sex, marital status, or restrictive attitudes toward sex. They even considered factors like similar childhood illnesses or sexual fantasies about someone other than their partners.

Most of the results fell into the expected logic: political stances don't exercise the slightest influence on orgasm frequency, and having a higher libido increases orgasms both during intercourse and during masturbation. The women who had more fantasies outside of their relationship experienced significantly less orgasms during coitus but many more masturbating, and having restrictive attitudes about sex didn't affect the orgasms with a partner, but did decrease the solo ones. I should point out that when I use the word "significantly" I am saying that the results indicated important statistical differences, enough to establish that there is a correlation between both factors. Social class had no affect on orgasms during coitus, but the higher it was, the greater the frequency of orgasms during masturbation. In fact, the biggest differences were found in educational level: women with higher levels of education had less than the average number of orgasms during coitus and more than the average when masturbating. Having a neurotic or psychotic personality did not play any role, and both extroversion and impulsivity slightly increased the frequency of orgasms in coitus. Married women also had more orgasms with their partners, and very slightly below the average with masturbation. The number of past sexual partners didn't affect sex with the current partner, but it was associated with many more individual orgasms. And, as we said earlier, later initiation in sex with penetration was related to less orgasms in adulthood both during intercourse and during masturbation. Undoubtedly, several of these factors are interrelated, but the researchers highlight in their conclusions that they are actually more independent than had been previously believed.

Other studies have analyzed specific factors separately, and associated less orgasmic function to various illnesses, some mental disorders, dissatisfaction, hormonal alterations, and aging. And among healthy people, aspects such as emotional intelligence, positive attitude toward sex, variety of sexual practices, and the use of vibrators and lubricants represent a clear and significant improvement. This is still very far from solving the mystery of the 25–30% of women like Megan who have problems reaching orgasm during penetration, or the 10% that almost never achieve it even with masturbation. Individual therapy continues to be the most effective way to treat each particular case, but science is still revealing data that, in the hands of experts, will undoubtedly contribute to a better description and diagnosis of sexual diversity, taking into account factors ranging from the cultural to the genetic.

I Was Multi-orgasmic and Didn't Know It

Before writing this book, when I would hear people who practice tantric sex assure me that all men can control their ejaculatory muscles and be multi-orgasmic, the truth is I was pretty skeptical. I envisioned someone making love, ejaculating and continuing as if nothing had happened, and honestly I found it hard to believe. So when I interviewed a couple of tantric teachers, Mark Michaels and Patricia Johnson, one of the questions I wanted to discuss was that mystery of the multi-orgasmic male.

I broached the subject carefully, since I knew that tantric practitioners aren't thrilled about everybody being so interested in that particular aspect of their sexuality, which they actually consider secondary. In fact, sex is a minor facet of this philosophy born in India circa the fifth century, which has close links to Taoism, Hinduism, Buddhism and practices like meditation and yoga, and whose objective with partners is working on the emotional, physical and spiritual connection. In fact, in terms of the strictly sexual aspect, the goal of their long sessions of caresses and physical intimacy is not having a lot of orgasms, but accumulating a large amount of sexual energy that, according to them, invigorates the body and mind. They gather this concentration of sexual energy by remaining as long as possible in that incredibly intense state of arousal and extreme preorgasmic sensitivity, which seems to be a point of no return, but which they can control and expand so that when they do finally climax, the emission of energy is experienced as a release. Even if you decide to avoid orgasm, you can retain that sexual energy inside your body and harness it to feel much more vital. Interesting, very interesting. But, who am I kidding? What I wanted was to learn how to be multi-orgasmic.

To my shock I found out that I already was multi-orgasmic and didn't know it! Honestly, this was one of the great revelations I had while researching this book. I had been multi-orgasmic in the past, I just hadn't realized it. Allow me to explain.

The key is in that point of no return, that moment in men when we have an increase in genital sensitivity, we feel the immediate arrival of our orgasm, and we know that we will ejaculate automatically in a few seconds, even if we stop all stimulation. Yes, it is just that point that sometimes shows up by surprise sooner than we predicted and can be a real bummer. It has happened to pretty much all of us at some point.

Over time we learn to control rhythms, to know in what positions the friction doesn't excite us "too much," to use resources like thinking about the refrigerator, our boss, or recreating in our minds any image that decreases our excitation (yes, ladies, as pathetic as it may sound, many men resort to inhibiting thoughts to try—normally, unsuccessfully—to retard ejaculation). When all this fails and we feel we are approaching that point of no return, another option is to come up with a way to pull out, stop for a few seconds using some excuse like changing positions, constrict our pelvic muscles, feel some relatively pleasurable pulsations in the penis that may be accompanied by the release of a few drops of liquid, note a slight relaxation afterward, and return to coitus without losing our erection almost as if we were starting over again.

Well, that is being multi-orgasmic. The enormous difference between my pseudo-orgasmic pulsations and those of tantric practitioners is that they knew how to recognize them, appreciate them and enjoy the moment. During my covert orgasms I was struggling to hold back my pleasure and avoid ejaculation. They, on the other hand, had learned to regulate that process, to control the pelvic musculature responsible for ejaculation and be capable of living the moment more intensely, distributing the sexual tension throughout their body in what they call a "corporal orgasm." And obviously continue if they so desire. The difference between one experience and the other was vast, and this is what is so complicated to achieve and what requires training. It turns out being multi-orgasmic is not the hard part, but rather experiencing it as something extremely pleasurable without triggering ejaculation. I'm not here to talk about my life, but the day when after several attempts I suddenly felt those pulsations escape my genitals and travel up my body giving me shivers even in my arms, I thought I was feeling something that I had heard some women describe.

The tantrics talk about directing sexual energy toward the inside of the organism, distributing it through all the chakras of the body and impregnating their spirit. Which is fine, everyone uses their own language, and in the end we are all referring to the same thing. But what is clear is that we need mental control to focus our attention and sensitivity in other areas that are not the genital region, which we can train ourselves to do, and that there's nothing mystical about it. Orgasms without ejaculation are clearly more intense, but the body really remains in a state of greater sensitivity, you don't lose even a hint of your erection, you can continue with renewed energy, and the final orgasm in which you release everything is much more pleasurable. It's impossible to compare, but I have the impression that it must not be so different from the female multiple orgasm.

I have asked researchers and sought out scientific books and articles, but no one has been able to explain the physiological mechanism through which an orgasm after a marathon session of stimulation is felt much more intensely than a 4-minute quickie. The greater sensation of sexual satiety could be due to more release of prolactin after orgasm, and who knows if the intensity of the pleasure is because of a greater accumulation of dopamine, more extreme activation of the sympathetic nervous system, or some psychological effect such as suggestion. No scientist has ever thought to study what happens to tantrics during a 2-hour-long sexual encounter, or using them to investigate the separation between orgasm and ejaculation. They could surely offer interesting information, and it was a surprise to see that normative science had never even considered it. Actually, the science of sex is pretty conservative, and despite claims to the contrary, it is too influenced by the obsolete concept of "normality." Members of different sexual communities have more interest in science than the other way around.

I saw Mark and Patricia several more times at the Pleasure Salon that they organized the last Thursday of every month at the Happy Ending Lounge on the Lower East Side of New York. There, all sorts of sex-positive people gathered to share experiences and converse, from polyamorous couples to sadomasochists, burlesque performers, writers of erotic columns, people with diverse sexual orientations, and others interested in Tantra or other sexual expressions, or the wildly entertaining retired porn actor Big Joe, with his cane and his raspy voice telling the most hilarious anecdotes in the vein of "porn isn't what it used to be," because according to him before they didn't take pills and they didn't edit the images and the interaction with the actresses was more authentic. A real riot. But beyond that, I was met with absolute interest in my scientific contributions. I confess that I was expecting more negative reactions of the sort "What would science know about this!" but it was the exact opposite. Mark, in particular, had a profound interest in neuroscience and, in general, everyone felt that scientific information could complement some of their experiences. Also, understanding diversity can be a path to accepting it and not having such a closed-minded and normative view of sexuality.

The "Coolidge Effect" and My Envy of Men Without a Refractory Period

They say that in the late 1920s, the first lady of the United States, wife of Calvin Coolidge, was strolling through the gardens of her country home, and when she saw a couple of chickens going at it she asked the farmer who was with her: "How many times a day do they mate?" "Dozens!" replied the farmer, to which Mrs. Coolidge asked, "Is that so? Could you please inform Mr. Coolidge of that fact?" When the farmer told the president, he replied: "Do they always copulate with the same female?" "No, sir, they switch," answered the farmer, to which the president responded: "Is that so? Can you please inform Mrs. Coolidge of that fact?"

The "Coolidge effect" got its named from this anecdote. It refers to the fact that the refractory period of some males (the time needed between ejaculation and further sexual interest and erection) diminishes considerably when there is a new sexual partner involved. The Coolidge effect is very well proven in various animal species. If we leave a male rat in a cell with four females in heat, before long he will have gotten with them all and we will find him in a corner, trying to avoid any physical contact despite the females' constant insinuations. For a while he won't want to repeat the sex act with any of them, but if we suddenly put a new female rat in heat in front of him, you can be sure he will pounce eagerly on her. Obviously, there are no similar studies with humans (scientists say it wouldn't be ethical), and there aren't many men who can give an opinion based on their personal experience. Mark is one of the few who can. When I met him at a swinger party he told me that he always came with curious girlfriends, after warning them that he would give free rein to his desires. I talked to him about the Coolidge effect and when I asked him if he had found that to be the case for him, he thought about it and said that he could recall situations where he had had sex with a gorgeous woman who had wanted to go for another round and he had ask for some time to get it up again, but when a new woman—who wasn't even as attractive—showed up just then, he was able to get aroused immediately.

In order to defend the existence of the Coolidge effect in humans, researchers use indirect arguments such as showing men female faces and proving that they prefer novelty, often citing the loss of sexual interest in long-term marriages, and noting that in men that has more of an inhibitory effect than in women. They are all refutable arguments, but overall it does seem that the Coolidge effect does exist in humans, despite it never having been scientifically proven.

In addition, evolutionary psychologists say that both the refractory period and the Coolidge effect make perfect sense from the perspective of evolution. The real goal of sex is not pleasure and, under that premise, a male rat has no need to immediately repeat intercourse with the same female. It makes a lot of sense that natural selection has favored a refractory period that keeps males from wasting energy and sperm reserves uselessly, but at the same time they can get excited again if a different female in heat shows up. Making one of those non-experimental extrapolations that evolutionary psychologists are so fond of, the Coolidge effect could be evidence of males' innate tendency toward sexual promiscuity.

And it does seem to be neurochemically coded. Remember that the basic physiological functioning of the sexual response is not very different between rats and humans, and with rats we can do experiments where we inject them with different hormones in order to observe how the duration of their refractory period changes. One of the first hypotheses was that serotonin could be involved. Knowing that antidepressants that act by increasing levels of serotonin create erection problems, prolong orgasm and increase the time needed to have another erection after ejaculation, they carried out studies with rats that proved that decreasing serotonin could shorten the refractory period and vice versa. But that's not all. They also suspected dopamine and adrenergic hormones could be implicated. Adrenaline is the hormone that activates our metabolisms and vital signs, and dopamine does the same with our minds. Dopamine is the hormone of motivation and euphoria that we secrete when we suddenly see or imagine a stimulus we want to obtain, and it is perfectly possible that it could be secreted at the sight of a new female, contributing to a shortened refractory period. Again, experiments with rats confirm that the increase in both dopamine levels and noradrenalin reduce the time needed between erections. OK, but that still doesn't explain the existence of the annoying refractory period and that mysterious sudden loss of sexual interest and arousal after male orgasm (and in some case female orgasm too, as we will see a few paragraphs later).

Here is where prolactin comes into play. Aware that in situations of hyperprolactinemia there is a significant loss of sexual desire, and that prolactin is secreted in enormous quantities following male and female orgasm, several researchers thought that this hormone could play a crucial role in the regulation of the refractory period. In fact, German scientists experimentally proved that, an hour after sexual activity, both men and women had very high levels of prolactin only if they had reached orgasm, but following the same sexual activity without climax, their levels were unchanged. In order to prove this hypothesis, German researchers headed up by Doctor Tillmann Krüger asked a 25-year-old boy without a refractory period to masturbate several times in a row in order to analyze his prolactin fluctuations.

Because yes, some men have no refractory period. Allow me to digress here for a moment. There have been several points during the making of this book where I felt like Kinsey and Pomeroy, comparing my sexual discoveries with friends and even new acquaintances of both sexes, and taking note of their experiences. I've been surprised by the number of women who have come across micropenises, who have a female friend who has never had an orgasm, and the very small proportion of people who have never been unfaithful. But without a doubt the most unexpected moment of them all came one morning when I was with some good friends, the kind you share so much with that you think you know most all of their secrets. I was talking to them about the scientific article on the German guy who didn't lose his erection after ejaculation, when one of my friends shocked us by confessing that the same thing happened to him. If he wanted to, after ejaculation he could pull out, calmly take off the condom, go find another one, keep going until the next ejaculation and then repeat if he felt like it. If he stopped after ejaculation his member flagged like most men's, but if he wanted to he could maintain his erection with no problem. In fact, he sometimes ejaculated during coitus yet continued until his partner reached orgasm and then he would usually decide to stop. We were all impressed and we started treating him like our new idol. He assured us it wasn't a big deal and that for a long time he thought that it was normal, until one ex-girlfriend told him that he had some sort of a "gift." Putting aside petty male envy, I found, after discussing this with more people, that several told me they had come across men with this type of multi-orgasmic capability. There are no statistics on this topic, but it seems it's not that uncommon.

Going back to the study with the young German man published in 2001, in his case he only needed three short minutes after masturbation to obtain a full erection, and he said he didn't usually lose his sexual appetite. Scientists at the University of Essen wanted to investigate what was different in his organism, and they recruited nine volunteers of the same age and physical conditions, but who reported an average of 19 minutes between erections.

The procedure was very simple. With the help of erotic films, the ten participants masturbated several times at different intervals while the levels of prolactin in their blood was continuously measured. The results were conclusive. Not only did the young man without a refractory period masturbate more times (he managed to have two orgasms with ejaculation separated by only 2 min), the levels of prolactin in his blood were unchanged. However in the other volunteers, their prolactin levels increased substantially after the first orgasm, decreased a little bit 20 minutes later and rose again after the second orgasm. The researchers concluded that the secretion of prolactin induced by orgasm was one of the mechanisms involved in the refractory period. But there were motives for thinking that it couldn't be something so simple.

In an extensive review of scientific bibliography on the refractory period published in 2009 in *The Journal of Sexual Medicine*, Roy Levin argues that hyperprolactinemia doesn't always cause lack of desire, and that while the reports from doctors about individual cases are informative, it is risky to draw conclusions based on them. Levin believes that the role of oxytocin needs further investigation, and that the model should include a look at the loss of activity in areas of the brain such as the amygdala. He was surprised to find that no one had ever investigated the refractory period with fMRI scanners, comparing which specific areas activated during initial stimulation but not with identical stimulation following orgasm. I sent Levin's article to Barry Komisaruk, at the University of Rutgers, and it may be that as you read this lines I am volunteering as a guinea pig for that type of research.

The Refractory Period Also Exists in Women

But what surprised Levin even more was the complete lack of research on the refractory period in women. Of course there are many multi-orgasmic women whose excitement barely diminishes after climax, and if the proper stimulation continues they can soon have another orgasm. But there are also many women who after orgasm feel an uncomfortable hypersensitivity of the clitoris similar to that of a man's glans, and ask their partners not to go on.

In fact, a study by Canadian researchers published in 2009 distributed a questionnaire to 174 university students (of an average age of 25) with various questions on their sexual behavior including: (1) "Does your clitoris become more sensitive when you have an orgasm? (*a*—Yes, at the moment of orgasm; *b*—Yes, but not until after orgasm; *c*—No)"; (2) "After having an orgasm, do you wish to continue direct clitoral stimulation? (*a*—Yes, right away; *b*—Yes, but I focus on my partner; *c*—Yes, but I need some time; *d*—No)." 96% of the women answered "*b*—Yes, but not until after orgasm" and the other 4%, "*c*—No." In response to the second question about their desire to continue clitoral stimulation, 86.2% of the women responded "*d*—No." 11.5% answered "*c*—Yes, but I need some time," 1.7% "*b*—Yes, but I focus on my partner," and only .6% said "*a*—Yes, right away." These results suggest that hypersensitivity of the clitoris following orgasm is very frequent in women.

We should note here that we are talking about direct stimulation of the clitoris, not about continuing intercourse or gentler stimulation. In fact, one of the other questions was "How often do you have multiple orgasms?" and 6.9% answered "always," 10.3% "only when masturbating," 8.6% "only during intercourse," 25.3% "occasionally," 21.8% "rarely" and 27% "never." In response to the question "How many orgasms do you usually have during intercourse?" 50.6% said "one," 13.8% "more than one" and 29.9% answered "none." Which is to say, at least among Canadian co-eds, it is more common to not orgasm during intercourse than to come several times.

Out of all participants, the researchers selected a group of 11 volunteers to do a more exhaustive follow-up on clitoral hypersensitivity, and they proved that hypersensitivity can last anywhere from a few seconds to several minutes, that in most women the tip of the clitoris is the most hypersensitive spot, and that the bothersome sensation is less intense during masturbation than with a partner because the women themselves can regulate the stimulation better than their lovers. The study does not indicate that the female refractory period is identical to the male, but it does show that, unlike what was established by Masters and Johnson, one does also exist in women.

And that makes a lot of sense. We have already explained that, on a physiological level, the clitoris and the penis are not so different, and that various mechanisms of sexual function could coincide. We also know that women's orgasms are more varied, that they occasionally include the ejaculation of liquids similar to prostatic fluid, and that many of these ejaculations can go unnoticed because are retrograde, meaning they squirt backwards into the vagina. All of this gives rise to the speculation that women can have some sort of refractory period similar to men's with very intense orgasms that are accompanied by ejaculation. For the time being this is all just hypothesis, impossible to test on rats, but completely coherent with another important aspect of sexuality: the refractory period and loss of desire are produced after ejaculation, not orgasm. Two closely related, but independent, processes.

Masturbation and Its Disadvantages As Compared to Intercourse

An extensive British study titled *Prevalence of Masturbation and Associated Factors* established a very curious difference between men and women with regard to masturbation: when asking about sexual practices in the last 4 weeks, it was seen that in the case of men, the less intercourse they had, the more they masturbated. Yet with women it was the opposite. Obviously these are general tendencies and do not apply to every particular case, but the authors suggest that for men masturbation usually acts as a substitute, while for women it is more a reflection of a wider repertoire of sexual behaviors. Who knows, really, there are many possible interpretations.

Masturbation is difficult to analyze scientifically because of its enormous diversity, but there are some interesting studies. For example, in surveys that ask why we masturbate, pleasure is always the top reason in men but boredom and tension relief also usually show up, and in a study of almost 1,900 women, 10% said they masturbated to reduce menstrual pain, 39% to relax and 32% to fall off to sleep.

There are those who say that in Western societies women masturbate more now than in previous time periods, but the truth is that is impossible to corroborate empirically. The first general surveys of masturbation frequency weren't done until the late 1940s, by Alfred Kinsey, who published that 95% of men and 40% of women had masturbated at some point in their lives. In a macrosurvey at the start of the twenty-first century, the statistics were 95 and 71% respectively, which seems to indicate that female masturbation is indeed more frequent now. In fact, that same survey established that 73% of men and 37% of women had masturbated in the last 4 weeks, and that 27% of men and 8% of women stated doing it at least once every 7 days. Obviously it varied depending on the age ranges surveyed, and the National Survey of Sexual Health and Behavior published in 2010 in the United States revealed that almost 50% of women in their 20s had masturbated at least once in the last month, 38% of women in their 30s, also 38% of those in their 40s, 28% of those in their 50s, 21% of those in their 60s and 11% of those over 70. In every age range from the 20s to the 50s, men were above 60%, dropping to 55% from 50 to 59, 42% from 60 to 69, and 28% for those over 70. Even when separated by age groups, the truth is that such generic statistics give us little information and are basically useful as a curiosity, or to evaluate tendencies and compare between societies.

As for individual diversity, obviously external and psychosocial determinants have an influence, but surely the endogenous ones do as well: an adolescent will masturbate more because his spiking testosterone levels make him constantly fantasize about sex, and it has been proven that women without partners masturbate more frequently during ovulation and right at the end of the follicular phase, a few days before ovulating. Even the fact that many girls begin to rub themselves and feel genital pleasure before puberty, which as adults they will interpret as some sort of childhood masturbation, could be conditioned by a developmental phase called adrenarche around 6–8 years old, when there is an increase in androgens.

This book does not cover the fascinating topic of the development of childhood sexuality, partly because so few researchers have dared to study or give an opinion on it. But any pediatrician can cite cases of mothers who come to them concerned after discovering their daughters (it is much more common in girls than in boys) rubbing themselves with objects or moving their legs and blushing as they feel more relaxed. For example, in 2012 *The Journal of Sexual Medicine* published a case of two twin girls that literally said:

Two monozygotic twin sisters who were 11 months old were presented with atypical behavior that worried their parents. It was learned that they pulled their legs toward their abdomens while sitting, rubbed their own legs together, and groaned, and in the meanwhile, they had facial flushing and sweating, whereas afterward, they calmed down. Physical examination of the twins was normal. The twins were diagnosed with infantile masturbation (IM). IM is a behavioral idiosyncrasy rather than a disease. Behavioral therapy and advices were recommended.

While 11 months may seem very young to us, remember that a review of the scientific literature published in the journal *Pediatrics* in 2005 established that "gratifying behaviors" could easily begin at 2 months, and their frequency peaking, according to the authors, at 4 years old. Obviously this is not equivalent to what we understand as masturbation, and perhaps we shouldn't even call it that, but, again, we act as if sexuality began in puberty, when it is extremely interesting to investigate its true beginning, development and determinants in childhood.

Masters and Johnson, for example, documented an enormous range of masturbation methods, especially in women, and they linked some of them with first experiences in the discovery of sexuality. They described boys who, from a very young age, learned to give themselves pleasure by rubbing on the floor and would repeat the action whenever they could, or women who stimulated their clitoris on the armrest of their favorite chair or with their pillow, because that was how they first felt sexual pleasure as girls and even as adults it was the most effective way for them to reach orgasm. Beyond the fact that during masturbation some women include vaginal penetration and others don't, there are countless varied techniques that go beyond the aims of this book, such as the use of pelvic muscles with crossed legs or standing and rubbing up against the edge of a door.

Perhaps pleasurable rubbing that happens unconsciously in childhood can be associated with some instinctual behaviors we observe in the animal kingdom. Beyond the typical images of primates masturbating and dogs and cats in heat rubbing their genitals up against everything, if we were to start talking about nature, we would come up with infinite curious examples. One of the most peculiar cases is that of the promiscuous squirrel in Namibia. Many biologists establish that certain animals masturbate to improve the parameters of their semen before possible mating encounters, but American researchers, analyzing the behavior of this African squirrel, saw that the males usually masturbated right after coitus. Their hypothesis was that since the females in heat had sex with some ten different males during a period of 3 hours, the masturbatory reflex could be some sort of cleaning to avoid infections. In fact, they associated it with the fact that during some minor genital infections, men feel a certain itchiness linked to the desire to masturbate.

Speaking of how masturbation affects the quality of semen, the question that always comes up about whether it's "a waste" or, quite the opposite, it's "fresher" may seem ridiculous, but it turns out that this has been well studied by sperm bank clinics. A retrospective study with data from 6,000 donors analyzed their sperm's total volume, concentration, mobility and morphology, and concluded that the optimal time for donation is 1 day after the last ejaculation. On one hand, masturbation clearly reduces the concentration of spermatozoa in semen but, on the other, it is a way to eliminate those that decrease in quality over time. Since the clinics can choose the spermatozoa with greater mobility, they usually request 3 days of abstinence.

It may seem anachronistic to say this, but I will anyway. Masturbation does not cause any of those supposed health problems that some religious orders have used as a strategy to strike fear into the hearts of teenagers. The origin of that myth dates back to 1758, when the Swiss doctor Simon Auguste David Tissot (1728–1797) published *L'Onanisme, Dissertation sur les Maladies produites par la Masturbation* ("Onanism, a treatise on the diseases produced by masturbation"). In it, he enumerated several supposed physical problems associated with masturbation, which of course science has ruled out, but which unfortunately continue to be bandied about, usually by religious factions. The strange thing about this is that it can prove counterproductive, because there are signs that suppression is behind more mental obsessions, and an increase in fantasies and frustration than when masturbation is accepted as a natural act.

I don't want this to be interpreted as an ode to masturbation. I vividly recall the therapist Richard Krueguer, a specialist in cases of hypersexuality, telling me about patients who masturbate more than ten times a day, which is clearly a sign of psychiatric disorder and obsessive compulsive behavior. But he also told me that there were men who came to him feeling tortured because they felt the desire to masturbate several times a week, and he had to help them to understand that it was absolutely normal.

It is also possible that masturbation may have slightly pernicious effects on a psychological level. One of the more original theses I've ever read was written by the controversial Scottish sociologist Stuart Brody, who carried out many investigations comparing different aspects of masturbation and intercourse, before concluding that intercourse has much more positive effects—both physical and mental—than masturbation.

Stuart Brody is a strange case in the community of researchers on human sexuality. He is a prolific author who has published influential studies on cultural factors and risk of HIV infection, but at the same time it seems he is carrying out parallel investigations that don't seem as rigorous. Brody has published texts asserting that you can predict whether a woman has vaginal orgasms from the way she walks, that women whose upper lip tilts skyward have more orgasms, that sex without a condom has more positive effects on mental health and that the vaginal orgasm is more beneficial than the clitoral. These articles have gotten a lot of media attention, but they are highly criticized by most of the sex researchers I have consulted. They say that Brody's methodology is sorely lacking, that his conclusions are exaggerated, and they accuse him of attention seeking. However, the articles have all passed the filter of peer review, most of them published in *The Journal of Sexual Medicine*, which has the most impact on the field. I mention this because I even personally asked Irwin Goldstein, the editor of *JSM*, why he published such criticized articles. His reply was simple: "I know, but as an editor it isn't my decision. Brody's articles were reviewed by peers and passed." It's a delicate subject. My reflection is that sexual medicine still has a long road ahead before establishing itself as solid science and that, in many aspects, preconceived ideas and favoritism still have more weight than experimental data.

But going beyond this reflection, which I consider an essential one, Brody's work comparing masturbation and vaginal intercourse and analyzing the importance of context and the subjective experience in which an orgasm takes place, has some very interesting points. In a first study published in 2006, Brody saw that the release of prolactin (the hormone related to satiety and satisfaction) was up to four times greater after intercourse than following masturbation, both in men and in women; according to him, this explained the greater physical satisfaction after sex with a partner as opposed to on your own. But Brody insists that it is not only a chemical measurement; when he distributed surveys with the orgasmic rating scale he observed that coital orgasms were always evaluated higher than those produced through masturbation. Brody argues that sexual investigation has not sufficiently considered the importance of the context in which the orgasm takes place. Referring to the statistics that open this chapter, he states that when asked about their motives for masturbation, people with steady partners often respond that they do it for pleasure or to satisfy their partner while singletons respond that they often masturbate to release tension or out of boredom.

In his most controversial work, Brody associated better mental health with greater frequency of penile vaginal intercourse (PVI) but not with more masturbation or anal sex. In a review published in 2010 with the title "The

Relative Health Benefits of Different Sexual Activities," Brody argues that frequent masturbation is associated with higher indexes of depression, less happiness and less love in partner relationships, and he states that women who experience "vaginal orgasms (defined quite conservatively as having "had an orgasm solely through the movement of the penis in the vagina") were more satisfied with their mental health than the minority of women who had only experienced orgasms through direct clitoral manipulation." In a dangerous line, Brody mentions the immaturity of psychological defense mechanisms. He talks about vaginal stimulation having greater analgesic properties than clitoral, which help to keep the pelvic muscles in shape, and that "presence of seminal component prostaglandin PGE1 in the vagina after ejaculation might maintain vaginal oxygenation and blood flow. Improving blood flow could be expected to support sexual response and vaginal health (and perhaps general health). Using condoms deprives women of many benefits." He also claims that the quality and volume of semen ejaculated is greater after intercourse than masturbation, and that vaginal sex is better for cardiovascular health.

This last conclusion is reinforced by a study published in 2012 by Brody himself, in which he measured heart rate at rest over 5 minutes in 143 men and women, before asking them about the frequency with which they had practiced various sexual behaviors in the prior month. Heart rate variability at rest is a marker of health and longevity, and the only improvements observed were in frequent intercourse in men (not masturbation) and in vaginal orgasms in women, either through coitus or masturbation, but not with exclusively clitoral stimulation. Actually the thesis Brady was defending is that vaginal intercourse is the only sexual behavior associated with physical and psychological benefits, and that masturbation and anal sex oscillate from neutral to problematic.

Who knows, perhaps intercourse and masturbation are much more different than they may seem on the surface. Brody's ideas are controversial, but they have some experimental data to support them. However in science hypotheses are only accepted through independent studies that mange to replicate results, and it seems that in this case we will still have to wait for that.

Vibrators, Lubricants and Aphrodisiacs to Increase Sexual Pleasure

One morning, Pablo was going at it with a girlfriend when all of a sudden his cell phone, which was on the bed, started to vibrate. "Don't worry about it," said Helen. But Pablo stopped and made an impish face. "What's up?" asked Helen. "Nothing…" replied Pablo. "Tell me!" demanded Helen with a smile.

After a slight pause, Pablo picked up his vibrating phone and placed it on Helen's clitoris while he tried to continue having sex. Helen burst into laughter, squirmed and they had to stop for a few minutes until their laughing died down. After that Helen had the fastest and most intense orgasm that Pablo could recall. It's a shame they never hooked up again.

I explain this anecdote to illustrate two things: laughter makes sex easier because it is a source of muscular, nervous and emotional relaxation that also predisposes us mentally for enjoyment, and incorporating a little vibrator into coitus can be very useful for women who need clitoral stimulation to reach orgasm. Obviously certain sexual positions can serve the same function, but a vibrator ring placed around a man's penis or a small device in a woman's hand can enormously increase sexual pleasure. In fact, in a study published in 2009 by researchers from the University of Indiana and the Kinsey Institute with data from 3,800 women between 18 and 60 years old concluded that 52.5% of women had used vibrators, and most of those both for masturbation and sex with a partner. The study proved that the use of vibrators is clearly on the rise, and associated that with higher levels of arousal, lubrication, desire and orgasm. If you are already using vibrators, none of this will surprise you, but if you are among the other 47.5% of women you might want to give them a try.

Actually the idea behind vibrators is quite simple, and even though many women say they are magic and that "unlike men, they never let you down," there really isn't much magic or even science involved. In the sexual medicine conference I asked the experts at Lily if they took physiological aspects into consideration when designing them and they told me their choices were more aesthetic.

Perhaps the most curious thing about vibrators is that they were first conceived as a medical tool to cure hysteria. From an extremely sexist perspective, in the late nineteenth century doctors assumed that women didn't masturbate, and that if they weren't married they had no release for their sexual tension, which could accumulate and cause hysteria, so they designed some machines to liberate tension through orgasm. Those were the first vibrators.

Later ones were devices not actually designed for that purpose. The most classic example is the Hitachi Magic Wand, a long back massager with a ball at the end that was created so that you could give yourself a back massage, but many women decided to explore using it on other parts of their body to pleasing results. Decades later, the classic model of the Hitachi Magic Wand is still sold in sex shops and specialized stores. If you go to one of those shops, or take a look online, you will see the wide range of shapes, sizes, textures, materials and intensities for vibrators and sex toys that there are on the market, both for men and women, heterosexuals and homosexuals, to enjoy alone or with a partner.

Perhaps one of the most famous is the "rabbit," popularized by the series *Sex and The City*, which tickles the clitoris with bunny ears on a dildo that spins inside the vagina at different speeds. It is a way to put pressure on the G-spot with an intensity that a woman might not be able to reach on her own or with penile penetration. Stimulation of the G-spot, generally located 3–4 cm from the entrance in the upper wall of the vagina, should begin gentle like a caress, but to increase pleasure the intensity should gradually increase, and that is much easier to do with a dildo than with your fingers or penis. In fact, sexologists state that one function of sex toys and masturbation is to learn about your own body's reactions, and that exploring what produces more pleasure is essential to a healthy sex life and taking the initiative in a relationship.

There are many women who, as intercourse progresses, need the movement and intensity to be very precise in order to reach orgasm, and it works very well if at that point they take control of the rhythm and position. But in order to do that they have to have knowledge of their own genital response. According to a broad study published in 2009, when asked their motives for starting to use vibrators, 55% of women said it was just out of curiosity, a third said it was to facilitate orgasm, and 27% because their partners asked them to, but the vast majority agreed that they had helped improve their sex life.

Something similar happens with lubricants. In a study by Indiana University published in 2011 and headed by Debby Herbenick, researchers chose 2,453 female volunteers and assigned them six different types of lubricants to try in masturbation and intercourse (anal and vaginal) over 5 weeks. This was the largest study of lubricants to date, and it proved that their use greatly improved sexual experiences not only for women who suffered from vaginal dryness or dyspareunia (pain during penetration), but in almost every one whose level of lubrication was not very abundant. It was also proven that, although overall there were few cases, the water-based lubricants caused less vaginal symptoms than the silicone-based ones. The silicone-based lubes are retained for longer inside the genitals because the water-based ones are more easily reabsorbed, but the authors of the study suggest that each person should experiment with the type, quantity and moment in which to use external lubrication. There are also options of liquids that generate warmth in the genitals, and others that claim to increase blood flow, arousal and sensitivity. According to experts, using lubricants is the easiest thing you can do to improve sexual pleasure.

The concept "aphrodisiac" is a very broad one, it goes beyond remedies for erectile issues, and the truth is that it makes sense that certain natural substances can have positive effects on sexual response. Obviously there are foods and natural compounds that can cause vasodilation, increase blood flow and facilitate genital excitation. Yohimbe and the amino acid L-arginine, for example, have passed the test of clinic studies comparing them to placebos. It is also possible that there are substances that increase levels of testosterone or dopamine and favor sexual desire, and others that momentarily boost our energy and vitality making us feel more in the mood. If we use a wider definition, we could consider alcohol and marijuana aphrodisiacs because they lower inhibitions and, in the case of the latter, also increase sensitivity.

Of course, the company and conversation during a romantic dinner can affect the sexual appetite more than the food itself, but it is perfectly logical to think that some specific foods and traditional medicines can contribute chemically inside our bodies to increasing vasodilation, improving our mood and boosting sexual desire, either by lessening inhibition or increasing arousal. What happens though is that in the herbalist's shops and on the Internet there are many products that make big claims, but very, very few scientific studies have analyzed their side effects or confirmed their efficacy. To top it all off, an investigation published in April of 2012 analyzed nine herbal dietary supplements sold as treatments for erectile dysfunction, and it determined that four of them were adulterated with versions of sildenafil (Viagra) and three with other drugs prescribed for erectile problems. So, yes, natural products work, but so do the synthetic ones, and perhaps even better.

In one of the few clinical trials carried out, prominent researcher Cindy Meston proved that the alkaloid Yohimbine—extracted from a tree called Corynanthe yohimbe—does significantly increase vaginal flow in women shown erotic images, compared to a placebo; however, *Ginkgo biloba* extract, which is also sold as a powerful aphrodisiac, did not improve female sexual response more than the placebo. The placebo effect is terrifically interesting, since Cindy Meston herself has proven that giving placebos without any active ingredient increases desire and arousal, and has very positive effects in certain cases of sexual dysfunction. Undoubtedly, the mind has an influence on the nervous system, relaxation and mood; as a result, even if by suggestion and the placebo effect, many natural products really can work in specific situations. The big risk is that yohimbine, for example, is toxic in large quantities, and the vast majority of natural products have not passed trials to evaluate their side effects in large doses.

It is also important to remember that the most common complaint in women is lack of sexual desire, not lubrication problems that can be easily solved with lube, nor blood flow to the genitals, which is not as essential as it is in men. In males, the opposite is true and the most frequent problem is not lack of desire, but rather insufficient arousal and erection. In those cases, vasodilating substances that contribute to the production of nitric oxide like arginine, ginseng or yohimbine, among many others, could work (surely not as effectively as Viagra or similar drugs), but their effect on the female sexual response would be much less relevant. Aphrodisiacs are not a myth, but classics like chocolate or a little glass of wine that boosts our mood and adds to a feeling of wellbeing will continue to be, along with various placebos, one of the best options for facilitating sexual encounters. And I am not simplifying here. In fact, we should be less concerned with the aphrodisiacal effects of certain foods than the antiaphrodisiacal effects of having the stomach too full and the mind clouded.

The Treacherous Effects of Alcohol in Arousal and Orgasm

Many women have been through this plenty of times. You have a drink, two, three or even four, and you are feeling euphoric, uninhibited, having spicy thoughts, your blood is flowing and you're incredibly attracted to that guy with the playful gaze who is trying to seduce you. One drink later you go to bed with him, and whether he's your boyfriend, husband, lover, friend or recent acquaintance, and despite being super-relaxed, sensitive, excited and engaged, you struggle to reach orgasm. Sometimes you can't get there no matter how hard you try. That is the trap of alcohol's misleading effects.

Alcohol is a powerful depressant to the central nervous system that interferes with receptors of the neurotransmitter GABA (gamma-aminobutyric acid) provoking generalized inhibition of neuronal excitability. In other words, it makes the entire brain "work slower." That means that if after drinking a little bit more than we should we feel a greater desire for intimate contact, it's not really because we are genitally or neurochemically more excited, but because we are mentally less inhibited. The decrease in brain activity reduces our self-control; however, even though it doesn't seem like it, it also lowers our physiological arousal response. It is true that the net effect after several drinks is a subjective feeling of more arousal, but if someone were to measure blood flow in vaginas—which has been done—they would see that it is significantly less than if they hadn't had any alcohol. Physical excitation is reduced, and along with the difficulty of activating the sympathetic nervous system when inebriated, it is much harder to reach orgasm. And the same thing happens to guys.

In moderate doses, alcohol is effectively one of the most infallible aphrodisiacs that exist. Having a couple of beers lowers inhibition, generates euphoria, relaxes physical and emotional tensions, and facilitates sexual enjoyment. But if we drink a little too much we reach that paradox where we feel very excited but have difficulty reaching climax.

The context of the drinking is important and not everyone reacts in the same way, but the effect has been proven many times over. In a review of the scientific bibliography, published in 2011, "Women's Sexual Arousal: Effects of High Alcohol Dosages and Self-control Instructions," researchers at the University of Washington and the Kinsey Institute concluded that of the eight most rigorous studies carried out on the effects of alcohol consumption on female sexual response, five observed that after moderate quantities of alcohol, lubrication and vaginal blood flow diminish. On the other hand, of the 15 studies analyzing the subjective perception of arousal, 13 concluded that alcohol increases it. Which is to say, the subjects were mentally more aroused and physically less so.

In addition to this bibliographical review, researchers made several of their own studies. In the first they selected 78 female college students between 21 and 35 years old who were occasional drinkers without any physical or mental problems, single but interested in men, and they proposed an odd experiment: they were to arrive at the laboratory without having had any alcohol in the previous 24 hours and without having eaten anything in the last three. After making sure they weren't pregnant or menstruating (in the latter case the experiment was conducted on a different day), they separated them into two groups. The first group would gradually drink a cocktail of one part vodka to four parts orange juice, until they reached a blood alcohol level (BAC) of .08%. That equals .82 ml of alcohol per kilo, or in more colloquial terms, a bit tipsy. The second group would drink the same quantity of orange juice, but "without additives." Then they would go into a room, sit with their legs raised in front of a television, and the researchers would insert a vaginal photoplethysmograph into them, a tampon-shaped device that, after sending out pulses of light and capturing their intensity reflected by the vaginal walls, would measure the changes in genital blood flow. Once they were comfortable and relaxed a 2-and-a-half-minute documentary about birds appeared the screen, and then 6 minutes of an erotic video that previous studies had shown to be arousing for women. They divided both the group that had drank alcohol and the one that drank juice into two subgroups. Of those, half of the women were supposed to try to avoid getting excited during the erotic film, and the other half were supposed to try to maximize their excitement. Various prior studies showed that women have better mental control over their genital response than men do, and the researchers wanted to see if the alcohol affected that ability. When the test was finished, they let them rest in a room until their blood alcohol level went down to .03%, gave them 15 dollars per hour and asked them for their subjective impressions of whether they had gotten more or less excited. They would then use that data to correlate their mental perception with their physiological response.

The researchers carried out another identical study with 64 other women, but the difference in their degree of intoxication was significant, reaching 1.25 ml of alcohol per kilo of body mass, a quantity that inhibits some functions of the central nervous system.

They came up with several conclusions from each of the two studies. On one hand, the women who were just tipsy were still almost as capable as the ones who drank juice to maximize or inhibit their genital response while watching the erotic images, but the ones who consumed higher levels of alcohol lost that ability. In fact, the difference in arousal between the drunk women from the second experiment who were asked to increase their desire and those who were asked to diminish it was tiny. As for the total level of physical excitation, it was observed that the women in the first study who had drank a little bit of alcohol and were instructed to maximize their desire reached the highest peak of excitation more quickly than the control group that hadn't drank any alcohol. Which is to say, that a little alcohol facilitated genital response. But in the group with a high dose this effect disappeared, and there was even an overall decrease in physical excitation. As for the differences between subjective perception and genital response, they were seen in most of the women in every group, but contrary to what we said before, the consumption of alcohol didn't seem to make them feel more excited than they actually were. The authors of the study justified this effect by saying that the laboratory context could be important and not representative of a real encounter. "Could be," they say...

A couple of years earlier, the same researchers carried out a similar experiment with young men. They selected 125 students, got half of them intoxicated to the point that their blood alcohol level (BAC) was .10 % (the limit for driving in Spain is .05 %), they gave orange juice to the other half, and with a plethysmograph on their penis (a cylinder that measures the change in thickness), they first showed them the bird documentary for 2 and a half minutes and then erotic videos. They also asked half of the men in each group, control and intoxicated, to try to maximize their excitement and the other half to try to put the brakes on.

For once, the male results were more interesting than the female ones: as was to be expected, the initial time required for the penis to go from the basal state to a 5 % growth (the start of an erection) was shorter among the sober than among the drunk. However, the final time for reaching the peak of erection was almost identical. It was only slightly slower in those who had had the laboratory cocktail. But also, in the sober men, the difference between trying to hold back their erection or stimulate it was very significant, and they had quite a bit of control over their penis response. On the other hand, among those that had been drinking there were practically no differences; in fact, the erection began earlier in the intoxicated who were trying to control themselves than in the sober who were trying the same thing. Interesting reflections.

The authors' other conclusions were that in healthy young men moderately high doses of alcohol didn't interfere in erections. Erection problems following large dosages of alcohol are very well documented, and due to cardiovascular effects in frequent drinkers; but if a man has trouble getting it up after moderate alcohol consumption it isn't due to a physiological effect, but rather some other lack of motivation. We can no longer use that as an excuse.

Motivations for Anal Sex

I had another shock during the sexual medicine conference in Chicago, when after a lecture I went over to a group of scientists and said: "I just came out of a very interesting session on anodyspareunia, pain from anal penetration." One of the scientists quipped: "Of course it hurts, that's not what the anus was designed for." The others smiled and I, after so much phallocentrism and normativity at the event, couldn't stop myself from replying: "Well, anal sex is a common practice among most gay men, its rates are skyrocketing in the heterosexual population, and almost half of women between 25 and 30 have had anal intercourse at some point. I think sexual medicine should consider it, despite the fact that some people may find it inappropriate." "Of course, of course…" was the not very convincing response I got from an American, in whose country up until 2002 there were ten states with laws against anal sex under any circumstance, and three (Kansas, Oklahoma and Texas) that banned it only between homosexual couples. Obviously those laws no longer apply, but in the 1990s there were still some notorious cases of gay couples being denounced for sodomy.

Digressions aside, the study presented at the conference analyzed the frequency of anal pain in a sample of 1,190 Belgian homosexuals. The results were that 41 % felt no pain at all, but 32.7 % of them did feel some mild pain, 17.2 % mild to moderate, 4 % moderate, and 1.8 % severe. Further analysis of the results showed that the more experience they had, the less pain, but there was a sector of the population which always suffered pain, and that often hindered their having satisfactory sexual relations. It was also documented that 28.2 % regularly practiced barebacking, or anal sex without a condom, despite the fact that risk of HIV contagion is 20 times higher through anal intercourse as opposed to vaginal.

These risky behaviors, and the growing cases of anal cancer from the human papilloma virus, are what most concern scientists investigating this type of practice between "men who have sex with men," the expression they use to include anal sex among the population that does not necessarily define itself as homosexual.

Only a few researchers are beginning to deal with aspects of anal sex in heterosexual couples. And that is strange, because the numbers show that it's not an isolated practice. According to Kinsey Institute data, 46% of American women between 25 and 29 have had anal sex at some point. If we look carefully at the age ranges, we see something curious: the percentage of women who have had anal sex at least once in their lives decreases to 35% in those between 50 and 59, 30% in those 60–69, and 21% for women 60 years and older. The interpretation is quite obvious: heterosexual anal sex is becoming more normal and much more frequent now than it was decades ago. That is if we trust the results, since a study published in 1999 analyzing how much people lie in surveys on thorny issues showed that the proportion of adult women who hide having anal sex is greater than those who hide having had abortions. It is certainly a taboo subject.

There is more interesting data: among people who have practiced anal sex in the last 12 months, 69% of men and 73% of women did so in the context of a stable relationship and, according to a 2003 survey, 58% of women did it for the first time because their partners requested it. Another study, of 2,357 heterosexuals who had practiced anal sex in the last 3 months, revealed that only 27% had used a condom every time, and 63% hadn't used one at all. This is something that is of serious concern to researchers, since according to other surveys almost half of students both male and female considered anal sex less dangerous than vaginal in terms of sexual disease transmission. And it is completely the opposite: the epithelial tissue, the higher frequency of injuries and the inflammatory responses produced during anal sex make it much more risky.

But why do heterosexual couples practice anal sex, and more now than in previous decades? Obviously pornography has played a very large role in the second part of the question. While there is no data to corroborate it, we are more exposed to images of anal sex and that makes more people interested in trying it. But besides curiosity, the classic wanting to remain a virgin and avoiding pregnancy continues to be a valid reason among some teenagers. In an American study published in 1999, 19% of college students thought that anal intercourse wasn't exactly sex, and that practicing it didn't mean losing your virginity.

Undoubtedly the other main reason why many couples practice anal intercourse is simply because they find it pleasurable. In what sense? There isn't much scientific bibliography on this subject, but Kim McBride, from Indiana

University, reviewed different studies to establish six large groups of reasons why people said they enjoyed anal sex. They were: (1) intimacy and trust (some people associate anal sex with a practice that they would only do with someone they really trust); (2) the search for different and new sensations; (3) linked to control and domination games; (4) the increased eroticism of breaking taboos, (5) the pain can be pleasurable (for many people in states of high sexual excitement a bit of pain feels good) and (6) it is just another practice in a couple's sexual relationship.

The most conservative view of anal sex among heterosexuals sees it as simply an atypical and more dangerous version of vaginal intercourse, but clearly for many people it is something more psychologically exciting; with all the preparation and trust it entails, it can be considered more intimate and it generates a different type of pleasure. The anus is an erogenous zone with independent nerve endings whose stimulation can be gratifying, and in men it also allows access to the prostate.

One of the few researchers who is studying heterosexual anal sex in depth is the Croatian sociologist Aleksander Stulhofer, whom I interviewed during the International Academy of Sexual Research meeting in Portugal, where he presented his fascinating studies on perceptions of anal sex in women. In a first study, Sasha surveyed more than 2,000 women from 18 to 30 years old, and saw that 63% had had some experience of anal sex. But Sasha found something peculiar: while half acknowledged that their first time they had to stop because of intense pain, most of them repeated the practice. In fact, of the 505 women who said they had had anal sex two or more times in the last year, 9% did it despite experiencing severe pain that could be classified as anodyspareunia. So why did they do it? Sasha wanted to investigate this aspect more deeply and he carried out another study asking specifically about pain during anal intercourse. Sasha explained to me that there were three large groups of women among those who had tried anal sex: approximately a fourth feel no pain at all and practice it with no problem, about 40% experience intense pain and avoid it, and for around a third it hurts but they like that because they eroticize the pain. This last group is the most interesting according to Sasha, and a very clear example of the fact that sexual pleasure does not just depend on physical sensations but also on the context in which it takes place. A straight man can want to have anal sex with his partner because it finds it kinky or to feel a different kind of pressure during penetration, and a woman can enjoy it also for pleasure or the erotic charge it has. It isn't simply a version of vaginal intercourse.

In fact, the physical pleasure is another interesting point. In the case of men, anal penetration with a penis, fingers or dildos is pleasurable because it internally stimulates the prostate, and that is true for both gay men and heterosexuals (a man can enjoy receiving anal penetration without being attracted to men in the slightest). For some, prostate stimulation is very pleasurable and can lead to orgasm and ejaculation, but why is it also pleasurable for some women, even in some cases to the point of orgasm? The eroticization is a factor, and the sensory fibers around the anus is another. But the latest studies with sonograms are being carried out to determine whether anal penetration can reach some internal parts of the clitoris. These studies are still underway, but if they confirm that it would explain why some women feel so much pleasure with anal sex while others can take it or leave it.

Who knows whether in addition to what we already know to be essential: lubrication (better water-based than silicone), relaxation (there are various muscles in the anus: some we can relax voluntarily and others that are only relaxed when the rest of the body is as well, breathe deep), preliminary stimulation with the fingers or toys (lubricated and with a condom), hygiene (yes, this is an anus we're talking about), condoms (as I've already explained, because of the type of tissue and the small lesions produced, HIV, human papilloma and other diseases are much more frequently spread) and never using any pressure or force, our physiology will offer some positions or practices that will make anal intercourse more satisfying, being that it is one of the sexual practices that is growing in popularity. More education—particularly in teenagers who are often led by highly unrealistic pornographic images—is needed.

In fact, I would like to wrap up here with another speculation/hypothesis/challenge for science that no researcher has yet been able to answer for me: just as the pelvic and pudendal nerves stimulate the penis, clitoris and the external part of the vagina, but some women have orgasms they describe as deeper or less localized through the stimulation of interior areas of the vagina via the hypogastric nerve, some men experience orgasm through prostate stimulation as more "full-body." And remembering that the physiology of men and women is actually much more similar that we believe it to be, perhaps the male orgasm induced by prostate stimulation is relatively similar to the "vaginal" orgasm in women.

Penis Size Does Matter, but Clitoral Size Doesn't

There is such an obsession with penis size, but it is in clitoral size where we see the most range. A study published in 2005 by British gynecologists recruited 50 premenopausal women who had not had surgery or hormone treatments, and determined that there was a range in clitoris length from .5 to 3.5 cm (differences of up to seven times, in a sample of only 50 women, and without seeing any extremely large cases, of between 6 and 7 cm, as have been described in medical literature). The average was 1.9 cm long and .5 cm thick, with a variability in width from 3 mm to 1 cm. This study confirms a few previous ones that showed that variations in the size of the female clitoris are much, much wider than those in the male penis. In fact, a substantial difference lies in the fact that the clitoris can significantly change size in adulthood. This is due to the fact that the clitoris has testosterone receptors, and the fluctuations in this hormone can make it grow or shrink. The effect is very obvious in transsexuals who start to take androgens, or in women who use testosterone creams topically as a supposed remedy for lack of desire, and they observe how in a relatively short time their clitoris grows considerably in size. This is very speculative and there has yet to be a study confirming it, but it is said that the fact that birth control pills reduce testosterone levels could make the clitoris also shrink slightly. That said, it is important to clarify that size does not in any way affect the sensitivity or sexual function. The quantity of nerve endings is the same in a small clitoris as in a big one, and there really is no difference.

The British study also found significant differences in the distance between the clitoris and the urethra (min: 1.6 cm; max: 4.5 cm), the color of external female genitals (most were darker than the skin around them but in 18% of women they were the same color) and in the size of the vaginal lips: the labia majora didn't vary too much, but the internal ones could fluctuate between 2 and 10 cm in length, and .7 and 5 cm in width.

This deserves attention, since very large inner lips can cause physical irritation and discomfort in some women, and the surgical operations to reduce them are very simple and are becoming more and more common. In a review of scientific literature on cosmetic genital surgery published in 2011 in *The Journal of Sexual Medicine*, it was established that the rates of satisfaction among patients following the operation were 90–95%, and that 80–85% noticed an improvement in their sexual relations. Labiaplasty is the most widespread, but there is also the reduction of the clitoral hood and perineoplasty to strengthen the vaginal muscles that age and births may have left flaccid. If to all this we add the diversity found in pubic hair and its modifications (a study by the Kinsey Institute concluded that 21% of Americans between 18 and 24 remove all of their pubic hair, and that this is associated—without establishing causality—to higher rates of sexual satisfaction), it is clear that there is much more diversity in female genitalia than in the male penis. But since phallocentric questions are still what dominate the conversation, let's have a look at the myths and realities surrounding penis size and its importance in sexual relations.

Width Has More Impact than Length

If you are one of the 45 % of men who don't feel entirely satisfied with the size of their penis, let's defuse the situation by comparing ourselves to gorillas. A male gorilla, with his 200 kg of muscle and daunting appearance, has a ridiculous 3–4 cm tool. It's enough for the little he uses it. Since it's not female gorillas who choose their mates but rather the males who duke it out for the rights to be head of the harem and keep the others from copulating with their concubines, it was evolutionarily better to develop muscle and avoid wasting energy unnecessarily: a gorilla's testicles are like two peanuts you can barely see.

Bonobos and chimpanzees, who are much more closely related to us, have larger penises and particularly bigger testicles, but the human phallus is the largest of all the primates. Why? There are several reasons: on one hand, walking upright and having it exposed turned it into a sign of sexual attraction similar to the female breasts (women are the only primates whose breasts are inflated when they aren't lactating, and evolutionary biology explains that they remained this way as a sexual inducement). Another reason is that with our species' promiscuous past, large penises and that peculiar mushroom shape of the glans were much more useful for removing from a woman's vagina the semen of the person who had fornicated with her previously. It is speculated that this function of removing others' sperm led our coitus to be longer and to require more ins and outs before reaching the ejaculatory reflex than the sad few seconds it takes gorillas. This is also why sperm composition is not uniform: the first drops of ejaculation contain spermicidal substances that attack the spermatozoa from previous encounters.

When you really think about it, maybe there is an evolutionary justification for why men are more obsessed than women with the size of their penises. But really, we do make too big a deal out of it. The earlier statistic of 45 % of men wanting to have a larger penis comes from an online survey of more than 52,000 heterosexual men and women, which also revealed that 85 % of women were perfectly satisfied with their partner's penis. It is an interesting discrepancy.

There are countless individual opinions. If you get together a group of women friends in a bar and you ask them about the importance of the penis size, some will tell you it's more important than others, but almost all of them will agree that it is not one of the most important factors in a sexual encounter.

Avoiding casuistry and instead recurring to larger samples, a Dutch study asked 365 women who were in the hospital after giving birth about various aspects of their sexuality. Only 1% classified the length of the penis as very important, 20% as important, 55% not important and 22% not important at all. They gave similar responses about the thickness, but in general most stated that the width was more important than the length.

Comparable results were found in a survey among American students, in which 45% of women said that thickness was more important than length, and only 5% disagreed. The rest didn't care either way. Masters and Johnson created the myth that size doesn't matter when they proved that the vagina adapts to the penis, but physiologically it seems that a thick penis can better stimulate the external parts of the vagina and clitoris. Although, obviously, the key factor involves more skillful movements and awareness of which positions best arouse your lover's genitals. So size matters, but not that much.

What Is the "Normal" Penis Size?

Historically, urologists have made three types of measurements: flaccid, erect and extended (when flaccid, pulling the tip out as far as possible). If you are curious, the extended and erect state are almost the same, but the flaccid can offer many surprises, since it isn't necessarily at all related to the size when excited. Penises that significantly change size when erect are called growers and are eight in ten, while two in every ten men have showers, or penises that are always fairly swollen and don't change increase much in size when erect.

The first study documenting penis size and establishing an average was published in 1899 by the German Heinrich Loeb, who measured 50 men and arrived at an average of 9.4 cm in the flaccid state. In 1942 Schonfeld located the mark at 13 cm extended, and comparing by age he observed that length didn't seem to change over time. In 1948 Alfred Kinsey examined an enormous sample of men, establishing 9.7 cm in the flaccid state and a very generous 16.74 cm average when extended.

As surprising as it may seem, it wasn't until 1992 that measurements of erect penises were published. One hundred and fifty European men agreed to be injected with papaverine and prostaglandin to cause erection, resulting in an average of 14.5 cm long and a width of 11.92 cm in circumference. The widest study to date came in 2001 when Italian investigators measured the members of 3,300 Italian men between 17 and 19 years old to establish an extended length of 12.3 cm, 9 cm in the flaccid state and a circumference of 10. One of the conclusions of this study was that penis size was slightly correlated to the man's height and weight. The taller and heavier men generally had somewhat larger penises. That means that motion with the thumb and index finger indicating an apparently inverse relationship between body height and penis length—sorry to tell you, guys—isn't endorsed by science.

Usually the penis is measured from the tip to the base on the abdominal wall, but there is a lot of irregularity in the methodology. A 2011 study with 2,276 Turks came up with an average extended penis size of 13.98, and a review of all the scientific bibliography done by American urologists concluded that a regular penis was between 12.8 and 14.5 cm when erect (less than the Italians) and 10–10.5 cm thick.

There are more peculiarities. In the 1980s, a study with a small sample group suggested that foot size was related to penis length, and despite it being refuted in 2002 by English researchers, the idea that some part of the body could be linked to penis length stuck in the popular imagination. No study has every established any relationship, except for the oft-mentioned link between the ring and index fingers as an indication of exposure to testosterone during pregnancy.[2] Korean researchers measured the penis and fingers of one hundred and forty-four 20-year-old men and published that the larger the ring finger as compared to the index, the longer the penis. This data has yet to be replicated.

There are also doubts about the honesty of 935 homosexual men who answered a survey in 1999, which when compared to 4,187 heterosexuals they came out ahead with an average of 16.4 as opposed to 15.6. They also claimed to have thicker penises. This study had much impact and is constantly cited, but it also received several methodological criticisms, which argued that the measurements that each man makes of his own penis are not very objective.

Another study of 6,200 young Bulgarians suggested that rural men are better endowed than urban ones, proposing that greater testosterone production during adolescence could favor this relationship. In 2009 an Egyptian study with 1,027 adults observed that men with erectile dysfunction had a slightly smaller penis (11.2 cm extended as compared to the average of 12.5 cm), speculating that problems of insecurity could be behind the association. Because really there are more myths than experimental data about penis size. Perhaps that is why Italian urologists prepared an article entitled "Penile length of most men who seek out enlargement techniques is normal."

---

[2] No one really knows why, but exposure to testosterone during pregnancy affects finger length. The more testosterone, the longer the ring finger is compared to the index. Since this relationship was discovered, researchers are using that measurement as an indication of prenatal exposure to androgens, and comparing that to a large range of physical and behavioral characteristics. There are those who dispute the validity of this association.

It is true that cutting ligaments in the base of the penis can add a couple of centimeters, and that there are other techniques both for lengthening as well as widening it, but all the urologists and sexologists I've consulted told me that they always discourage it, even in the case of micropenises. They assure me that most of the patients who visit them with this concern have normal penises, and the truth is, after reading about all the problems that can arise from these techniques, a psychotherapist seems to be the best treatment for penises that are considered small.

The problem of the micropenis is another subject: this means a penis that is less than 7 cm in the erect state. It is calculated that one in every thousand men is affected by this, and it is not the result of normal variety but rather congenital causes that impede the correct development of the penis, such as low secretion of the hormone gonadotropin in the hypothalamus or failures in the hormonal secretion of the testicles during embryonic development. I call this a problem because despite the look and function of the micropenis being conventional, and many men handling the situation well, it sometimes does cause difficulties for them and their partners. In any case, the reality or myth that is most often bandied about when talking about this matter is whether skin color is related to penis size. Conclusion: the color, in and of itself, is not.

The phrase "you can't generalize" is absolutely false. You can. In fact, statistically analyzing correlations in the population is one of the great tools of epidemiology in order to establish associations between, for example, one type of diet and diabetes, smoking and cancer, or childhood abuse and the appearance of adult sexual violence. These associations allow for the development of better-informed therapies and preventive measures, as well as help to investigate the causes behind these differences. You can generalize. What you can't do is individualize and say that one person in specific will be more violent because he or she suffered childhood abuses. That is something we cannot do. But knowing the correlations is definitely useful for confirming hypotheses or ruling out myths. Although the studies have to be done correctly.

When I used to work at the National Institutes of Health (NIH) in the United States, I remember attending an internal seminar on genetic studies and minorities in the American population. They were debating about the higher predisposition towards diabetes in Latinos, Tay-Sachs in Jews, glaucoma in African-Americans... when a petite, skinny black researcher stood up and said something along the lines of: "I've been saying this internally for a long time and nobody pays me any attention: my family comes from a region of Africa that has nothing to do with where the descendants of African-Americans who arrived here in the slave era came from. Yet, in the studies, I'm lumped together with them as black. It is a serious error."

I explain this anecdote to reflect what science says at the moment: that, having seen a correlation between height, body weight, and penis size, obviously an African-American who is 190 cm tall will have a larger penis than an Asian man who weighs half the size, but not because of his skin color or race. After living as a single woman for 8 years in China, my friend Laure doesn't agree and says that the popular legends are based on something real… And Anne says she prefers black men, shooting me a look that implies I'm jealous. But when I ask Anne if her black lovers were stocky or scrawny, she confirms that they were well built.

Despite that, it is only fair to acknowledge that there has yet to be any scientific investigation published—possibly because of how little use there would be for such a study—that statistically compares a wide range of sizes between different nationalities and races. A study using the Indian population came up with an average of 13 cm, significantly smaller than in other countries, and another in Nigeria concluded that the black population had slighter longer penises on average than the Caucasians. But both evaluated absolute measurements without including height or body constitution. Penis size will continue to be an unresolved subject for science, simply because there really is no justifiable reason to study it. Sexual medicine has much more serious concerns.

6

Sex in the Doctor's Office

It was 1983 and researcher Giles Brindley was giving a conference on erectile dysfunction at the meeting of the American Urological Association held in Las Vegas. Doctor Brindley had already published various studies suggesting that the direct injection of vasodilating substances into the penis could stimulate blood flow and generate a hard, lasting erection in seconds, but his work had been highly criticized, and there were rumors that the photos of more or less erect penises in his articles may have been rigged.

In the middle of his talk, suddenly Professor Brindley stepped away from the podium, reminded the audience that giving a conference was not a sexually stimulating activity and lowered his pants. Beneath his underwear he had an unmistakable erection. Brindley explained that minutes before his lecture he had injected himself with papaverine in his hotel room, and that this was the definitive proof that said vasodilating substance worked perfectly against erectile dysfunction. He didn't stop there, as various attendees recalled in the article "How (not) to Communicate New Scientific Information," he lowered his underwear too and advanced toward the front row offering people the chance to confirm the degree of tumescence. He wanted to convince everyone that it wasn't an implant, but the screams and shocked expressions made him dress again quickly and continue his talk.

This true story is more than an amusing anecdote. Injections of papaverine and other substances investigated by Dr. Brindley became the first pharmacological treatment for erectile dysfunction, and some of them are still used today when Viagra or other phosphodiesterase-5 inhibitors don't work.

Twenty-nine years after Brindley's famous lecture, I attended the International Society for Sexual Medicine's conference in Chicago and saw for myself that male erection continues to be the primary preoccupation of urologists and sexual medicine, which has historically been centered only on men. But I also saw that in recent years a rapid and important change is taking place. During a presentation, Irwin Goldstein, the editor of *The Journal of Sexual Medicine*, asked the therapists or gynecologists whose main focus is female sexuality to raise their hands. Approximately 20–25 % of the arms went up, eliciting a "wow" from Goldstein; he commented that there were many more than there had been in the past, and that the medical community's interest in female sexual dysfunctions was growing at a much higher rate than interest in the male issues. In fact, from a scientific perspective, it had already bypassed it. If we introduce the search terms "female sexual dysfunction" and "male sexual dysfunction," and we make specific searches by date, into the Pubmed database, which gathers all the scientific articles publishing in the realm of biomedicine and health, we see that 2012 is the first moment in history in which more scientific articles on female sexual dysfunction were published than on male. This change is very significant because some decades ago the difference favoring men was vast, but it actually is extremely logical for several reasons. One is that the arrival of oral drugs to produce erections brought about an enormous improvement in the primary male sexual dysfunction, and another is our society is changing and sexual pleasure is no longer understood as more important for men than for women, and thirdly because the percentage of women who feel they have some sexual problem (lack of desire is the most frequent) is significantly higher than in men (according to surveys carried out by Edward Laumann, from the University of Chicago, 43 % of adult women and 31 % of men said they had had some sort of sexual disorder in the last 12 months).

But one underlying reason for the shift in scientific interest toward female sexuality is the indisputable fact that it is much more complex than its male counterpart. And I'm not saying that to please my female readers, nor of course to downplay the various problems plaguing men, but because it is truly the majority opinion among sexologists, and because on an evolutionary level it makes all the sense in the world. In our species, like in the rest of the primates, the regulation of the reproductive cycle of females is infinitely more complex than that of males, and the appearance of sexual desire depends, out of evolutionary necessity, on many more physical and psychological factors. There are many more internal variables in women to study and that makes both the diagnosis and the sexual therapy itself more difficult than with men.

Again, I am not trying to take any importance away from male sexual dysfunctions. We also suffer countless disorders of psychic origin, which we will discuss further on, but in general our most common medical problems are better identified and usually have a greater degree of physiological origin to them. As for science and sexual medicine, the hottest topics right now are about female sexuality.

That said, in this chapter we will review the primary sexual dysfunctions that affect men and women from a medical perspective (not the sexological one used throughout most of this book, which involves a wider, multidisciplinary vision). But before I begin I want to make two important points. First, that this book does not attempt to be a guide to sexual medicine, and one of my main messages is that if there is something bothering us about our sexuality we shouldn't keep it hidden and it's worth checking with a doctor and/or a specialized therapist. Secondly, I insist on the need to have a biopsychosociological perspective of sexuality that brings together medical, psychological, relationship and sociocultural factors in both diagnostics and therapy. Sexologists of the future cannot leave out any of those aspects, and the big challenge will be how to combine, in each case, pharmacological treatments (for example hormonal supplements) with cognitive therapies (such as discussing fears or negative feelings both in relationships and outside of them) and behavioral ones (for example, modifying behavior patterns or exploring sexual stimulation exercises, both with a partner and alone).

In any case, without undercutting the intuitive component based on long experience, one of the comments I've heard most frequently while writing this book is that there are countless "sexologists" who tackle sexual problems with invented, antiscientific and sometimes downright idiotic pseudotherapies. We are not by any means defending here a mechanistic view of sex; we assume that sexology will always be one part art and one part science, and that therapies must be individualized. At the same time, we think that scientific information will increasingly contribute to building a medicine and a psychology of sexuality based on evidence and not beliefs or casuistry.

Expectations and Other Male Sexual Dysfunctions

Searching through the medical literature, we come across photographs of men with two penises, cases of paraphilias that we never could have imagined and descriptions of all sorts of rare physical and psychological disorders. We find quite a bit of discussion of hypersexuality and obsessions that we'll discuss in other chapters, many publications about medications' side effects and analyses of how certain diseases such as cancer or diabetes affect sexual function. The diversity is overwhelming but—to outline—the most common problems of male sexuality are:

Erectile Dysfunction

Erection problems can have physical or psychological causes, and as we have already seen in Chap. 2, specialists have various tools for identifying the origin of the dysfunction in each particular case.

The most frequent psychological causes are depression, stress, relationship problems and performance anxiety. This last phenomenon occurs in both adults and young people when, despite feeling mentally aroused, insecurities, nerves or fears provoked by prior traumatic experiences impede erection when they are about to engage in sex. When erectile dysfunction of psychological origin happens in the context of a stable relationship, the combined advice of a specialist and behavioral therapy should be enough to solve the problem. In the case of someone single the treatment is a bit harder, and it is based on eliminating fears, learning to take the lead in sexual encounters and occasionally using oral medication to facilitate erection and thus regain confidence. Among other reasons this is also important because men with erection difficulties try to avoid condoms in order to avoid losing stiffness when they put one on, and as a result several studies have associated erectile dysfunction with a higher rate of sexual transmitted infections and diseases.

As for the physical causes, we have already seen how in recent years a very important message has been spread: erectile problems can be a first symptom of cardiovascular diseases and could become a tremendously useful early detection tool. To give just one example, arteriosclerosis is the hardening of the arteries, which can cause coronary problems, but in initial stages can harden the small arteries in the penis and provoke erectile dysfunction. Diabetes or hypertension would have similar consequences. So, if erection problems show up without any apparent psychological motive, we should discuss it with our doctor because it could be a valuable first sign of something in our organism that is starting to fail.

Beyond that, everything that affects the circulatory system, such as obesity, smoking or excessive drinking also contributes to erectile dysfunction. Although it is not well known because of doctors neglecting to sufficiently inform their patients, erectile problems are a side effect of many prescription drugs (for example, those used to treat hypertension), and it is worth demanding information and alternatives from your doctor. And then, obviously, aging itself is a factor. We will all have erectile dysfunction at some point in our lives, and we should know that if it bothers us, there are solutions. Medications based on phosphodiesterase-5 inhibitors, such as the sildenafil in Viagra or the tadalafil in Cialis, increase blood flow to the genitals and, always under medical supervision, are the simplest remedy for erection problems. If they are contraindicated, you can resort to injections of vasodilating substances, vacuum pumps or implants in the case of permanent dysfunctions. A urologist will be able to advise you in each case. Although a multidisciplinary approach should be sought out anyway, since there are many situations in which physical and psychic factors reinforce each other. A middle-aged man can start to notice that his erections are not as immediate as they were in his youth, which is a normal process that merely requires trust and longer stimulation time, but it could lead to performance anxiety that worsens the issue.

Premature Ejaculation

I vividly recall the German sexologist and historian Erwin Haeberle explaining to me the absurdity of the term "premature ejaculation" and its conceptualization as a dysfunction. "It should be called unsatisfactory time of orgasm," wise Erwin told me, arguing that for nature reaching climax in 1 or 2 minutes is not abnormal in the least. The problem is that many men don't even make it to that first minute and some even ejaculate uncontrollably right at penetration. And along with the refractory period needed to have another erection, this *is* a problem that seriously affects the confidence and sexual life of individuals and couples. In those cases that combine quickness, lack of control, and suffering, premature ejaculation is clearly a dysfunction. In fact, among men younger than 40 it is the most frequent sexual disorder.

Many studies have tried to calculate the average time between first penetration and male orgasm, and even though it is probably around 5 minutes, the key is not in the time but rather the lack of control and whether it is a recurring problem. It is certainly not pathological if you are very excited and somewhat tense with a new partner and you feel a different friction or heat on the glans that provokes ejaculation sooner than desired. The problem is when it happens all the time. In premature ejaculation, the situational and behavioral causes are more frequent than the physical ones, but surely there are some cases with a physiological component. It is not a phenomenon that is entirely understood, but it has been associated with low levels of serotonin and genetic polymorphisms in the receptors of this neurotransmitter.

There are various types of treatments that range from cognitive therapy to reduce anxiety and learn to regulate your excitement, behavioral therapy with desensitizing exercises to do alone or with a partner, anesthetic creams for the glans that should be used with a condom so they don't affect our partner, and even medications that in clinical studies have proved themselves clearly more effective than the placebo. As I have already explained in earlier chapters, those treatments originally come from medications for depression based on selective serotonin reuptake inhibitors (SSRI), after observing that they delayed ejaculation and even provoked anorgasmia as a side effect. They started to use them successfully in smaller doses to treated premature ejaculation, and now they have already designed SSRIs specifically for this dysfunction.

It is worth noting that certain rhythms and positions during intercourse can affect the timing of ejaculation, but that the classic remedies of masturbating beforehand or trying to mentally distract yourself during penetration are not generally very effective. We should also note that often erectile dysfunction and premature ejaculation are linked by anxiety and psychosomatic effects, and on many occasions solving the erection problem helps to increase confidence and lengthen the time before ejaculation.

Delayed Ejaculation or Anorgasmia

It is calculated that between 3 and 5% of men can have serious difficulties reaching orgasm in some period of their lives. The disorder could be brought on by injuries to the nervous system, drug consumption, medications such as antidepressants or tranquilizers, or having psychological determinants that range from relationship crises to having learned to masturbate in a way that is very specific and—although many dispute this last part—too vigorous. When the cause is medication, searching for an alternative or counteracting it with another prescription drug (Viagra, for example, but always with medical supervision) can improve the situation. In other cases, vibrators or lubricant gels can help, although the best thing is to see an experienced therapist.

Lack of Desire

Lack of desire affects more women than men, but with social changes and the increase in life expectancy, many men refuse to accept that their libido decreases over time. Obviously mood, relationship problems and a large number of psychological aspects influence lack of desire, but it is also a normal consequence of the progressive loss of androgens or hypogonadism (colloquially called "andropause"). Those who conceive of their own aging as a process of deteriorating health that they can try to reverse argue that testosterone patches, creams and injections can help contribute to maintaining a younger body and spirit in several aspects, including sexual desire. Testosterone doesn't affect everyone the same way, it has real side effects, it is contraindicated due to higher cancer risk, and its use should be supervised by a doctor. Absolute levels are not very indicative, and comparing between individuals, we see men with low testosterone and high desire, and also vice versa. Researchers believe that the key is more in the quantity and type of cellular androgen receptors than in the blood levels, but clinical studies have clearly confirmed that supplying testosterone to adults with a deficit has a positive impact on sexual function, and could improve general health in some cases. Those who are reticent to use it argue that it is overrated—especially in the United States—since obesity, diabetes, alcohol consumption, diet and exercise are much more limiting factors.

Other Male Dysfunctions

Sexual pain is much less frequent in men than in women, but there are cases caused by infections, nervous or inflammatory problems, penile trauma, or Peyronie's disease, in which the buildup of scar tissue causes a strange and sometimes painful curvature of the penis. Priapism is caused when an obstruction in the blood flow inside the penis produces an involuntary erection that doesn't go down for at least 4 hours. It is a rare condition that requires immediate attention because of the problems of hypoxia and cellular death that can occur. There are also cases of micropenises, penises that grow inwards (buried penises), pain specific to ejaculation and, obviously, a countless number of psychological disorders, some of which we will discuss further on.

Female Concerns. The Key Is Satisfaction, Not Desire

Female sexual problems can be conceptualized and classified into four broad categories according to their causes: (*a*) sociocultural factors such as lack of sexual information, overwork or religious pressures; (*b*) relationship problems such as conflicts, different interests, partner's health or lack of communication; (*c*) psychological factors such as anxiety, fears, low self-esteem or past traumas; (*d*) medical problems such as pain during intercourse, infections or hormonal changes that can lead to lack of desire or of genital arousal. All these categories are interrelated, and a skillful sexologist should be able to identify the origin and consequence in each case, avoiding preconceived ideas that underestimate the physiological or the psychosocial component.

Before I go on to describing specific dysfunctions, I wanted to comment on a controversy that began as academic but moved into the medical realm in 2013.

Desire Is Not the Same as Arousal

Something that illustrates the higher complexity of the female sexual response as compared to the male is the intense debate that went on during the World Meeting on Sexual Medicine in Chicago, which even devoted a specific plenary session to it, entitled "Are desire and arousal in women distinct psychophysiological functions, and should their problems be managed as such?"

It isn't a trivial matter. As we explained in Chap. 2, at this point the DSM-IV (Diagnostic and Statistical Manual of Mental Disorders, created by the American Psychiatric Association) does distinguish between two different disorders called "hypoactive sexual desire disorder" or HSDD and "female sexual arousal disorder" or FSAD. The first refers to the continued absence of desire and sexual fantasies, and the second to problems of lubrication, vaginal blood flow and clitoral erection. Even though they are clearly correlated, for most experts the distinction between mental desire and physical arousal is a very useful one in their therapeutic practice, and they document cases in which clearly only one of the two components is failing.

However, it is highly likely that both disorders are joined together in the next edition of the DSM-V as a single disorder called "sexual interest/arousal disorder" or SI/AD. Why? For many this would be an error based more on academic grounds than on medical ones, and perhaps because of preferences from the pharmaceutical industry.

If we recapitulate a little bit, in 1966, Masters and Johnson published their famous linear model of human sexual response, made up of stages of excitement, plateau, orgasm and resolution. Later, in 1979, Helen S. Kaplan added the phase of mental desire as the beginning of the sexual response and, in 2001, Rosemary Basson described her circular model in which sometimes mental desire can precede physical excitement, but physical contact and excitement can generate sexual desire. Since then, various studies have demonstrated that an equal number of women identify with the linear models by Masters and Johnson and Kaplan as with Basson's circular model, but on a practical level the differentiation between mental desire and physical excitement remains in use.

Nevertheless, those in favor of joining them in the DSM-V under the umbrella of one single dysfunction argue that there is no solid scientific evidence for differentiating HSDD from FSAD, that both are constructs created based on the opinions of experts, and that the fact that Viagra doesn't work on women implies that desire and excitement are not as separate for them as they are for men. Those who defend keeping the disorders separate argue that desire and excitement are regulated by different hormones, that on a practical level it is very useful, and that Viagra does work in some concrete cases such as those women who have arousal problems caused by antidepressants.

The subject is controversial and not something that concerns us in our daily lives, but I mention it because it is truly one of the most intense discussions in sexual medicine and because everything indicates that, despite the discrepancies, HSDD and FSAD will be unified in the next DSM-V (the primary world reference in psychiatry) under the new sexual interest/arousal disorder or SIAD.

SIAD will be defined as the "lack of interest and sexual desire during a minimum of 6 months manifested as the absence or decrease of the frequency or intensity in at least three of these six categories: (1) interest in sexual activity, (2) erotic thoughts or fantasies, (3) initiation of sexual activity and lack of receptivity to partner initiation, (4) excitement/pleasure during the sex act in more than 75% of encounters, (5) interest or excitement in response to erotic or sexual stimuli and (6) genital and non-genital sensations during sexual activity in more than 75% of encounters." Important to note: this is only considered a dysfunction when "the change causes a clinically significant discomfort or impediment." This last point is essential.

Lack of Desire Is Not Necessarily a Problem

Medical concern for the lack of desire skyrocketed in 1999 when Laumann and Rosen published, in *JAMA*, data from the American health survey (NHSLS), according to which 43% of adult women between 18 and 59 years of age suffered some type of sexual dysfunction, the most prevalent being the lack of desire, which affected 33% of women in that age range.

This data generated a big stir in the world of sexology, and was used to defend the idea that medicine should start to look at female lack of desire as a serious health problem, that science should devote much more energy to investigating the biopsychosociological causes of this so-called dysfunction, and that pharmacology should search for possible treatments.

Over the following years the subject was much talked about, but in 2008 a new study lessened the supposed seriousness of the problem. The study, called PRESIDE (Prevalence of Female Sexual Problems Associated with Distress and Determinants), carried out a very large survey of 31,581 women in which it not only asked about their lack of desire, lubrication problems, anorgasmia, etcetera, but it added a question about whether these disorders generated distress. The results were impressive: when adding the question about distress, the percentages lowered significantly. Which is to say, the number of women who felt a lack of desire or arousal was still very high, but for many of them this was not a source of any discomfort or concern. Specifically, the percentages of women with some type of sexual dysfunction accompanied by distress were 10.8% of those between 18 and 44, 14.8% between 45 and 64, and 8.9% in women over 65, and in the particular case of lack of desire, 8.9, 12.3 and 7.4 respectively.

The conclusions were very obvious: lack of desire continues to be a very important sexual health problem for women, it should in no way be considered "normal," and psychotherapists as well as doctors and scientists should be attentive to it. But it also true that for many women the lack of sexual desire in no way interferes with their wellbeing. At the extreme end of this position, activists like Leonore Tiefer defend the idea that the concept of female sexual dysfunction is an artificial construct fostered by the pharmaceutical industry and a sexist society that demands higher than natural levels of libido in women. Although Tiefer's stance may seem excessive in some aspects, in many others it makes a lot of sense.

One of the focuses of Tiefer's attacks with her New View Campaign is the search for a "female Viagra" or medication to increase desire and arousal in women, which seems like it may be very close to reaching the market. In Chap. 2 we already mentioned that in 2010 the FDA did not approve the German manufacturer Boehringer's drug Flibanserin, arguing the lack of studies proving its efficacy and safety, but the company Sprout has acquired the patent and is carrying out those clinical studies with apparently good results. The new drug doesn't act the way Viagra does, by increasing blood flow, but rather it alters the levels of dopamine, serotonin and norepinephrine involved in the appearance of desire. Although the FDA approved the drug in 2015 after the publication of more studies underwritten by Sprout, it is too early to judge its possible efficacy, risks and the cases it may benefit. It will surely be much talked about.

The fact is that in the United States, despite it not being indicated for women, many clinics recommend testosterone supplements to augment female sexual desire. I have talked to both doctors who say they've gotten very positive results and experts reticent due to the possible long-term effects of breast cancer, insulin resistance or higher cardiovascular risk. The truth is that androgen levels in women decrease with age, and some studies show that—with vast individual variation—testosterone does, modestly but significantly, increase female libido in certain situations, especially in cases of androgen deficit. In 2008, for example, the *New England Journal of Medicine*, one of the journals with the most impact on the biomedical field, published a double blind clinical study in which 814 post-menopausal women with hypoactive sexual desire disorder (HSDD) who were not receiving estrogen replacement therapy were randomly assigned to a group that, over 52 weeks, received 150 µg of testosterone from a patch, or another that received 300 µg of testosterone and a third that received a placebo. The conclusions showed that those who received 300 µg of testosterone a day slightly improved their sexual function.

Obviously the effects of testosterone on breast cancer and cardiovascular diseases still has to be further investigated, but those who defend supplementary therapies also argue that androgen deficit is associated with loss of bone density and muscle mass, as well as deterioration of mood and physical wellbeing. One of the most heated debates surrounding this topic at the conference was whether the birth control pill could have negative effects because it lessens testosterone.

The Birth Control Pill Could Reduce Sexual Desire

Taking birth control pills reduces testosterone levels. It happens to all women, it is a chemical process. On one hand, oral contraceptives reduce the production of steroids in the ovaries, and on the other, they increase blood levels of a protein called SHBG or sex hormone binding globulin, which attaches to testosterone and estradiol and diminishes their bioavailability. When levels of SHBG are increased in the blood stream due to oral contraception, the testosterone levels go down, without exception.

This isn't alarming, and most likely it has no effect. However, according to a review of the scientific literature published by Andrew Goldstein in September of 2012, there are a small percentage of women whose sexual desire lessens after starting to take the pill. There are no studies yet with reliable estimations, but in the conference in Chicago they told me that it could affect some 10% of women. On one hand, what's really shocking is that this hasn't been investigated or explained to pill users much earlier, and it once again shows that sexual function is rarely considered as a side effect of medications. On the other, what caught my attention was the question of why only some women were affected.

Francisco Cabello is one of the few sexologists in Spain who combines excellent scientific productivity with long clinical experience. He says he manages it by not sleeping much. At the European Federation of Sexology Congress held in Madrid in September 2012 he confessed to me that they are lost when it comes to this issue of testosterone. Francisco knows about the studies unequivocally linking the increase in androgens with sexual desire, but in his practice he also finds countless patients in whom it has absolutely no effect, including pederasts or hypersexuals whose testosterone levels are chemically diminished with drugs and yet they don't lose any of their desire. According to Francisco Cabello, the relationship between testosterone and libido exists, but it is plagued by exceptions.

This enormous interpersonal variation makes investigators think that the determinant from a biological standpoint is not the levels of free testosterone in the blood, and they are beginning to suspect that the key could be in the cellular androgen receptors, which would make the cells more or less sensitive to shifting testosterone levels. In fact, Andrew Goldstein announced that they are analyzing genetic polymorphisms linked to those androgen receptors in order to find some relationship to whether increasing testosterone influences sexual desire or not. In the future, this could lead to personalized sexual medicine, which would serve to identify which types of women could benefit from androgens, and in what quantities. Perhaps even by analyzing genetic information, some women could be prescribed testosterone supplements along with their birth control pills.

We have talked a lot about chemistry when lack of desire is much more highly influenced by sociocultural, relationship and psychological factors. I'll justify that by saying that from the scientific perspective this is the most novel aspect, and that the big, controversial debate that will soon erupt is about the medicalization of female sexual health. It will be important to have voices speaking up to remind us that what's truly important is not desire so much as satisfaction.

Dysfunctions Due to Lack of Arousal and Lubrication

When a woman becomes aroused, blood flow in her vulva and vaginal walls increases, she begins to lubricate and her clitoris grows larger. When this doesn't happen, intercourse is less pleasurable and can even be painful. There are many physiological factors that intervene in physical arousal and in the feeling of greater genital sensitivity. For example, the loss of estrogens after menopause significantly lessens lubrication and some prescription drugs or cardiovascular problems that negatively affect male sexual response could also have an impact on female genital arousal. That is why at first they thought that Viagra could help in arousal dysfunctions, but several studies in the early 2000s confirmed that it was only effective in some very specific cases, and that in most it had no effect at all. Estrogen replacement therapies also improve sexual response, and if lubrication is the only problem, a good lube that doesn't irritate the vaginal tissue is more than enough. Some of those products include substances that can topically increase blood flow, but for the moment one of the best strategies for improving physical arousal is cognitive and behavioral therapy that includes relaxation practices and sexual exercises that should be done alone or with a partner.

In this sense, one of the consequences of the cyclical model of sexual response put forth by Rosemary Basson is that sexual desire doesn't necessarily have to be the first step to starting intimate contact. Many therapies are based, even with no desire, on doing subtle genital massaging or caressing that generates arousal and with it desire and further physical excitement. This cyclical process would work in cases in which the lack of desire and arousal are clearly linked and not so much when one exists without the other.

Some cases of anorgasmia could be related to the disorders of lack of arousal and be susceptible to improvement with the same therapeutic approach. According to the statistics, about 10% of women have never experienced an orgasm and a much higher percentage have trouble reaching climax during intercourse. Some of them experience this as more of a problem than others do, but if there is concern it is definitely worth seeing what medications could be influencing your sexual response (primarily antidepressants), and seeking out the help of a sexologist who can recommend masturbation techniques or sexual exercises that are effective in quite a few cases.

Persistent Genital Arousal Disorder

A radically opposed case is that of women who, without any desire at all, feel their genitals permanently excited and on the edge of orgasm. Some of them have to constantly be giving themselves orgasms in order to be able to relax for a few minutes. It is an extremely uncomfortable situation and not pleasurable in the least, even leading to some suicide attempts. Interestingly, doctors have tried blocking and even cutting the nerves that lead to the clitoris and the situation still continues. Persistent genital arousal disorder (PGDA) is a strange syndrome whose etiology is as of yet unknown, but I mention it because in September 2012 Barry Komisaruk published a possible explanation.

After learning about the case of a woman who had PGDA and Tarlov cysts where the pudendal nerve enters the spinal column, Barry contacted an association of women affected by PGDA and asked that they send him magnetic resonance images of their sacral spinal region. Eighteen women complied and 12 of them had between one and three Tarlov cysts pinching nerve endings. If these results were replicated with a larger number of patients, it would confirm that PGDA could be a result of the persistent arousal of the nerves at the height of the spine, which would explain why blocking the nerve endings at the height of the clitoris didn't help at all, and would suggest a possible treatment based on local anesthesia application to the area where the Tarlov cysts are, and this would be a fabulous example of how scientific knowledge can better our health and wellbeing.

Vaginismus and Pain During Intercourse

Pain during intercourse, or dyspareunia, can be caused by congenital malformations, infections, lack of arousal, inflammations, insufficient lubrication, allergic reactions, vaginal atrophy or a very peculiar condition called vaginismus.

In vaginismus, the muscles of the pelvic floor involuntarily contract, closing the vagina and impeding any type of penetration, making sexual encounters or even introducing a simple tampon impossible.

Conventional opinion is that vaginismus has psychological causes and stems from anxiety and fear of sex, perhaps as an unconscious reaction to traumatic experiences in the past or the anticipation of pain during intercourse. Sometimes it is understood as a "phobia of penetration," but many researchers believe that in certain cases it could be linked to muscular or nerve injuries, or be a reaction similar to "trying to stick a pen in your eye," and that on a diagnostic level, a physical examination is as necessary as a psychological one. It is a very active field of study with diverging opinions, well illustrated by the comment made by a gynecologist at the sexual medicine congress to a therapist who assured him that vaginismus was always psychological: "You shouldn't attribute dyspareunia to anxiety or depression just because you have no idea what is going on in your patient's vagina."

What is true is that psychotherapy to eliminate blocks and aversions works in many cases, while Botox injections to stop muscle spasms have also had positive results, and one of the most frequently used treatments is based on the patient periodically introducing a series of different sized dildos into her vagina until her muscles are used to it and then trying penetration with her partner (or as the shockingly arrogant scientist called it "Surrogate versus couple therapy in Vaginismus," with the therapist acting as a "surrogate partner").

Dyspareunia is one of the sexual problems with more possible different causes, and the message continues to be the same: if you feel continued pain during intercourse you should talk to a gynecologist about it because there are solutions.

The Art of Sexual Therapy

I started this section by saying that sociocultural, relationship and psychological factors are involved in sexual problems, and I realize that so far I've focused almost exclusively on the medical ones. This book deals with science as the unifying tool for the wide variety of theoretical approaches to sexuality, but recognizing from the very start that the scientific perspective is just a partial and therefore limited view of human sexual behavior. I am well aware that Johan and Darlane, a New York couple 43 and 39 years old respectively, with excellent physical and psychic health, an awesome son, a comfortable socio-economic position, a real feeling of mutual appreciation and an apparently happy life, have no medical or physiological reasons for not having made love in the last 6 months and not feeling sexual desire for one another. It would be foolish to think that testosterone could increase Darlane or Johan's arousal. In fact, he has no problem getting an erection when he masturbates watching porn on the Internet, nor does she have any desire problem when fantasizing about other men. Sexology will never be an exact science and will always require the art, intuition and individualized therapy of an experienced sexologist. We will describe many more psychological and sociocultural aspects in the next few pages of this book, but first we will finish this chapter devoted to sexual health with the unavoidable mention of the protective microorganisms and pathogens that live together in our genitals.

The Microorganisms That Cohabitate in or Invade Our Genitals

Even in healthy conditions, women do not all have the same bacterial communities in their vaginas. In fact, in 2012, researchers at the Human Microbiome Project established five different categories of healthy vaginal flora. Some communities make more acidic environments, some have more of an odor, but they all create essential balances to protect the genitals from other pathogenic microorganisms.

Likewise, circumcised men have different microorganisms in their penises than uncircumcised men. If we think about it, it makes all the microbiological sense in the world: in a circumcised penis the glans is exposed to an environment with oxygen that favors the growth of aerobic bacteria; but an uncircumcised glans spends most of its time covered by the foreskin so no air gets in, allowing anaerobic microorganisms, whose metabolism prefers that lack of oxygen, to thrive.

Different Bacteria in Circumcised and Uncircumcised Penises

I know what you're thinking: if you wash carefully down there, you won't have any bacteria. That's false, not even if you scrub so hard you get past the inner layers of the epithelium. Like in our eyes, mouths, intestines and nostrils, the deepest layers of the skin on our bodies are filled with bacteria that carry out basic functions we couldn't live without.

In fact, we have 10,000 bacteria on just a square centimeter of the skin on our arm. Most of it is washed away in the shower, but don't bother scrubbing too hard because they reappear quickly. If we dig a little bit on that same square centimeter, we will find 50,000 bacteria per square centimeter on the more superficial layers of the epidermis, and if we do a biopsy that reaches the hair follicles we would find a million. But it's not the amount that should impress us, given that since microbiologists have had powerful genetic sequencing techniques to analyze microbial ecosystems in a totally new way they've found that just on our skin there are a hundred times more types of bacteria than previously known. This represents a paradigm shift for some aspects of medicine, including gynecology, and for microbiology itself.

Just think that, not long ago, if a conventional microbiologist wanted to find out what kind of microorganisms were living in a pond, our mouth or our vagina, he would take a sample, spread it out into different culture mediums, wait a few days for various colonies of bacteria or fungi to grow and then analyze their characteristics. The process was slow, but what was really tricky was when they started to carry out genetic analyses, the microbiologists saw that in those culture plates only an infinitesimal part of all the bacterial diversity of the original environment would grow. Bewildered by that, in 2003 the US National Institutes of Health began the Human Microbiome project, whose objective is to study all the diversity of characteristics and functions of the microorganisms in our body. To do so, they took samples from 18 areas in the female body and 15 in the male from a total of 242 healthy volunteers between the ages of 18 and 40, and they are analyzing them with new and powerful techniques of metagenomics: taking a sample, directly sequencing all the different genes it contains and, based on that, using tools from computational biology to discern the bacterial species. In addition to allowing us to see the broad hidden diversity that doesn't grow in culture plates, since it is faster and the sample can be analyzed directly in the collection site and not from a colony, it allows for more frequent analyses, observing their evolution and understanding more functions. Metagenomics is for microbiology what the fMRI is for neuroscience and the LHC is for particle physics. And there are already important results.

To finish up the matter of the penis before getting into vaginal flora, Jacques Ravel, at the School of Medicine at the University of Maryland, tells me: "In fact, the bacteria in uncircumcised penises is more similar to the anaerobic bacteria that grows deep inside vaginas, and that in certain cases could cause less alterations to the vaginal flora and less likelihood of vaginosis." Two important clarifications show up soon after those words: the first is that in monogamous relationships, the partners' microorganisms end up being very similar, whether there is circumcision or not, and therefore circumcision is irrelevant in stable partners. The second clarification is that in casual relationships, since in theory we should be using a condom, that exchange of bacteria from the glans is practically nonexistent. That said, in the case of new relationships without protection, circumcision—which in an earlier era of less personal hygiene was promoted as a supposed protection—does indeed now pose a possible problem for certain categories of vaginal flora, especially when there is ejaculation and the semen increases the vaginal pH from 3.5 to 7, leaving the bacterial flora temporarily vulnerable to invaders that can start growing unchecked.

There are many more factors to explain. In 2010, an American study compared the microbiota of 12 Ugandans before and after circumcision, and observed that two families of anaerobic bacteria (Clostridiales and Prevotellaceae) practically disappeared after circumcision. That was important because some bacteria in those families are linked to inflammation processes that activate the Langerhans cells that transmit HIV to the CD4 lymphocytes. In laymen's terms, what the investigators concluded is that in the case of protection from AIDS, circumcision can offer some protection because of the change in microbiota it entails. Obviously, shedding light on this point could have important implications on the public health level, and also for the care of the vaginal ecosystem.

Odor and Excess Hygiene in the Five Categories of Vaginal Flora

Jacques Ravel is one of the world leaders in the study of vaginal microbiota. As he explains to me, "Of course bacterial flora changes during the menstrual cycle, with age and hormone levels, but we already knew that. What's new is that we've established five categories of microbial flora, all of them healthy, but with different properties. That means that treatments and recommendations should be personalized for each woman according to their vaginal flora."

And we really do have to visualize the vagina as an ecosystem. In the same way that a forest is an ecosystem with different species in a balance that can be broken if external invaders come in to hunt, chop down trees or introduce new species, bacterial flora act in exactly the same way. The different species of Lactobacillus try to maintain the balance and protect it from the invaders by generating an acidic pH of 3.5. But just like a forest in the Pyrenees doesn't have the same trees and plants as one in the Mediterranean, not all healthy vaginas have the same microorganisms nor are they equally susceptible to various aggressions. "For example, there are vaginal flora that take longer to restore their balance after intercourse or a douche."

This would imply that, if out of an excessive idea of cleanliness, a woman douches before going out with a potential new lover and eliminates part of her own bacterial flora, she is really making herself more vulnerable to the bacteria the man could transmit to her. Although it all depends on the type of vaginal flora she has, which can also change over time.

This data is from 2012 and surely it will be expanded and modified but, according to Jacques Ravel, categories I, II and II are dominated by Lactobacillus crispatus, gasseri and iners, respectively. These Lactobacillus produce lactic acid, but each one allows other species around them. Categories IV-A and IV-B are more heterogeneous and have much less Lactobacillus and a much higher pH. In IV-A, strict anaerobic bacteria such as *Anaerococcus*, *Corynebacterium*, *Finegoldia* and *Streptococcus* are found, and in IV-B, large amounts of the genus *Atopobium*, along with species of *Prevotella*, *Parvimonas*, *Sneathia*, *Gardnerella*, *Mobiluncus* and *Peptoniphilus*. They are all healthy, stable communities; yet, the latter, for example, entails more risks and in a promiscuous phase has to be much more cautious. We assume the use of condoms during coitus, but even receiving oral sex could be more delicate depending on the vaginal flora at that moment.

Jacques Ravel has genetically analyzed the evolution of the bacterial flora of a wide population of women, and has observed that in African-Americans category IV is much more common than in Caucasians, and that women with type II don't often transition to other types, but that those with III and IV-B frequently switch back and forth. He doesn't yet have conclusive data, but it seems that differences between the categories can also affect fertility or predispose someone to more bacterial infections such as chlamydia or viral infections such as HIV. "At this point gynecologists continue to treat all women the same way without taking their vaginal flora into consideration, the challenge now is to reach the medical community so that they can begin to personalize their work more," explains Ravel, who in his article published in March of 2012 in *Science* says that "now, clinical examinations look at pH, the presence of odor and characteristics of vaginal fluid to establish whether a flora is normal or abnormal. But this is very limited because we see that it changes over time and that not all healthy women are the same." A curious example is a study by researchers in Houston that compared the vaginal microbiota of 24 pregnant women and 60 who weren't pregnant, and observed that, just as they prepared for birth, the bacterial flora changed dramatically, the diversity of species suddenly lessening, possibly so that the newborn—a sponge for bacteria—finds a less aggressive environment.

The case of odor is paradigmatic. It is often associated with infections, but in some women it is something constant and normal. Why do some vaginas have that characteristic smell and others don't? It's not necessarily a question of hygiene, but rather one of being in a category that allows for the growth of bacteria that produce amine compounds or not. For many women it could be uncomfortable for them or their partners and gynecologists can treat it, but according to Ravel in the future they will even be able to use probiotics to completely modify vaginal microbiota. Don't be surprised if, within a few years, 2 weeks before going to a checkup at the gynecologist's, women receive a kit in the mail with various vials to take different samples from their vagina. They would then take them to their appointment, and right there, in a question of minutes, they would analyze them to establish their bacterial flora category, and depending on that they would be prescribed a treatment and given specific recommendations on their sexual conduct, which could be very different from those offered to the next patient.

Sexually Transmitted Diseases

I wouldn't want to simplify information on sexually transmitted diseases. If you think you have been exposed to risk or you notice any irregularity, consult your doctor or seek out information from reliable sources. There are many good places online, including Wikipedia. But be careful, the symptoms of some genital infections can overlap and be confusing. Get regular checkups, because most infections are easily treatable but if they aren't dealt with they can cause sterility or other problems. That said, I will give you an overview here of the most common microorganisms that can cause from passing discomfort to very serious diseases.

Bacteria

Chlamydiasis is one of the most common bacterial infections. Caused by different strains of the bacteria *Chlamydia trachomatis*, it can be contracted through vaginal, oral or anal sex, or through contact with areas of the skin that contain the bacteria. It is calculated that in the Untied States there are three million infections per year, of which only a little over a million are diagnosed by a doctor. That is because in 75% of women and 50% of men chlamydia doesn't show any symptoms, or they are very minor and disappear over time. But the fact that it is easy to treat should not be overestimated; it can sometimes remain latent and manifest after various sexual encounters that destabilize the bacterial protection. The most common symptoms show up 5 days after infection and are a burning sensation in the genitals, pain during intercourse, the appearance of a yellowish liquid in women and a sort of pus in men. There can be little pimples or small ulcers. If you notice symptoms after sleeping with someone, it doesn't necessarily mean "they gave it to you." You may have already had it in your body and the sexual activity favored its proliferation. As with everything explained here, let your doctor make the diagnosis.

Neisseria gonorrhoeae is a bacteria that can grow in the damp parts of the genitals, mouth, anus and even the eyes or throat. When it manages to set up shop in your nether regions and multiply uncontrollably it causes gonorrhea, a symptomless disease in 80% of women and 10% of men. Pain during urination—more in men than in women—is a clear symptom, along with unusual secretions or even vaginal bleeding.

Syphilis is caused by the bacteria Treponema pallidum. It grows in the genitals, mouth and anus and is transmitted by direct contact with open wounds or the typical sores produced by the disease itself. It is much less common nowadays. The appearance of sores is a clear way to easily identify that there is a problem. If it goes untreated it can lead to fevers, serious neurological problems and even death.

Bacterial vaginosis is the most common infection, although there is some dispute as to whether it should be classified as a sexually transmitted disease. Its origin is an imbalance in the vaginal flora that reduces the number of acid-producing bacteria. The pH increases and other microorganisms, such as Mycoplasma, grow, causing abnormal whitish or grayish discharge, and a very strong odor that is reminiscent of fish. Normally vaginosis occurs after a series of frequent sexual contacts, especially with new partners, or after too much douching. But the causes are still not well known. The use of a condom doesn't prevent it, and it can occur even without sexual contact. Occasionally antibiotics help, but there are also natural remedies, and the system can rebalance itself on its own. The frequency of reinfection is high.

Fungi

Seventy-five percent of women will have at least one fungal infection at some point in their lives. Candidiasis caused by the *Candida albicans* fungus is the most common. The symptoms are serious itching—more intense in the vagina than in the penis—and reddening and irritation of the genitals. A man's glans can ooze a sort of solid white pus with a texture similar to that of cottage cheese, generally odorless. Personal hygiene is basic for prevention. Antifungal creams or pills help, but a medical analysis is essential to ruling out other diseases that sometimes accompany it. In fact, although sex can irritate the genitals and increase the odds of passing it to your partner, candidiasis is not really a sexually transmitted disease, but rather the result of an imbalance in our community of microorganisms.

Virus

We aren't going to say anything new about HIV or hepatitis B here. Both can be caught through the exchange of fluids, there are some sexual practices that are much riskier than others, and in a sexually promiscuous life you can never be careful enough. Most people are vaccinated against hepatitis, but no such shot yet exists for AIDS.

Herpes is a much, much less serious virus, but it can stay with us all our lives in a latent form and awaken in different periods with annoying itching and sores. Its symptoms are very varied, and your doctor will tell you about how you can check regularly for them, which is something you should be doing if you aren't (yes, guys, I'm talking to you).

The human papilloma virus (HPV) perhaps deserves a little more attention because of recent data linking it to so many cases of cervical cancer, and the relatively recent appearance of vaccines against it. There are more than 150 different subtypes, of which 40 are easily transmittable through cutaneous contact and between mucous membranes. In fact it is considered the most frequent sexually transmitted infection in Western societies. Most of the people infected with HPV are unaware of it, never have symptoms, and only some ever get little warts on their genitals and anus, but it is clearly linked to cancer of the cervix, to a lesser extent to cancer of the penis, is the main cause of the strange but increasingly frequent anal cancer, and is also tied to the continued rise of oral and buccopharyngeal cancers. The vaccine doesn't immunize against all the subtypes of HPV, but it does cover the most frequent ones, and is recommended for teenage girls and young women.

And while we're talking about microorganisms, we will continue to see how they were the ones who invented sex in nature, and we'll discover the fascinating evolution of sexual reproduction from bacteria, amoebas, plants, arthropods, reptiles, birds and mammals until we get to the primates, to see if that can help us to understand some of the irrational behaviors of those bizarre Homo sapiens.

7

Sex in Nature

If we think human sexual behavior is diverse, let's take a look at nature and be humbled. No erotic ritual matches the sophistication and ingenuity of the hermaphroditic flatworms' penis fencing.

Planarians are a species of 2-inch-long flatworms that can reproduce sexually or asexually. Asexual reproduction is simpler: like starfish or geraniums, they cut off a little piece of their bodies and from that a new organism is created, which is genetically identical to the "parent." Flatworms have such an extraordinary ability to regenerate that, even if we slice off just a bit of a tail, it will become an entire body, head included. Which is why biologists have been studying them for more than a century.

Some planarians also reproduce asexually through germination, like yeasts and very simple marine animals: one cell from their bodies takes on a life of its own and begins to multiply on the flatworm's skin. It grows and differentiates itself until it is a new organism and, once the process is complete, it separates from its identical progenitor.

Asexual reproduction is very practical, but it has the disadvantage that it leads to very little genetic diversity in the species. That doesn't bother bacteria because they have other ways of exchanging genes between them, but for an animal this lack of diversity is a big risk in the face of possible environmental change. That's why if the conditions are right, flatworms prefer to develop female and male gametes and reproduce sexually, so they can combine the genes of two different individuals so that the resulting newborn is completely genetically new.

But since they are hermaphrodites (each individual flatworm produces both ova and sperm), which one is the mother and which is the father?

The truth is that all the flatworms prefer to be dads. If you think about it, it's a much more comfortable role: the worm who takes on the male role just has to leave his gametes in the body of the "female" one, and she'll have to suffer through all the energy expense of gestation. And in the end they'll both have equally contributed 50 % of the genes. It's unfair, but that's how nature is. How do the various species of flatworms resolve this conflict? Well, through brute force: fighting over who sticks their penis into whom.

The images are spectacular. Two purple flatworms face each other and arch their bodies, lifting their trunks threateningly and challenging each other with some sort of white cone located at what would be our abdomen. Those are their penises, and on their tips they have a thin needle they each try to use to inject semen into their rival's body. The battle begins: the flatworms take position, they approach each other, sizing each other up, and then suddenly one will pounce on the other, who swerves to dodge the insemination attempt. He counterattacks but is also blocked. Soon their bodies join, they twist and fold, until one of them manages to prick the other's skin with his penis, marking the end of the battle. The semen will travel subcutaneously through the losing flatworm's body until reaching its female gametes and starting gestation.

There are many curious examples of sex in nature, and there is no need to look for it in rare animals. The seventh arm of the male octopus is adapted to transport semen from his body to the reproductive organ of the female. If you ever see an octopus with its arm extended to caress the nether regions of another one, well, let's just say he's not tickling her.

And scientists are always pointing this stuff out. For example, there was an article published in 2009 in the scientific journal *PLoS* by Chinese and English researchers entitled "Fellatio by Fruit Bats Prolongs Copulation Time." Oral sex is quite rare in the animal world. Beyond a few licks without any sexual intent, it's only really bonobos and us humans who use it as part of our stimulation process. But the article about bats says: "Our observations are the first to show regular fellatio in adult animals other than humans," and describes observing "the female lowers her head to lick the shaft or the base of the male's penis but does not lick the glans penis which has already penetrated the vagina. Males never withdrew their penis when it was licked by the mating partner. A positive relationship exists between the length of time that the female licked the male's penis during copulation and the duration of copulation." In fact, in the article they assert that for each second of fellatio, the copulation time is lengthened by 6 seconds. They speculate that this peculiar case of contortionism can increase reproductive success and avoid disease. Another example of how fascinating sex's evolutionary history is.

Why Do Ducks Have Penises and Roosters Don't?

At first glance, ducks and chickens don't look that different. They are birds of a similar size, they lay eggs, we see the chicks and the ducklings following along behind their mothers… so if we told you that, evolutionarily speaking, they are very closely related, I suppose you wouldn't be particularly surprised. But no, they aren't, and there is one very peculiar difference between the two species: ducks have a penis and roosters don't.

In fact, only 3% of bird species have penises. Why? What evolutionary puzzle gave ducks a penis and not roosters? This question was bothering me for months. I didn't find anything on the Internet to explain it, and none of the evolutionary biologists I consulted could help me either. I finally got some clues to the mystery from Patricia Brennan, a Colombian biologist at the University of Massachusetts. She was studying how the sexual conflict between genders had molded the shape of duck penises and vaginas into corkscrews. To sum up what she explained, the females try to make it difficult to enter their vaginas to avoid being fertilized by just anybody without their permission, and in turn natural selection reacted by favoring penis shapes that make it more difficult to disengage once they've been introduced. Leading to the duck's corkscrew shaped penis. But, in fact, this is the exception, since the first ancestors of the birds evolved from dinosaurs did have penises, but something happened to make evolution gradually faze them out in most bird species. Why?

If you were born in the city and are wondering how a rooster procreates without a penis, the answer is very simple: both males and females have an orifice called a cloaca that serves as both an excretory and reproductive apparatus. If you recall the first few chapters, during an embryonic stage we too had a single cloaca/hole that later divided into anus, urethra and genital conduits. When a rooster wants to fertilize a hen, if she's up for it, she will get into a position that exposes her cloaca and opens up the hole. Then the male places himself on top of her and in a matter of 2 seconds drops into her the semen stored in some glands located very close to his orifice. The female closes her cloaca and stores the semen inside her oviducts. There, that semen will fertilize several eggs over the following days.

But, again: if evolution created the penis as a very useful organ for joining gametes, why is it that 97% of bird species don't have them? In natural history, chance can explain some isolated cases, like the few snake species that have two penises, but undoubtedly this consistence in birds must be based in some very important evolutionary justification. Patricia told me that she isn't positive, but that there are several published hypotheses that attempt to explain the disappearance of the bird penis.

The first hypothesis is that it could be a way to reduce weight for better flying. If that bit of flesh is not essential, they can travel more lightly. Patricia doesn't subscribe to that explanation because ducks have long curved penises and they are migratory birds that fly very long distances. A second hypothesis is that it disappeared in an ancestor common to all bird species, and it only remained in a few waterfowl, like ducks, since they needed a penis to ensure that their semen didn't scatter and get lost in the water. This isn't a "watertight" theory either, since the ostrich lives on land and has a penis. Patricia tends more toward the third hypothesis: being that the cloaca is both anus and reproductive organ, avoiding the introduction of a penis would reduce the risk of infection. Although there is insufficient data to corroborate it, it's not a preposterous possibility. Patricia is studying whether the length of the penis in different duck species is correlated with the number of infections per sexual encounter.

Nevertheless, the fourth hypothesis is the most widely accepted: the penis was lost over time to favor the females having control and being able to choose whom they wished to mate with. Without a penis, no bird could force himself on anyone, and the female would only be fertilized by the male she chose. And in an animal group where the female invests so many resources into caring for her young, this is fundamental. The problem, again, is that confirming this hypothesis experimentally is impossible, researchers can only carry out phylogenetic studies to see if it adds up. Patricia explained that there are investigations underway and data that suggests that the loss of the penis happened independently in various groups of birds, such as the Galliformes. This convergence would imply that the lack of a penis did indeed suppose an evolutionary advantage in birds, and the feminine selection theory would be the most plausible. My goodness, when bacterial reproduction is so simple…

The Origin of Sex in Bacteria, Amoebas and Sea Sponges

We often say that we humans are the first species to ever separate sex from its reproductive function. But strictly speaking, bacteria did that billions of years ago when they started to share genetic material in a controlled way. Bacteria reproduce asexually: one bacterial cell has all the DNA scattered throughout inside of it, and at a certain moment it compresses it into chromosomes, makes those chromosomes duplicate, sends a copy to each side of its cellular body, and then the bacteria splits in half. This creates two identical organisms, which actually should be considered just one. In fact, which is the original, the left one or the right? It's really both.

This reproduction through bipartition worked pretty well, but soon bacteria invented something called conjugation, which consisted of copying just a small piece of DNA, shaping it into a circular plasmid and passing it to another bacteria through a thin little tube that connects them for a few seconds. This process recombines genetic material, and even though there is no reproduction properly speaking, it could be considered the first form of sex on the face of the earth.

It's more a question of terminology, since really the early stages of what we understand as sexual reproduction wouldn't occur until billions of years later with the appearance of single-celled eukaryotic organisms.[1]

Let's use the amoeba as an example. Originally those single-celled beings continued to duplicate their genetic material and divide themselves asexually into two identical organisms. But at some point in their evolution—still unclear to biologists—these microorganisms developed the ability to divide their DNA in half, package it and combine it with some other amoeba's half-DNA, thus creating a new, unique being. It was the origin of the gametes and sexual reproduction. Today, many protists alternate phases of asexual reproduction with ones of sexual reproduction. The question is obvious: why complicate life so much if you could already make countless copies of yourself on your own, and even feel the thrill of immortality? We sketched out the answer earlier: variation and the incorporation of new genes are positive things for natural selection, since they contribute a diversity that makes those organisms more resistant to environmental changes.

This generation of gametes was the evolutionary basis on which, later on, multicellular organisms would develop sexual reproduction with the ova and spermatozoa we are familiar with. But let's take it one step at a time, and the first step is recognizing that asexual reproduction is still very common in animals and higher plants. We talked about starfish and flatworms that can individually divide themselves, but there is another, very particular, type of asexual reproduction: parthenogenesis. A classic case is that of a certain species of lizards, of which there are only females. What they do is quite

---

[1] A eukaryotic organism is one whose DNA is contained in a well-defined cellular nucleus, instead of scattered everywhere like in the case of bacteria.

simple: instead of having an egg with half of the chromosomes waiting for a sperm cell to show up with the other half of the chromosomes, when these lizards want to reproduce they duplicate the genetic material in the egg and make it start to develop as if it had been fertilized. It is really a simple process and can be reproduced in the laboratory. In fact, it's been used to clone several mammals.

Parthenogenesis continues to have the problem of little diversity. Which is why the famous Komodo dragon and certain species of snakes prefer sexual reproduction, and the females resort to parthenogenesis only when there are no available males.

But we are gradually approaching the sexual union of female and male gametes, which nature designed three main ways of combining. The first and most basic was external reproduction: at a romantic full moon, sea sponges send out into the aquatic environment the equivalent of sperms and eggs, and there, floating in warm water, their love is consummated extracorporeally. Without pregnancy or anything like that.

Most female and male fish also expel their gametes into the water so that fertilization occurs completely externally, but there is a variant of this technique: when the females retain their eggs inside their bodies, the males release their sperm and it flows into the female bodies. We can even find examples of this curious way of reproducing outside of the aquatic environment, such as some male scorpions who leave their semen on the ground inside some sort of droplet, which the female then picks up and introduces into her abdomen herself. Somehow, this second method represents an intermediate step in the evolutionary appearance of the penis, an organ designed to allow the male to directly insert his gametes into the female's body. That is internal fertilization, and what superficially allows us to distinguish between male and female, overlooking the fact that this classification is much more fluid than we imagine.

Hermaphroditic Potatoes and Animal Sex Changes

Potatoes are hermaphrodites, like most flowering plants and trees. Perhaps you've never noticed, but there are male flowers with structures called stamen that produce pollen, and female flowers with pistils that are fertilized and end up creating fruits with seeds in them. There are even flowers like the pumpkin or potato flower that have both female and male organs. The male flowers of the potato plant secrete pollen and the female flowers allow themselves to be fertilized by the pollen that reaches them. Can these plants fertilize themselves? Perhaps in some species but nature prefers to design ways to avoid that and thus preserve the richness created by the combination of different genomes. One of the techniques for hindering self-fertilization is as simple as producing the female and male gametes at different times. Another is making the complimentary gametes of a plant or flower chemically incompatible. And a third is just that the reproductive organs of some hermaphroditic species are far enough away from each other to hinder self-fertilization. This is the case with earthworms and snails.

Many species of snails are hermaphrodites. The same snail will have female organs that produce ova and male organs that produce sperm, and can act as a male or as a female. However, snails cannot self-fertilize because of purely physical limitations and they require copulation between two individuals, which—by the way—can last more than 4 hours.

Another case that may surprise our anthropomorphic view of nature is that of animals whose sex doesn't depend on chromosomes, but rather on temperature. The vast majority of male turtles do not have a Y chromosome. A turtle egg will give rise to a male or female individual depending on the temperature of its incubation. If it is a low temperature, the turtle will develop as a male, and if it is high, as a female. In other reptiles the exact opposite occurs, and in the peculiar case of the crocodile, when temperatures are very high or very low, females are born and if they are more average, males result.

A question that comes to mind is: if gender is not determined by chromosomes, could it change over the course of an organism's life? Not in the case of turtles or crocodiles, but in some fish, yes. As we've seen in earlier chapters, the reproductive organs of men and women have a common origin and are more similar than we tend to think. A germ cell doesn't much care whether it turns into an ovum or a spermatozoon, and it has the genetic information to do either. It's a question of the chemical environment it finds itself in, which is obviously very different within an ovary than in a testicle.

That is why there are quite a few fish species that can modify their gender if certain environmental circumstances demand that, most frequently the relative number of males and females in the group. For example, all clownfish are born male but during their development a few change into females and create a hierarchy dominated by a female. If she dies, an adult male will turn female to take her place. Other fish species do the opposite, forming harems and, when the male dies, a female will convert into the phenotypic male gender. But, wait, there's more.

Sexual Dimorphism: You've Only Ever Eaten Female Monkfish

Every time we have a nice monkfish fillet, it is a female. Mostly because male monkfish are tiny little things, 10 cm at most, and at that size would only be good for making some broth at best. And for fertilizing females, of course.

It turns out that the idea that the males are strong, well built and display dominant behavior is pretty exclusive to mammals. In this case the male bodies do have to grow strong in order to compete amongst each other and see who is going to get more females. But the norm in the animal world is that the female is more powerful in every sense than the male. In species of insects and other animal groups in which the female doesn't need any useless male to offer her protection or nourishment, she is the large, strong one because she is the main one responsible for the continued survival of the species. The male spider only has to deposit a few spermatozoa in the female's body; she, on the other hand, will have to carry out the entire gestation and will require many more energy sources during procreation. Female spiders are the ones who prepare the web, hunt, lay eggs, protect them, and gather food for when the babies are born. The males are just insignificant little bugs in comparison. Honestly, people who are arachnophobic are scared of female spiders. If you are ever walking through the jungle and get spooked by a tarantula, you can rest assured it is a female, because of the accentuated dimorphism of the species, and because the males usually die young after having copulated. They aren't good for anything else. Like the classic case of the praying mantis, when in the middle of copulation the female begins devouring the male in one of the most peculiar cases of sexual cannibalism that exists. You've undoubtedly already heard about the praying mantis' cannibalism, but what you might not know is that in some species, when the males approach with reproductive intent, they bring the female a little bit of food to distract them so they might have a chance to escape. Or that some male spiders like the black widow try to immobilize their partner with web before copulation (it's the evolutionary origin of bondage!). Aren't nature and its instincts strange…

But the most interesting dimorphism isn't on a physical level but on a behavioral one. Really size is something relatively secondary, and the "survival of the fittest" is a viewpoint restricted to Darwin's natural selection. What is truly important for evolution is not survival but leaving descendants, passing genes along to the next generation. Otherwise how do we explain, for example, that some male fish risk their lives by approaching other larger fish that could eat them? It doesn't make sense, but by observing them carefully, scientists have proven that those brave fish only do that when there are females present. And that later the female chooses the male who was able to get closest without being hunted, since he has demonstrated having the best genes. Perhaps it is an evolutionary vestige, but a study by English researchers showed that men are more likely to cross the street into traffic if there are women standing on the sidewalk.

Before we get into behavior, why would the male peacock drag around a heavy tail that makes him more vulnerable to predators if not to show that he is healthy and managed to survive despite it, due to the strength of his genes? Reproduction is more important than survival in the end. The male praying mantis would live longer if he didn't mate, but in that case his existence would be meaningless. Bright colors make a frog or a fish more visible to predators in the forest or the coral reef, but also more attractive to their partners.

Moose antlers are another classic example of sexual dimorphism. Useless and heavy for females, only the males have them, as a way to decide based on head butting who will dominate the harem. The same thing happens with the sea lion's fat accumulation and aggressive behavior, as his is a species in which only 10% of the males—the most powerful—manage to have descendants. The females, on the other hand, have no need to compete, because in most cases they will all end up fertilized by the promiscuous dominant males who just go from one female to the next in order to fulfill their evolutionary mandate: the more offspring, the better.

Something very different happens in species where the male, in addition to reproducing, also has the obligation to act as a father to the children. Here things get more complicated, the requirements are more sophisticated, and real competition between females and controversial words like monogamy appear. The basic idea is that natural and sexual selection have not only shaped animals' bodies, but also their brains and behavior. This is what most interests us here in this book because of all the clues it provides on the innate predispositions of our sexuality. Clichés about the abundant sexual signs in women's bodies, like puffy lips and voluptuous breasts and hips, are already much discussed. Let's go beyond that and dare to interpret (which is different than explaining) the evolutionary conditions of the highly singular sexual behavior and conjugal lives of the hominids.

8

Sex in Evolution

There are two great mysteries in women's sexual nature, at least for evolutionary psychologists who believe that our behavior is conditioned by instincts that favor our chances of survival and of leaving descendants. The first mystery is why women feel desire to mate outside of their ovulation periods. For pleasure, obviously, but no other species does this. If you think about it, natural selection shouldn't allow women to feel eager to invest so much energy and risk when there is no reproductive possibility. A bit further on we will look at the evolutionary motives that made hominids an exception within the animal kingdom.

The second mystery is even more intriguing: why are women the only primates who don't know how to clearly distinguish when they are in their fertile phase? Obviously they can keep count of the days, or sense some signals in their body and temperament, but compared to the rest of higher animals, those signals are incredibly subtle. Female baboons' genitals swell up to show all the males that they are unequivocally receptive. Female parrots feel an impulse to shriek much louder when they are in heat. And female lemurs secrete an unmistakable pheromone perfume that drives all the males wild. Signs of ovulation are exhibited by all females, yet in humans there is not a single physical indicator that shows without a doubt when women are in their fertile days and it is the best moment to concentrate all efforts on reproduction.

From the evolutionary perspective it doesn't seem to make any sense. In theory, knowing and clearly showing the specific days of highest fertility would be very advantageous and would avoid wasting energy; yet, at some point in our evolutionary past, Homo sapiens began to hide their ovulation.

And this is not something affected by culture. In the debate over nature or nurture we can discuss to what point the attraction to voluptuous breasts outside of lactation remained codified when our species began to walk upright with our torsos exposed, or whether it is a preference that has been exacerbated by cultural influences in Western societies. Undoubtedly, many evolutionary psychologists and people in the media take evolutionary logic too far when interpreting our behavior. But the signs of ovulation are a fundamental physical trait for the very basic function of reproduction, and the fact that our species has hidden them cannot be an evolutionary accident.

It is true that during the process of natural selection certain traits often show up randomly with no adaptive function whatsoever, or one appears that only comes into play under certain conditions, and since they are harmless and don't bother anyone, there is no motive for them to disappear. But, again, we are not talking about an adaptation like being more or less tall or having a particular hair color, but rather about a characteristic essential to the survival of our genes. While a women's body can show subtle signs during ovulation that increase her attractiveness, and they could feel subconsciously more receptive to men with more masculinized traits, the differences between the moment of ovulation and other stages of the cycle are not nearly as obvious as in the rest of mammals. And this is something universal, that affects all the women in our species. It must be something intrinsic and favored by natural selection. Why?

The Trap of Hidden Ovulation in Women

Well, let's travel back in time a few centuries and imagine for a moment what would happen if this weren't the case. If the hominid male knew ahead of time when the female was in her fertile period, and she had no interest in having sex with him during the 3 weeks following it, what would happen? Well, like the rest of his primate relatives, the male would leave her in a corner and go in search of other females who were displaying sexual receptiveness and a predisposition to carry his genes. It is important to note that of course he would go out looking to score because of his instinct to maximize his number of descendants, but also with the peace of mind that no other hominid would impregnate his partner. Let's consider the most dramatic thing that can happen to a male in terms of natural selection—and it happens much more often than is believed: raising another man's children thinking they are his own. This is the worst thing because he would be committing all his survival resources to someone else's genes. The most brutal display of this selfish instinct is the infanticide committed by the males of some species when they mate with a new female and kill all of her prior offspring so she can focus exclusively on procreating with them.

Speaking in strictly evolutionary terms, we men are predetermined to feel jealous at the threat of possible sexual infidelity. We cannot tolerate a casual encounter with another male that could leave our partner pregnant, because we would have to take care of some other man's genes. On the other hand, a women leaving us for someone else isn't so bad, since at least we aren't working in vain. For women, however, the greatest risk is—I repeat, in strictly evolutionary terms—their men changing partners when they are pregnant or with young children who still require paternal care. The real concern is maintaining male collaboration in the family unit, but the male having a sporadic sexual encounter with a passing female isn't really that bad, even if he knocks her up.

If exclusively inherited instincts determined our behavior, men would be jealous of the attractive male who is only looking for casual sex and women would be jealous of the female their mate might fall in love with.

And this is related to another essential nuance that brings us closer to an answer about hidden ovulation: evolutionary proximity is a fairy tale. In certain aspects we are closer to birds than to chimpanzees. One basic example is that, like chicks, our children are also born defenseless and require the collaboration of both parents in order to survive. Generally the mother protects them while the father goes in search of food. And this conditions the fact that both birds and Homo sapiens have the instinct to band together in family groups that include socially monogamous couples.

But hold on a minute, because social monogamy is not the same as sexual monogamy. There was one study that sterilized several male birds of monogamous species and, mysteriously, the females continued to have children. In fact, other studies with birds (by the way, not even in the case of birds are the females sexually receptive outside of ovulation) show that the males that distance themselves more from their partners usually end up taking care of a greater number of offspring that are not genetically theirs. Which is to say, despite how happily mated some families of monogamous birds may seem, if while out hunting the male comes across a receptive mating candidate, he doesn't hesitate to run around, while his beloved may be doing the same back in the nest.

Taking all these considerations into account, the evolutionary goal of hiding ovulation and feeling sexual desire throughout the cycle seems very obvious: keeping the male close, afraid that his female could be fertilized by another man at any moment, and forcing him to collaborate in caring for offspring that he trusts are his. Being receptive to sex without showing signs of ovulation, the male will have incentives to stay and constantly seek out reproduction, as well as ensure that no other primate enters the scene. So natural selection has us profiled as unfaithful and distrusting monogamists.

Monogamy Is Natural, Faithfulness Isn't

Surely you have pondered this dichotomy at some point: is it natural to sexually desire another person despite being happy in love? Or is having exclusive sexual relations with the same partner what's unnatural? Well, at least in the animal world, monogamy and infidelity are not mutually exclusive.

I remember being in the remote scientific station of Tiputini, in the middle of the Ecuadorian Amazon, interviewing a group of primatologists who were spending 13 hours a day analyzing the behavior of different monkey species in that jungle. There were no tourists or indigenous population, so the animal conduct was in no way influenced by human presence. The primatologists could investigate all types of social behaviors, eating, sexual roles and conduct, and compare them to other species that lived in the same habitat. Sara, for example, studied spider monkeys, which are quite similar to chimpanzees, and like them show a lot of aggressiveness and establish relationships between several males and females. Which is to say they are polygamous and promiscuous; they all mate with each other. Amy, however, was studying some species of smaller monkeys called Titis and Sakis. Despite living in the same environment and forming groups of similar size and characteristics as the spider monkeys, the Titis and Sakis were monogamous species: they formed stable family units in which the males helped to care for the young. And it wasn't a casual arrangement. It was documented that spider monkeys that lived in other jungles were also polygamous, and Titis and Sakis, monogamous. Monogamy was imprinted in their genes. Although when it occurred to me to ask Amy if her Titis and Sakis had "extramarital relationships," she answered "of course!" Again, social monogamy is one thing and sexual monogamy quite another. The first is understood by nature as the instinct to establish a family unit and defend a territory in those species whose young require the care of both parents. It has been observed in birds such as swans, mammals such as bats and even some fish in coral reefs. On the other hand, sexual monogamy entails remaining sexually faithful and passing up the opportunity to procreate with a receptive female or a male with better genes. This sexual monogamy doesn't make much evolutionary sense, and in fact it is very, very rare in the animal world. Of course, no primate except humans practices it. If a female is in striking distance of a male who is better genetically endowed than her partner she doesn't hesitate to mate with him and go back to her partner as if nothing had happened. And if, while searching for food for his chicks, the papa bird finds a female in heat, he won't think twice before trying to fertilize her. Natural selection approves, because that way they each optimize their possibilities of having the most and best descendants possible. Faithfulness and sexual monogamy is much harder to justify, evolutionarily speaking, and seems to be limited to some Homo sapiens.

The evolutionary mandate is clear: males should try to maximize the number of children they father and females should be tremendously selective and always seek out the best available genes, whether they belong to their monogamous partners or not. It is obvious that both men and women carry promiscuous instructions in our genes, that nature perfectly allows the detachment of love from reproduction, and that our most basic instincts may not be at all in conflict with the fact that we feel strongly tied to our partner while wanting to procreate with another. This are the most primitive directives inserted into our brains, in both men and women.

But wait! Don't even think about using that as an excuse! First of all, just because something is most coherent with natural law doesn't mean it is best, or most forgivable. Dictatorships and xenophobia are also more natural than democracy and integration. As a society and as individuals we establish the moral norms we deem best, independently of whether they contradict our innate instincts or not. In philosophical terms, the argument defending what's "natural" as if it were the best or most permissible is called the naturalistic fallacy, and it is a fallacious, simplistic reasoning often used to justify homophobia or sexism. It is cultural and not biological evolution that must dictate what we want to be as a species. And secondly, because letting oneself be carried away by gut instincts in the case of internal conflict is a sign of a poorly trained prefrontal cortex. Especially because our behavior is much more flexible than what so many evolutionary psychologists have been promoting in recent decades with their abuse of behavioral genetic determinism. Allow me to clarify: it is obvious that not only our bodies but also our behaviors have been molded by natural and sexual selection, and there is no doubt that we are predisposed to a certain type of conduct. Of course, we cannot leave out evolutionary biology when studying human sexuality, but discoveries like the enormous plasticity of the developing brain, the epigenetic changes over the course of a lifetime, our vast capacity for learning compared to other animals and the need to adapt ourselves to such changing environments force us to think that evolution also designed us to be extremely flexible. In fact, I defend the idea that knowing the genetic determinants we are born with is very useful to knowing what innate tendencies—that flow naturally from childhood—we should foster and which we should actively try to mold or correct. But interpreting our current behavior depending on the life conditions that our ancestors overcame is more fiction than science, and extrapolating our behavior from primates' occasionally borders the absurd. If you don't yet agree, let's take a look at the bonobos and the chimpanzees.

"Bonobos' Way of Life," Are You More of a Bonobo or a Chimpanzee?

Honestly, don't ever let anyone try to explain the origins of our aggressiveness or sexual behavior by comparing it to chimpanzees, because evolutionarily we are just as close to bonobos and they are completely different.

When you hear someone justifying their greed by explaining that we come from chimpanzees who, when they find a pile of food, beat each other up to see who can get away with the most, tell them that we don't come from chimps. Explain to them that we separated 7 million years ago from a common ancestor who would later diversify into chimpanzees and bonobos, and that when bonobos come across a bunch of food they jump for joy, start copulating with each other, and after the party they share the food.

Analyzing chimpanzee and bonobo behavior is really interesting. Physically they are very similar species that only evolutionarily separated a million years ago when the Congo River formed and isolated the two primate communities. Those to the north evolved into the chimpanzees we now know, and those to the south, into bonobos. The two species are almost identical genetically and share very similar environments; however, their behavior is radically opposed.

Bonobos perfectly represent the hippy maxim "make love not war" from the sexual revolution of the 1960s. Unlike the chimpanzees, they never engage in aggressive competitions, their societies show a high degree of cooperation, they play together and enjoy recreational sex just for the fun of it. It isn't rare to see a female get on top of another female in the missionary position and rub her swollen genitals from side to side. In bonobos, oral sex is practiced both between males and females, they copulate facing each other and with a very wide range of positions, they have sex all the time whether they're ovulating or not and only stop when they're menstruating, and if a problem arises in the group it is resolved in an enviable way: copulating to relieve tension. The phrase "chimpanzees resolve their sexual conflicts with power and bonobos resolve their power conflicts with sex" perfectly sums up the difference between these two species, with both of whom we share a common ancestor.

There is an important first reflection to be made here: if differences in surroundings and culture have made chimpanzees and bonobos so different in just a million years, imagine how much we hominids have changed over time and how limited the relevant comparisons with primates are. But given that we don't want to entirely repudiate our evolutionary past, and accepting that the most fundamental basic instincts do have an important innate significance, and that some primates could have evolved more similarly in certain aspects than others, can science answer whether Homo sapiens are more like bonobos or chimpanzees? You bet it can, and this is also a nice way to visualize different scientific methodologies.

The first step in science is observing and constructing a hypothesis to describe nature, and that is what primatologists like the Chilean Isabel Behncke do expertly. She combines her research at Oxford University with long stays in the Congo studying the behavior of bonobos in the wild. I first met Isabel at the City of Ideas in Puebla, Mexico, and then at the "Creative Beings" festival in Madrid, and I was fascinated by her work. She explained to me that bonobos form matriarchies and that, in the case of chimpanzees, it is the males who are dominant. That chimps hunt, are violent with other groups and practice infanticide, while the bonobos don't hunt or use violence of any sort. Bonobos have neoteny, retaining childlike behaviors such as play and, of course, their peculiar and intense sexual activity which some humans see as something to emulate. Observing them, Isabel tells me that our exacerbated sexual desire makes us more similar to bonobos than chimpanzees, but in science there is an important difference between observing and measuring. The former offers us coherent hypotheses, but the latter allows us to put them to the test via experimentation.

Are there scientists researching genetic, hormonal, and cerebral factors, or analyzing behavioral patterns in the lab in order to respond to the dilemma of the vast sexual differences between bonobos and chimpanzees, the two species closest to humans? There sure are. To begin to tackle this quandary, one of the first questions we have to ask ourselves is whether a primate's sexual behavior is genetically determined or learned according to a societal model. Investigators at Duke University carefully observed behavior in chimpanzees and bonobos that had been born in the wild but captured by poachers at very young ages before any type of natural socialization. Despite having grown up in captivity, the bonobos soon began to display sociosexual behaviors in the presence of food. The chimpanzees raised in the same conditions never developed that type of conduct. An article published in 2011 confirmed that the bonobo's sexual nature was indeed particular to their species and installed in their DNA.

Primatologist Victoria Wobber is highly engaged with trying to understand the pacificism of the bonobos. When I visited her at her office at Harvard she explained one of her experiments: knowing that, in moments of aggressiveness, there is a marked increase in levels of testosterone and cortisol (the stress hormone) in the blood, Victoria and her team placed some food in front of the cell of two chimpanzees who had not eaten for some time, and measured the levels of testosterone and cortisol in their saliva. The chimpanzees knew that when the cell door was opened they would fight for the food and even before the conflict their testosterone levels had gone up. When they repeated the experiment with bonobos, whose natural predisposition is to negotiate, their testosterone levels remained stable while their cortisol increased. That indicates that from the very beginning, the two species perceive the situation differently. Even though this may confuse the matter even further, in equivalent situations, we humans experience rises in our blood levels of both testosterone and cortisol.

Is something different going on in the brains of each of the primate species? Again in 2011, researchers at Emory University in Atlanta published the first results on the neurobiological differences between bonobos and chimpanzees. Their results showed that the bonobos had more connection between the amygdala and an area in the cortex involved in controlling aggressiveness, and more activity in brain regions implicated in the perception of suffering, both one's own and others'. According to the authors, these changes are related to the bonobo's more social and empathetic behavior as compared to the chimpanzees. And not only that, in early 2012, a mixed group of American and French researchers analyzed the cranial growth of both species and showed for the first time that the bonobos' brains grow slower than chimpanzees', and this means that their offspring reach maturity later. This could explain their predisposition to play, to not competing over food and other aspects of their temperament. Temperament? That is what Max Planck's team put to the test when analyzing the reaction to new stimuli in 2-and-a-half-year-old chimpanzees, bonobos, gorillas and Homo sapiens. The results showed that both the kids and the bonobos showed more rejection of new things than gorillas and chimpanzees. This is consistent with another study that showed riskier behaviors in chimpanzees than in bonobos, and another that proved that the bonobos' cognitive abilities developed more slowly. Are there genetic differences? Little has been analyzed thus far, but a study observed more changes in specific regions of the Y chromosome in chimpanzees than in humans and bonobos. According to the study's authors, this reorganization could make redundant the sperm competition that leads female chimpanzees to mate with a large number of males only when ovulation makes their genitals swell. Since it is the female bonobos who choose their mates, that sperm competition isn't so essential, and even though the question is still far from being answered, Victoria Wobber is of the opinion that, once we have all the data on the table, we will find we are more similar to bonobos than to chimpanzees. Luckily.

I'm More Interested in Primatologists than Primates

I was only half-joking when, after meeting the primatologists at the Tiputini station in Ecuador and seeing that they spend 13 hours a day isolated in the middle of the jungle taking notes on the behavior of their monkey groups, I said, "They should be studying you instead of you studying them." I said something similar to Isabel Behncke when she described the harsh conditions, risks and loneliness of her expeditions to the Congo. It wasn't an innocent comment, deep down it was a statement that natural selection could never even come close to making sense of these primatologists' behavior, and that indirectly weakened many of the parallels that they themselves were making about primate and human conduct. And it is the same with aspects of sexuality. If we review the literature on science and sex aimed at a wide audience, we see that most of it deals with the subject from an evolutionary perspective. That's okay, because obviously we are born with genetic determinants, but from the perspective of a popular science writer who has spent much time explaining this material I have to admit that it has been extremely exaggerated. Evolutionist logic is filled with curiosities and provokes amused "aha" responses from readers, but some of its affirmations are closer to psychoanalysis than science, and it is highly limited as a tool for understanding the diversity of human sexual conduct in developed societies. It can explain instincts and desires, but only the most basic ones. Besides, it doesn't tell us anything about whether we "should" behave one way or another, since our morality is above nature's determinants. While 6 million years are nothing in geological time, this fact is usually used in deceptive ways. Actually a lot has shifted from the Australopithecus to us. Evolutionary theory is very useful as a reference and as a conceptual framework, but I still believe it is much more useful and direct to study primatologists than to study primates.

9

Sex in Bars

Social psychologist Viren Swami, from Westminster University, selected two homogenous groups of young men for his experiment. He asked one group to show up to the lab on an empty stomach and the other to arrive nice and full. Then he separately showed them the same series of photographs of women, all of them pretty and fairly similar, but some of them a little more filled out than others. The volunteers were just asked to rate the attraction they felt for each image, and the results were clear: those in the hungry group significantly preferred the women with more body mass.

The conclusion is obvious and corroborates that of many other studies: our internal physiological state conditions our perception of beauty and whom we consider more attractive in each moment. And not only that, our mood also has an influence. David Perret, of the University of St. Andrews, is one of the main experts on the investigation of factors that make a face more attractive. In one of his many studies he selected two equivalent groups of women, showing the first a series of very pretty young women and the second, photographs of women who were much less attractive and elegant. In this phase, Perret only wanted to unconsciously decrease the confidence of the first group and increase the self-esteem of the second. Then he showed both groups a series of male faces with varying degrees of masculinity, and the group of women who'd had their confidence temporarily reduced chose faces with less masculine features than the group who had had their self-esteem reinforced. The unconscious in action.

Others' Beauty Depends on Ours

What these studies really suggest is that discussing attractiveness with such generic and clichéd affirmations such as "men prefer large breasts and women wide shoulders because evolutionarily they reflect energy resources and strength for transporting food" is terribly limited. And besides they aren't entirely true. Virem Swami himself explained to me that, in cultures where women don't cover their breasts, men are pretty much indifferent to their size. Swami led a very extensive study in which he compared ideals of body weight and aesthetic preferences in ten different cultures around the world and he is emphatic: "This idea that the ideal proportion between the waist and the hips is .7 is a myth. Perhaps in Western society, but there is not a shred of empirical evidence that shows this to be a universal trait." Evolutionist logic establishes coherent hypotheses, but in and of itself it proofs nothing.

Of course there are a few traits that are universally associated with beauty, like symmetry and youthful features. But they are subject to all sorts of cultural determinants, life experiences, socioeconomic status and physiological and emotional states. Asians and Africans may prefer lighter skin and have a different ideal of beauty than ours, higher socioeconomic classes prefer thinness without minding a sickly appearance, and at a party a woman's instinct will be unconsciously drawn to a more or less masculine man depending on how many pretty women are near him. Social experiments have proven that a radiant woman who feels like the most fabulous in the room will direct her gaze and smiles at the most attractive, most masculine man there. But if she arrives at the party and finds herself surrounded by models, she will focus on the more conventional men and not even consider the macho man with the arrogant stance so handsome.

The attraction puzzle is enormously complex, but despite that fact scientists try to unravel and understand all the factors that influence our perception of beauty with the goal of showing that there is a logic to taste. Let's take it step by step.

In terms of universal beauty features, the most prevalent in all cultures is symmetry, and that does have evolutionary relevance. In theory, the external appearance of our face and body should be perfectly symmetrical, but a higher rate of inherited genetic mutations, infections during embryonic development, childhood illnesses or exposure to toxins and high levels of stress can provoke slight distortions in development resulting in subtle loss of symmetry. Inadequate nutrition in a pregnant mother, for example, affects the repair mechanisms in the fetus' DNA. So an adult with a perfectly symmetrical face indicates high-quality genes, a healthy childhood and a strong immune system. Clearly, we would want their genes to accompany ours in our offspring. In fact, several studies have demonstrated that symmetry is an external marker of good genes and it is established that symmetrical men are more intelligent, run faster, have more coordinated movements and even produce more sperm. There's no doubt about it. Perret explained to me that every time he uses his computer programs to correct the subtle asymmetries in a face until the right side is a mirror image of the left, all observers deem it more attractive even when they are unable to tell exactly what has changed.

Another universal characteristic of beauty is averageness: that a face's proportions are not too far from the mean. Here it is worth mentioning that, in the case of very attractive people, some more pronounced features can highlight their distinctiveness and further their beauty, but when we are speaking about more conventional faces, studies done by David Perret (and many others) using distorted photographs have confirmed that bringing all the traits closer to the average increases their attractiveness. This trend has been proven in different populations around the world, and Perret interprets it by saying that, unconsciously, strange faces could indicate genetic alterations or disease, and that the average represents what natural selection has chosen over time as the best adapted. Again, there are exceptions yet, in general, averageness is a universal characteristic of beauty.

And the third universal factor in attractiveness is those traits that reveal sexual dimorphism: men search for signs of youth and femininity, like a small jaw accompanied by thick lips, and women seek out testosterone, like prominent cheekbones and less facial fat. But here things already start to get complicated. For example, when shown the same series of modified photographs, men will always prefer signs of greater femininity, but women don't always choose the most virile face as the most attractive.

David Perret is the author of those famous experiments that prove that during ovulation women prefer more masculine men and aggressive behaviors that reflect testosterone, health and fertility. The key is that said features are also unconsciously associated with a higher risk of rejection and abandonment. This isn't so important when what you are looking for are good genes, but it is the reason why at other points in the menstrual cycle women prefer less marked traits of masculinity. Analyzing by age range, it has also been seen that women prefer more masculine faces when they are in their reproductive years than during puberty and post-menopause. Of course there are exceptions, and you probably know men who are attracted to less feminine women, but as a universal rule it does play out that men look for signs of estrogens and women for testosterone.

But there is also something new: Perret explained to me that he had been realizing for a while that something didn't quite fit in the faces that he was using for his experiments. At the time of this writing his results hadn't yet been published, but he has preliminary data that includes the stress hormone—cortisol—in the equation.

During periods of stress, testosterone levels also increase. This doesn't change the size of a man's chin from 1 day to the next but it does slightly alter our expression and the distribution of facial fat, making us appear more masculine in theory. But Perret suggests that when our high levels of testosterone are due to stress we can forget about looking more attractive. High levels of cortisol create a more sickly skin tone, negative expression and a different distribution of facial fat that eliminates the correlation between testosterone and beauty.

The Power of the Unconscious in Physical Attraction

Taking a big leap, we could describe countless cultural and social influences that condition our perception of beauty. Viren Swami, for example, just published a series of works on factors that affect our preferences in body mass index. In addition to the fact that being hungry makes men prefer more voluptuous women, an article published in August of 2012 in the scientific journal *PLoS* says that we also choose rounder bodies when we are stressed. In comparisons between different cultures, it has been observed that in poor societies fuller bodies are seen as more attractive, and even in the same social population, individuals with less resources usually prefer partners with higher body mass. In fact, the predilection for skinniness at a level below what our evolutionary instinct interprets as healthy is exclusive to people with a high socioeconomic status. Swami states that in this case advertising and the media are clearly responsible, but he points out that "men are more consistent with their instincts when judging a model as not sexually attractive because she is too thin. On the other hand, in women beauty ideals are much more upset by media influences."

We all usually overestimate what the opposite sex finds attractive. If we take the image of a guy in a bathing suit, modify it several times, adding each time a little more muscle mass to have a range from weak to very well-built, then ask a group of men to rate which they think women will like more before asking a group of women which they find most attractive, we find that men choose slightly bigger muscles than what women actually prefer. But if we repeat the same experiment with a girl in lingerie and different degrees of thinness, the discrepancies between male and female responses are much greater. The distortion of reality in terms of which bodies are more attractive affects women more than men.

Studies analyzing facial fat show that men continue to judge slightly heavier women as more attractive, because we interpret it as a sign of health. However, health and attractiveness are not as correlated in women. They prefer thin faces when evaluating potential partners, and when comparing themselves to other women they deem svelte figures the sexiest. That's really not the way it is. The following isn't an academic investigation, but neuroscientist Ogi Ogas analyzed millions of sexual searches for his book *A Billion Wicked Thoughts* and was emphatic when I interviewed him: "If you analyze the key words that men usually introduce in their searches for erotic websites, many of them search for specific videos of plumper women, but almost no one puts "skinny" into their search criteria." Ogi's work is not methodologically solid but his reflections are very interesting. He believes that the media doesn't really condition us as much as we usually think, because if that were the case we wouldn't find such a high proportion of searches for transsexuals, older people or scenes of dominance and submission. This is really a threat to the validity of more formal studies done by Swami and Perret, since it could insinuate that the responses obtained when asking about the attractiveness of bodies and faces are based more on purely aesthetic criteria and potential stable partners than on sexual desire in moments of arousal. It could be that, in terms of sex, in contemporary society curiosity and the appeal of the new outweigh all the criteria that make us see a person as good-looking from an evolutionary and reproductive point of view.

But getting back to David Perret's research, he explained to me that he was most intrigued about the preference for lighter skin in Africa and Asia, since it totally breaks with the search for similarity and characteristic averageness of a given group. He explained that in Africa it could be due to messages from Western society and an association between light skin and more wealth, but that doesn't follow in Asia. On the other hand, one of his most curious discoveries was how the increase in carotenoids from the consumption of vegetables modifies skin tone and increases beauty. Based on studies published in 2011, Perret saw that the tinted skin tone caused by foods like the tomato transmits the idea of a better state of health and makes a person more attractive.

In any case, the key message is that what is truly interesting from a scientific perspective is not whether evolution has conditioned us to prefer some traits or others, but how our surroundings and our own physical and cognitive state influence our preferences in a partner. In fact, knowing that could be quite useful. For example, if you go to a party hoping to score, whether you are a man or a woman, you'll have better luck if you bring a woman with you.

If You Want to Score, Bring a Girlfriend with You

Renowned evolutionary psychologist David Buss carried out a large study with 848 participants of both genders, and confirmed that men judge women less attractive if they are accompanied by a man instead of a woman. That might seem logical, but the strange thing is that the opposite was true of women: they also considered men accompanied by women more attractive than those who were hanging out with a male buddy. The conclusion is obvious: if we want to score, we should ask a female coworker to join us, whether we are male or female.

Something important to clarify in the case of women: in order to maximize your chances of success, your girlfriend should be similar to you but ever so slightly less attractive, so that way your beauty will be reinforced in other people's unconscious. Dan Ariely, an expert in behavioral economics, put this hypothesis to the test by showing a large group of students two photographs of equally attractive guys next to an image of the first one doctored to look uglier. When asked which they found handsomest, most chose the undoctored version of the first guy. When they repeated the experiment with a doctored image of the second guy, the unconscious comparison made the majority choose the undoctored image of the second one.

And another bit of advice: don't play the hard-to-get introvert, it's better if you transmit a sense of wellbeing. Viren Swami told me than in another of his studies they separated 2,157 students into ten groups and gave the same photographs to them all, but each group received different information about the personality of the guy or gal in the images. They insisted that all the students should evaluate them exclusively based on their physical appearance, but when the images were associated with words like neurotic, introverted, selfish or sad, the margin of acceptance of body mass was much smaller than with the extroverted, emotionally stable, generous or happy personalities. Obviously, red lipstick will fool our primate brain by associating the contrast with higher blood flow, levels of estrogens, sexual arousal, youth and heart health, but the same thing seems to happen with information on personality. If we like someone's character, we find them more physically attractive. Conclusion: we mustn't forget to smile and make essential eye contact.

It seems that it is also shown that the more we drink, the better looking the person in front of us seems. In a study in which they showed photos of men and women to students who'd had a couple of drinks, it was proven that their visual acuity diminished, they perceived less physical imperfections and they evaluated the images as more attractive than the control group of students who hadn't been drinking. And the most fascinating part of it was that this effect on our vision lasts for several hours and that is why when we go out the next morning after drinking we find people of our preferred gender slightly better looking. We are in the hands of our unconscious.

More People Stop for Hitchhiking Blondes

While we're on the subject of strange observational studies, French sociologist Nicholas Guéguen asked five young women between 20 and 22 years old to hitchhike on a highway in southern France. They were wearing the same clothes and had similar bodies, but two were blondes, two had black hair and one had brown. You can guess what happened: of the more than 2,000 drivers who passed, more men stopped for the blondes than for the women with black or brown hair. Female motorists, however, didn't discriminate based on hair color and stopped equally. In a similar study, Guéguen asked the same 20-year-old girl to try her luck at hitchhiking several times, but each time with a more stuffed bra. Again, her breast size was positively correlated with the number of men who stopped, but not with women.

Here comes a controversial bit of data: Viren Swami carried out a comparable study, seating a woman in an English pub on different nights, with her hair dyed blonde, brown or red. He observed that when she was blonde many more men approached her, but Swami wasn't convinced that they found her more attractive, since according to his data, in the United Kingdom being blonde isn't a sign of more beauty. So he designed another study asking the men to more completely evaluate various images of women seated in a bar. The results were that the men classified the brunettes as more attractive and genial, and the blondes as easier and more needy. This fit with two of Swami's ideas: that men really like brunettes better and that the media has deeply influenced our stereotypes about sexuality.

Siblings Can Condition Our Preferences in a Partner

One of the most surprising effects on the search for a partner is that we feel particular attraction for people with a range of beauty we consider similar to our own. Several experiments have shown that the effect is less in men, who usually prefer their women as good-looking as possible, but in general women look for men based on their perception of their own beauty. The consequence is that in the end there is usually a correlation between the physical attractiveness of the two members of the couple.

So does that mean that when picking someone out of "the crowd" at a party there is cognitive dissonance? By which I mean, do our brains trick us by telling us that person is better looking than they actually are?

Several authors have suggested that our brains lie to us to make us happy and to get us to accept going out with people who are objectively less attractive; however, a study from Columbia University entitled *If I'm Not Hot, Are You Hot Or Not?* in which interviews were conducted with members of an online dating site concluded that no, that self-deception does not exist in most people, and that both men and women admit that their partners are not the most attractive and that they selected them based on their own looks.

With so many options, online dating sites are an interesting tool in the analysis of preferences when choosing a partner, and something that they are clearly proving is our exaggerated predilection for similarity in certain features. When searching for a partner, investigators differentiate between vertical preferences and horizontal ones. Vertical preferences would be factors such as income, occupation or body mass, which statistically do not present a problem if they are different. Horizontal preferences would be those where people generally do look for similarities, like religious beliefs, certain cultural aspects, age, race and height. But the biggest revelation of these studies analyzing Internet dating is that when we have many more potential candidates and search criteria than we would have in a bar, we show a much more pronounced tendency toward similarity than we would think.

I remember that when David Perret assured me that we usually prefer people physically similar to ourselves I had the typical ascientific reaction of focusing on an exception to try to contradict his statistics. I told him that my Italian girlfriend is blonde, with blue eyes and fair skin, but that she's always told me she likes dark men with dark eyes. Then Dave asked me, "Does she have younger siblings who look like her?" I said yes, and David continued: "What we've seen is that—especially if they get along well with their parents—women—and not men—prefer similarity in their partner except when they have a younger sibling that looks like them. If that is the case, it generates certain aversion and they seek out someone totally different who doesn't unconsciously remind them of their brother." Thanks for existing, Guido!

The Internet Has Only Revolutionized the First Few Steps in the Dating Process

If one afternoon we go out for a stroll and happen to walk into a shoe store that has a display window that catches our eye, we may find several pairs we like. But the day we head out of the house with a very specific image of a shoe in our heads, we are most likely to go from store to store without finding any that meet our expectations. The same thing can happen when looking for a partner at online dating sites, according to the authors of the article "Online Dating: A Critical Analysis From the Perspective of Psychological Science," published in 2012 in the journal of the Association for Psychological Science. Researchers call this phenomenon relationshopping, and it consists of logging in to a dating site, browsing and analyzing profiles as if they were products in a shop, and ending up with the feeling that there isn't a single one you like. Especially in the case of women, it's very common for all the candidates to seem the same or weird and for none of them to catch your interest.

There is no doubt that the Internet and online dating sites are very successful and have been revolutionary, especially for minorities and people with atypical taste, but this feeling of overwhelm from too many options is one of their dangers. Another is their homogamy: several studies have shown that users usually search for people who are very much like them in tastes, characteristics, hobbies, political orientation and life goals. Even though this could seem like a positive thing, since it allows us to find users who in theory will be a very good fit, the authors of the article suggest that exaggerating this homogamy blocks the incentive for complementarity and the spark of discovering new things in a future mate. In fact, one of the study's most relevant conclusions was that the algorithms used by different dating sites to suggest people who fit our profile are really of no use at all. Sharing similar hobbies or expectations doesn't necessarily contribute to the success of a potential relationship. Among other things, because another of the dangers identified by the overview article is that users are not entirely honest when they describe themselves in their profiles. A study randomly selected 80 online dating site users to compare the age, weight and height they claimed with their real measurements, and concluded that eight out of every ten users had altered one of the values. Sixty percent lied about their weight, 48% about their height and 19% about their age (researchers point out that the sample was of relatively young people and that in older ones this last figure could be higher). Another investigation at Duke University compared the measurements of

21,745 users with the national average for people with similar characteristics, and discovered that women between 20 and 29 years old who were looking for partners online weighed 2.5 kg less than the average, those 50–59 weighed 10 kg less, and that the men were 3.5 cm taller. The interpretation of these results is obvious: people are less than sincere when they describe themselves online and this, according to the authors, is counterproductive, since it comes back to haunt you when you meet face to face. The smart way to proceed is just the opposite: even though it could mean less initial candidates, we should seek to cause a positive surprise upon meeting instead of a negative one. It's like leaving the theater more satisfied after watching a movie we weren't expecting to be a masterpiece. It's the same thing with the asymmetry of information on the Internet: exaggerating the positive and hiding the negative is really a waste of time for both parties, the one looking for a date and the person they end up meeting.

The authors also observed that exchanging messages is positive but that in excess it ends up diminishing interest, and that the best thing to do is limit the number of profiles you look at, avoid predetermined ideas, don't rule out candidates over little details, be honest in your description but highlight something that can make you stand out from the rest, don't be slow to meet up once promising communication is established and don't pay any attention to the suggestion the sites make about whom you should contact.

But, beyond these generic tips, which may be more or less on target in concrete cases, the underlying academic discussion is whether virtual contacts have actually revolutionized the way in which people find partners. Apparently, the data suggests they have. According to a study published in 2010, 22% of heterosexual couples formed between 2007 and 2009 in the United States met on the web, which represents the second most common way to meet, topped only by meeting through friends. On the other hand, the perception of skepticism of years past has improved quickly and in 2006 a study concluded that 44% of people thought that through the Internet was a very good way to find a partner. Meet-up sites have become a multi-million dollar industry, and in April 2011 alone, 25 million people registered for dating services online. Those are big figures, but anthropologist Helen Fisher tells me that it really isn't as big a paradigm shift as we think. According to her, Internet obviously represents a big advantage for minorities and older people because it allows them to reach many more candidates than they would be able to find in their surroundings, but that, actually, after this initial phase the definitive factor continues to be the chemistry that sparks—or doesn't—during the face to face, and that "our brains continue to fall in love the same way they have for thousands of years." Besides, according to her, having a stranger's basic info before meeting is not that different from the conventional date with a coworker or with someone recommended by a friend. Helen Fisher believes that "what's really strange is establishing a relationship with a complete stranger in a bar, but every first date has always followed a pattern of getting together to have a drink with some prior information, whether it comes from the Internet or from a mutual acquaintance."

Non-verbal Signs of Seduction

You send a polite message to that guy or gal who gave you their number at the birthday party of a mutual friend. All you write is: "I enjoyed talking to you, here's my number, see you soon," but you hesitate over signing off with "best," "hugs" or "xx" because you know that last one would change the whole meaning and intent perceived in the sentence before it. The signals of seduction and courtship have evolved, and we can learn to interpret them and use them in our favor, but they continue to reveal our more unconscious thoughts.

The dating scene in the United States is superfun, at least in a big city like New York, where the difficulty establishing strong friendships is supplanted by a constant search for sporadic relationships. Dates are not necessarily sexual encounters, but rather something between a one-night-stand and the concept of seeing someone. Single people date more than one person at a time, and it's understood more as a social act than a search for a partner. Of course it is not exclusive to the United States, but there it is quite a phenomenon and follows some truly hilarious rules. There are even classes to teach you how to do it.

Ron was 32 years old, had a good job as a consultant, had moved to Manhattan 6 months earlier, broken up with his girlfriend 3 months ago, and just finished a course to learn how to pick up women. They taught him about showing confidence, how to dress depending on the type of relationship he was looking for, what attitude to have in different situations, each and every one of the detailed steps to follow during a date… and of course, how to master non-verbal language. Being in control of the non-verbal language of seduction is knowing how to send messages through body language that reach the unconscious of the person we are trying to pick up, but above all it is knowing how to interpret the signs of acceptance, rejection, interest or boredom that they express during the date while hardly realizing it. If you see them tilt their head, things are going well, you can order another drink. But if you notice that they are gradually leaning slightly backward, it's better to react fast and suggest moving to another bar.

These things are both nonsense and not. Psychologist Monica Moore, whom I spoke with to put together this chapter, has been investigating the non-verbal bases of seduction since the early 1980s. If you have watched a documentary about signals of acceptance and rejection in courtship, it is very likely based on her seminal work in which she methodically analyzed non-verbal expressions of seduction between strangers in bars, parties and public places. In 2010 she published a review of all the scientific literature to date, in which she discerned between the anecdotal, which was more appropriate to teen magazines, and those behaviors that were truly consistent in flirtation in Western societies. "It is true that now texting, Facebook and other social networks have interfered in the first phases of courtship, but the in-person part continues to show the same patterns, as if they were etched into our nature," she explains, emphasizing that all animals have specific courtship signs, and that humans are no exception. From her apartment on New York's Upper East Side, anthropologist Helen Fisher agrees: "The arrival of the Internet has undoubtedly created a different environment, but our brain hasn't changed in millennia and the reproductive instinct is one of our most solid. We may find a profile on an online dating site that rationally seems fine, but the emotions of the face-to-face meeting are always the deciding factor," she tells me, convinced that the patterns of seduction haven't changed as much as we often think.

Cynthia also participated in a course similar to Ron's, but obviously neither of them confessed to that when they met at a barbecue organized by a mutual friend. They hit it off, exchanged telephone numbers, and both felt willing to put their new knowledge to the test. According to the classic steps of dating, the objective is to get a kiss on the first date, copping a feel on the second, get into bed on the third, and see what happens next. Any slight precipitation might indicate that he was a horndog or she was easy, and a delay could generate insecurities on both sides. Pathetic, but at the same time really funny.

Ron did what they'd taught him and waited 3 days before sending Cynthia a message so he wouldn't look desperate. She received his SMS while she was bored in a café but she also took a few hours to respond, not wanting Ron to think she was waiting around or too happy to get his text. She didn't have any plans that night or the next, but she suggested Thursday to make it seem like her schedule was booked up. Ron, in turn, took 70 min to reply, which he spent weighing up the different options for the plan he was going to propose: not dinner on the first date, because that's very formal, too risky, and could be quite expensive (in the United States, the guy always pays, no matter how much *Sex in the City* they watch). A coffee is good for sounding things out, but if you're feeling confident the best thing to do is suggest having a drink in some place nice. Not too close to home because that seems too obvious, better to save that for the second or third date.

I know that this story of Ron and Cynthia is a bit of an absurd generalization, but it really does work to illustrate Monica Moore's studies on non-verbal language.

Let's imagine that Ron is having a cocktail with Cynthia in the East Village, and let's interpret the signals. Ron was clever and chose to sit on stools, which allow proximity, ease of movement and the opportunity for the classic "hand on the knee," one of the standard codes of American dating. During a date, looks can be confusing, but if in an animated moment of the conversation one of the two people casually puts a hand on the other's knee for a few seconds, that person will be explicitly showing interest. It is usually the man who takes that first step, but if things are going well and she senses that he's a bit insecure, it's not strange at all for the woman to be the one to take the initiative and give that first sign of approval. There are more options: if instead of on stools you are seated at a table, even though you don't need to use the bathroom, get up at a high point and lean your hand on their shoulder oh so casually as you pass by. He or she will understand.

Either way, the desire to continue moving forward to the next steps has been clearly demonstrated, and now it is time to start the serious interpretation of their body movements and reactions of acceptance or rejection. Obviously, if the woman doesn't meet his eye or turns down another drink saying she needs to get home soon, that's a bad sign, but there are many other more subtle signals. Verbal language doesn't lie.

A hug or "I'm so glad to see you!" when meeting up with an old friend can be faked, but not the accompanying expression or gleam in the eyes. And the same thing happens when we sense someone is nervous at work, or if when having a drink at a bar we notice that the conversation flags, or someone's eyes are less expressive, their body language suggests they're uncomfortable, they are glancing at their watch: and that's when the time has come to say: "Well… it's getting late, isn't it? Should we go?" Even though excessive politeness and our companion's verbal language responds with a "I'm fine… whatever you want," we know—or we should know—that non-verbal language is what counts.

Monica Moore's first big work on courtship was published in 1985 when, after following 200 women for more than a hundred hours in bars and other social contexts, she established a catalogue of 52 female behaviors that showed interest, including looking directly into the eyes, unconsciously smoothing their hair, smiling, tilting their head, reflexively touching their neck or lips, asking for help or leaning forward. Doctor Moore saw that the signals were consistent, and she prepared an experiment in which she trained external observers to see if they could predict the success or failure of the interactions between men and women in a bar just by observing their gestures and movements during conversation. In 1989 she published a study showing they'd had a high index of correct answers, but also revealing an unexpected result: what really predicted men's approaching a woman and the success of the encounter was not the woman's beauty, but the number of signals she gave off. Monica was emphatic on this point: "We have seen it repeatedly: men take interest in the women who give off more signals, not the ones they initially find more attractive." Which is to say, if you want a man you've never met to approach you, a smile or a direct look is much more efficient than a plunging neckline or beautiful features.

Courtship Is Started by Women

In fact, one of the conclusions of the overview published by Dr. Moore is that in two thirds of occasions it is clearly the woman who gives the man clues that lead him to come over and strike up a conversation. She is always the one to give the signal. They have recorded situations in clubs, parks and laboratories, and it has repeatedly been observed that when the guy takes a step it is almost always preceded by non-verbal invitations from the woman. For example, a man starts a conversation after a woman looks at him, or runs her arm over his shoulder when she approaches, non-verbally soliciting an embrace. Obviously this isn't always the case, but Dr. Moore insists that "even though the man seems to be taking the initiative, the first non-verbal steps are always taken by the woman. Men react when they perceive an unconscious invitation."

The unconscious is revealing. If you are at a party and suddenly exchange glances with someone the most normal thing is that you will both look away suddenly, but if 2 seconds later you instinctively turn and your gazes meet again, you can trust that it wasn't a coincidence but a sign of interest. If you hold each other's gaze for a little while, one of you sketches a half-smile and gets one in return, that's the first phase of non-verbal courtship. A long look is still the main element in non-verbal communication, both voluntarily and involuntarily. In a Dutch study where they asked young men and women how they showed interest in someone they found attractive, the men leaned toward going over to talk, but most of the women said that they used visual contact as their primary vehicle of initiation.

In any case, a curious study published in 1992 based on the observation of 500 couples in public spaces concluded that during courtship stages it is the man who usually initiates physical contact or caresses, and in married couples it is the woman.

Returning to the non-verbal signs of seduction, experts observe countless subtleties, like sucking in your stomach and keeping your back straight, adjusting your shirt or skirt, and—especially in men—positioning the body in such a way as to close off the woman's visual field. If you see a man leaning toward a woman as he speaks, gesticulating vehemently, moving his hands a lot and nodding his head almost exaggeratedly, that means he is clearly interested. And if the woman licks her lips involuntarily or turns her neck a little as she stares into his eyes with dilated pupils, they should just quit talking, and instead smile and kiss each other. But the signals of rejection are just as important as those of attraction. Spontaneous yawning is very obvious, and don't even think of checking your phone no matter how addicted to email you are, because it represents one of the clearest signs of lack of interest. Wait to go to the bathroom to check your messages, as everyone on a date does instinctively. In fact, in one of her latest studies, Dr. Moore classified 17 attitudes linked to rejection, most of them opposites of the courtship signals. For example, if a woman is interested, she will lean her body forward, and if not, she will pull back. If she crosses her arms it's a bad sign, and if she touches her fingernails or moves her teeth, it's not that she's nervous but that she feels uncomfortable and wants to escape. If you keep watching, you'll soon see her gaze drift, her legs fidget and before long she won't be smiling as easily as before. And if at one point she starts to diverge on banal subjects instead of agreeing, we can interpret that as another of the clearest signs of rejection. Obviously, we are talking about in the context of one of the first few dates.

It's really quite obvious, but Monica Moore assures me that "the most successful guys are those who know how to best decipher the non-verbal signs of attraction or rejection in their potential partners." Even in a study where they gave a test to volunteers attending workshops to improve their people-meeting skills, they saw that those who were least successful identified significantly less courtship signals than a control group.

The Magic of Kissing

Cynthia was nervous. Her first two dates with Ron had gone very well, but then he went on a trip for 3 weeks and she was afraid that the incipient relationship had cooled off. In New York, everything happens very fast.

Ron was smiling when he showed up, but gave her a kiss on the cheek. "Bad sign," thought Cynthia. The conversation flowed in fits and starts, touching on trivial topics; the noisy coffee shop was uncomfortable and there were many glances that were hard to interpret. They actually really liked each other a lot, but they were both unsure about the other's feelings and were trying to conceal their insecurity best they could.

They left the coffee shop and began to walk through Central Park. There they were more relaxed, an ugly dog made them laugh, and they finally established direct, transparent visual contact. Ron sat in front of Cynthia, brought his body close, tilted his head to the right (like 70% of couples that British researchers had been watching kiss in public over months do), and he kissed Cynthia's unconsciously swollen lips. He started out gently, but they didn't pull apart until an indeterminate and indeterminable number of minutes passed. Cynthia's cortisol level lowered so much that her stress was reduced to the point that her knees grew weak. Perhaps she would have relaxed with a suggestive embrace, but neuroscientists at Lafayette College confirmed that loving kisses have more physiological impact on the organism than caresses do.

Investigators asked 15 couples to alternately kiss and affectionately hold hands, measuring their hormonal levels before and after each action. Clearly, levels of the stress hormone—cortisol—lessen significantly after kissing, and in the case of men it raised their oxytocin—or attachment hormone—level. But the chemical magic of the kiss is not limited to that feeling of wellbeing. The pleasurable endorphins secreted by the hypothalamus and the pineal gland skyrocket, and exciting adrenalin gradually rises, increasing blood pressure, dilating the pupils, accelerating heart rate and breathing, intensifying the oxygen level in the blood and making us feel much more energetic. Men's saliva contains testosterone, and there is also evidence that a long, passionate kiss could increase desire in women, but the key factor is the secretion of dopamine. Higher levels of this hormone involved in pleasure, motivation and the search for novelty generate eagerness and a desire for more frequent kisses. In fact, Ron's repeated kisses (according to a peculiar study entitled "Sex differences in post-coital behaviors in long- and short-term mating," the man is usually the one who initiates kissing before the sex act and the woman afterwards) make Cynthia move from a romantic state to a dopaminergic motivation for sexual arousal.

There are doubts over whether pheromones play a relevant role in human behavior, but it is clear that mouth movements sharpen our sense of smell and increase our ability to perceive odors and chemical substances.[1] In fact, this could

---

[1] Chemosensation is the ability to detect chemical substances present in the environment that bring about an automatic response in our bodies. It is very frequent in all types of living beings, even bacteria. In mammals the olfactory neuroepithelium has developed specific areas for chemosensory function, like the vomeronasal organ responsible for perceiving pheromones. What's important is that they aren't "conscious" smells that are interpreted as more or less pleasing, but chemical substances that are perceived through the nose and can, for example, increase sexual desire, or in one of the most famous cases, coordinate the menstrual cycles of women who live together.

be the evolutionary origin of the kiss: identifying a good candidate through the intensification of olfactory function.

The kiss is actually very strange in nature. Many species lick and sniff, but only we and bonobos practice kissing with amorous intent. For some anthropologists, kisses are simply a relatively modern cultural habit; for others, they derived from sharing food mouth to mouth between mothers and children, the way that some animals and certain ancestral tribes do. But the combined facts that our lips are faced outwards, are proportionately thicker than in the rest of animals, have one of the highest concentrations of nerve endings in the body, their representation occupies so much space in the brain's sensory cortex, and that we instinctively consider fleshier ones more attractive, all suggest that the kiss had an evolutionary role.

Some hypotheses propose that we like them because of their resemblance to genitals and as a remembrance of breastfeeding. But for the moment the most plausible hypothesis is that kissing is a behavior evolved from olfaction; a more sophisticated way to know that everything is okay and that this is a good candidate to procreate with. The kiss wouldn't so much be used to generate excitement as to eliminate bad candidates, ones who were sick or too genetically similar to us. That is why there are kisses and people that—without consciously knowing why—arouse no chemistry despite our good expectations. The kiss is really a critical moment in the start of a romantic story.

Other investigations suggest that through smelling while kissing we could detect potentially more genetically compatible partners. The major histocompatibility complex (MHC) is a family of genes implicated in the immunological system, and various studies have shown that women instinctively prefer kisses from people whose histocompatibility complex is more different from theirs, because that way their offspring will have a more diverse and thus stronger immune system. The first studies to analyze this unusual hypothesis were Claus Wedekind's famous ones, in which he asked a group of volunteers to smell the sweat on different men's tee shirts and observed that their preferences coincided with greater diversity in the MHC complex. The proposed mechanism is that different combinations in MHC would relate to subtle nuances in body odor that would be recognized by females. It is a controversial topic because some later investigations have not replicated the results, while others only did so in very specific cases. I don't think that it would be approved by ethics committees, but perhaps the definitive experiment would be to have some women blindly kiss different volunteers and, unbeknownst to them, slip their brothers in and see if their histocompatibility complex is clever enough to reveal with whom they should or shouldn't genetically mate.

Aside from chemistry, what is clear is that kisses act as a fundamental first filter to trigger or inhibit the later sexual encounter. And at least in Cynthia's case, that didn't depend so much on the size of Ron's shoulders or on the brand of pants he was wearing. It was something much more magical and visceral.

Sex Without Commitment and "Hookup Culture"

In 1978 the psychologists Russell Clark and Elain Hatfield carried out a peculiar experiment at the University of Florida. Their objective was to analyze the differences in predisposition to casual sex between men and women, so they asked a student of average attractiveness to stroll through the campus, approach women in a friendly way and always say the same thing: "I've seen you around campus, and I find you very attractive," and then ask some of them "would you go out with me tonight?" and others "would you come to my apartment tonight?" and yet others "would you go to bed with me tonight?" Approximately half of the women that he asked to go out on a date said yes, but not a single one said yes to his proposal of going to bed with him. On the other hand, when investigators repeated the same experiment with a woman as the one approaching strangers on campus saying "I've seen you around campus, and I find you very attractive, would you go to bed with me tonight?" 75% of the men said yes they would, and some even insinuated there was no need to wait for nighttime. The "I find you very attractive" experiment has been replicated several times with similar results, and it is frequently cited to justify that women have a much lower predisposition than men to casual sex with strangers.

However, there's something about this experiment that wasn't sitting right with me. My intuition told me that a lot had changed in the last three decades and that the atmosphere on college campuses was nothing like the bars of New York City. So I convinced a good female friend of mine, we screwed up our courage, and I decided to try out my own version of the "I find you very attractive" experiment. A huge party in Brooklyn offered us the perfect scenario. The atmosphere was casual, there was good music, plenty of alcohol, and around one in the morning, when the party was at its height, my friend and I, separately, spent an hour approaching strangers who seemed to be single, smiling at them and saying in a friendly tone: "I've been watching you and I find you very attractive. [pause of varying duration] Would you go to bed with me tonight?" I have to admit that I somewhat lost track, that there were many ambiguous answers, that I only approached women who seemed receptive and always after making visual contact, that there was absolutely no methodological rigor to the pseudostudy, and that giving numerical percentages would be pretentious and scientifically foolish. But the feeling we got was clear: my friend said that approximately three out of every four guys responded with a direct "yes," although she suspected that some of them would have backed out when push came to shove. As for me, I got a significant collection of confused looks and firm "no" answers, but also quite a few women said "maybe" and despite their ambiguity indicated a very high predisposition and contradicted the absolute negativity of Clark and Hatfield's study. Again, the study was full of holes and in no way was an attempt to invalidate the prior academic data from campuses, but it does reinforce the idea that women's predisposition to casual sex or hookups has changed quite a bit in recent years.

The term "hookup" refers to the classic unpremeditated one night encounter without either partner having the slightest intention of starting a romantic relationship. It could end in intercourse or go no further than some heavy petting, but in general it is conceived of as a relatively spontaneous encounter whose objective is short-lived sexual fun, and whose most characteristic image would be that of waking up next to someone whose name we can't or don't even want to remember.

Of course this isn't a new phenomenon, but researcher Justin Garcia from the Kinsey Institute assures us that the normalization of casual sex among young people is one of the most notorious recent changes in sexual behavior in Western societies. And he has the figures to back it up. Justin has been studying the hookup phenomenon among college-age people for several years and, just a few weeks after our extremely interesting meeting at the Kinsey Institute, he published the scientific article "Sexual Hookup Culture: a Review," which is the most extensive biographical review to date of all the studies and sociological works on the prevalence, characteristics and effects of casual sex during adolescence and early adulthood.

We Can Fall in Love with Sex Without Commitment

There are four very clear messages that come out of Justin's work: hookups or casual sex with semi-strangers is increasing among American young people. Male and female attitudes are increasingly similar; really it's not just sex they are looking for, and in general the encounters are experienced as positive but often entail risks and psychological upset that should be taken into consideration.

Americans give everything names and amusing acronyms, and perhaps before we continue we should mention some of the terms that have gradually been incorporated into popular culture. A hookup is the equivalent of a one-night stand. It's fortuitous, usually unplanned, and, at least in the academic definition, does not refer to infidelities or encounters within the dating context. Often colloquially the term NSA or "no strings attached" is added to specify from the start that there should be no expectations of commitment. If it later turns out that the sexual chemistry works and both parties decide they want to sporadically repeat, but without even a friendship, they agree to be "fuck buddies" and are open to "booty calls" (a phone call to get it on when the desire arises). When there is a friendship in addition to casual sex it's called FWB or "friends with benefits," which has its own established limits. The review by Justin Garcia establishes that about 60% of college students have had a FWB relationship at some point in their lives, that 36% of them continued to be friends with their last FWB after they stopped having sex, and that 29% ended both the sex and the friendship (the rest still have those "benefits" or have started a romantic relationship). There are many other terms related to casual sex, but perhaps one of the funniest is the "walk of shame," which refers to someone returning home dressed in the clothes they were wearing the night before, showing that they slept at someone else's house. Although maybe the name should be changed because some people experience it more as a "walk of pride."

But getting back to the main objectives of Justin Garcia's work, the first was determining whether hookups were really on the rise and becoming normal among young people. A first consideration is that the frontiers of casual sex had already widened since the arrival of the birth control pill in the 1950s, the near eradication of syphilis thanks to penicillin, the easy access to condoms, and the resulting sexual revolution of the 1960s, which liberated an entire generation from fears and secrecy. Then AIDS erupted in the 1980s, lessening predispositions to sex with strangers and encouraging more cautious attitudes, but Justin points out two phenomena that seem to be generating a new sexual revolution among the young. Firstly, the time between puberty and the age that men and women say they feel ready to settle into a long-term romantic commitment has lengthened considerably. For many reasons, including socioeconomic ones, there are now many more young men and women who prefer to wait to search for the perfect person with whom to build a stable relationship, without that meaning they want to renounce enjoying sex in meantime, during their years of highest energy.

Even though this delay in the age of parenthood and marriage or equivalent commitment contributes to the rise of hookup culture, for Justin Garcia the truly decisive factor is the greater acceptance of casual sex in the media and popular culture. American society is often deemed puritanical because it doesn't allow nudity on television, but that is less than accurate. Contradictory would be a more fitting adjective. If we look closely we see that while there is rarely a breast shown on their TV series or reality shows, there is a total acceptance of casual, uncommitted sex as something normal, positive and desirable, and this is a significant change with respect to prior decades.

The messages that young men get continue to be pretty conventional, but now it is being suggested to young women that they can act freely without the sexist restrictions of yesteryear, and that casual sex is a totally valid option they have no need to be ashamed about. Without getting into the moral question, Justin Garcia observes that these media messages generate some confusion when they overlap with other more conservative ones, but in general have made casual sex something absolutely normal among most young people.

A curious study published in September of 2012, and not included in Justin's review, selected a hundred and sixty female college students and gave half of them explicit texts about sexual relationships taken from the women's magazine *Cosmopolitan*, while the other half read texts on different subjects. Some time later, they were given a test on their attitudes regarding sexuality and the ones who had been exposed to the contents of *Cosmopolitan* displayed more of a predisposition to defend that women should search for sexual satisfaction to their own desires and they saw less risks in sporadic sexual encounters. The authors conclude that "the complex and sometimes conflicting representations of female sexuality in the mass media and popular culture could potentially have both empowering and problematic effects on women's developing sexual identities." Along with those effects, the key point of Justin's work is that positive messages about casual sex are now much more present in mass media than they were a couple of decades ago.

Absolute figures are too generic and miss cultural nuances, but the most recent data establishes that between 70 and 80% of university students have had some experience with casual sex during their lives. According to one study, 67% of hookups happen at college parties and only 10% in bars. Spring break seems like it was designed for blowing off sexual steam. Another study indicates that condoms are used in less than half of the hookups involving coitus and oral sex, and that 64% of young women having casual sex did so after having consumed alcohol.

The reality of the rise in hookup culture in the United States reflects that society's prevailing superficiality, individualism, consumerism and loneliness. Perhaps that's why another of the most interesting conclusions of Justin's study is that people are looking for something more than sex in casual sex. In a study with 681 adults of college age, 63% of the men and 83 of the women declared that at this point in their lives they would prefer a traditional romantic relationship to a sexual one with no commitment. In another study, 65% of women and 45% of men admitted that after their last hookup they were hoping that encounter would lead to a committed relationship (51% of women and 42% of men even discussed that possibility after a hookup). In fact, in an investigation published in 2008, Justin Garcia himself observed that when he asked about the reasons behind hooking up, 89% of men and women responded physical gratification, but 54% also answered emotional gratification and 51% the desire to start a romantic relationship. According to Justin Garcia, the most curious aspect was that he didn't find significant differences between the genders, and that was in addition to much more data from surveys that indicated that men's and women's attitudes about casual sex were more similar than we thought: there are increasingly more women unabashedly looking for uncommitted sexual satisfaction, while more men admit to wanting an emotional component along with casual sex.

While recognizing the tremendous diversity among individuals and cultures, the data is actually consistent with the conclusion we came to in the last chapter: both men and women desire both romanticism and sexual diversity. There is little data compiled on homosexual sex, but Justin indicates that while hookups are much more frequent in the male gay community, there is also a clearly observed high level of desire for commitment and emotional satisfaction that perfectly reflects the coexistence of romantic and promiscuous interests.

Justin is an anthropologist and evolutionary biologist, and he defends the idea that our sexual instincts are strongly conditioned by natural selection, but in his article he recognizes that evolutionary logic doesn't fully explain the diversity and complexity of sexual behavior among young people in Western societies. According to him, the growing trend for men and women to hold increasingly similar viewpoints is due to cultural forces, and the now very clear distinction between sex and reproduction makes sociocultural pressures carry much less weight than genetic pressures when it comes time to shaping the type of sexual relationships we will have in the future.

As for the risks, the negative consequences of hookups include regret and emotional upset, cases of sexual abuse, infections and unwanted pregnancies. Even though it seems inconceivable, various studies have shown that there continues to be incredibly little knowledge about sexually transmitted diseases even among young people, and that condom use is still alarmingly low. In a study with 1468 college students, only 46% said they had used a condom in their last casual sex encounter. An important factor for understanding this data is the prevalence of hookups under the influence of alcohol, which is tremendously high, particularly in the United States. As a result, a study published in 2005 established that 72% of college-age people admitted having regretted hooking up on at least one occasion. Another study that focused exclusively on young women found that 74% stated they'd had regrets at least once, 23% never and 3% many times.

In general the effects of the hookup seem to be more positive than negative, since it increases wellbeing and satisfaction, but the emotions around it are confusing. Another study by Justin Garcia determined that on the morning after a hookup, 82% of men and 57% of women felt fully satisfied with their sexual adventure, but another published by Owen in 2010 with 832 students concluded that 49% of young women and 26% of young men reported a negative emotional reaction related to their last hookup, while the reaction of 26% of females and 50% of males was positive. The same author wanted to explore the reasons behind these differences; in order to do so he followed 394 young people over the course of a semester and observed something curious: in those with more depressive symptoms and feelings of loneliness, casual sex improved their emotional state, but in those with a richer social life and lower depression levels, the hookups made things worse. This result doesn't quite match with other studies that relate less self-esteem with a greater number of hookups and associate a positive attitude towards casual sex with more beneficial effects.

In light of all the data and studies in Justin's review, it seems obvious that individual diversity is vast and that generalizations only serve to observe trends, but it does seem clear that hookups are becoming more normative and socially accepted, that they are generally satisfying although sometimes there are regrets, and that both men and women are actually searching for romanticism and an emotional link in addition to sex. Justin Garcia told me that literally one out of every three hookups turns into something more serious even though that wasn't the stated intent, that "the culture of dating and the hookup are dramatically different among today's young people," and that it will be interesting to observe whether this is a passing thing during their youth or whether it will somehow affect their coupling models once they've reached maturity.

Cultural changes are slow, but it seems clear that with growing social and media acceptance of sexual permissiveness in Western societies we are moving toward a greater normalization of the separation between sexual pleasure and love. Possibly the combination of sex and love with trust and devotion will always endow human relationships with more magical fullness, but every one is free—or should be—to decide individually and with their partner how to balance and express their romantic and sexual emotions.

Educating is much better than restricting, and with current methods of protection and our cultural growth we should promote an education that fosters the healthy development of sexuality based on a knowledge of its diversity, risks and richness.

For the early chapters of this book I conducted interviews with researchers and reviewed scientific studies as my initial sources, searching to then relate that knowledge to everyday life. In the following chapters I took a different strategy: I set out to get to know different aspects of human sexual behavior and I then explored science to see what in it could relate to them. I have to admit that in many cases I learned more from individual testimonies, life experiences and conversations with experts on various forms of sexual expression than I did from visiting academics and laboratories. But I still find it marvelous that science allows me to understand what is going on inside my organism when, for example, during a tantric sex workshop I tried to have an orgasm with only the power of my mind.

10

Having an Orgasm with the Power of the Mind

I am sitting with my legs crossed and my back straight on the floor of the main room of the Atmananda yoga center, in the very heart of Manhattan. Barbara Carrellas asks us—the 5 men and 16 women attending her workshop "Erotic Meditation and Sexual Ecstasy: a tantric approach to love and life"—to spend 4 min breathing as quickly and deeply as we can, focusing as if each intake of air could reach the base of our genitals and caress them. If we do it conscientiously, we can feel a slight tickle and have the sensation that it is actually happening.

It was three in the afternoon, and the high point of the workshop was approaching: trying to have an orgasm with just the power of our minds. We had spent the whole morning reviewing tantric sex concepts, talking about different erotic practices to increase arousal and pleasure and doing exercises for sexual communication in couples. I was the only man who had come without a partner. When I arrived I surreptitiously sat next to Dana, the most attractive of the unaccompanied women. Dana must have been about 45 years old. She was tall, svelte, and blonde with a probing gaze, and she told me that she had been practicing tantric sex for 6 months. Having her before me staring into my eyes and giving me details about how she likes to be stroked and what fantasies she finds most stimulating, even though I knew it was just a communication exercise, was exciting in the most conventional way for a non-tantric like me. Actually that was the point. Unblocking negative feelings towards sex, opening up the imagination, allowing repressed desires and your own sensuality to flow. Highly recommended. I suggest you try it with your partner or friends, even if you don't think you need it.

There were many people in that room who, because of romantic failures, traumatic experiences, inhibitions and anxiety, needed that release. In fact, that was what they expressed openly when it came time to explain to the others what they were expecting from the workshop: most of them said they were there to overcome a block. The few folks who didn't acknowledge having any problem simply wanted to explore new ways to improve their sex lives. That was the case of three heterosexual couples, a lesbian couple, Dana, another woman and, in theory, me. The rest, who were aged between 25 and 60 years old and seemed to have a cultural and socioeconomic situation somewhere between medium and high, did feel that—some knowing why and others not—something was impeding their sex lives from being as satisfying as they would have liked. There were many tears shed.

We were sitting in a circle breathing as fast as we could. After 4 minutes of hyperventilation, Barbara Carrellas asked us to relax, spread out through the large room and lie down on the floor. Thirty minutes earlier, in front of everyone, she had had an orgasm induced only with her breathing and mind, and now it was our turn. When a week earlier I interviewed her at the Vapiano restaurant on Fifth Avenue in New York, Barbara explained the scientific studies they had done on her at Rutgers University. It wasn't a hoax. We can moan all we want to fake an orgasm, but we can't just decide to have a spike in blood pressure over 200 mmHg, or change our skin's conductance or dilate our pupils at will. Barbara Carrellas had become famous for mastering a technique that allowed her to have orgasms with her mind. She assured me that I could do it as well and she invited me to attend the workshop she was going to hold a week later.

I was lying down on a rug in a corner of the room, completely predisposed to letting myself get carried away. Goodbye rational constraints and prejudices, hello wide-open chakras. I was convinced I could use my mind to direct all my energy toward a state of extreme sensitivity that could lead me to sexual arousal and maybe even orgasm.

Barbara asked us to bend our knees 45° and leave them partially open so that we were comfortable, and to start breathing deeply again. We didn't have to breathe as quickly as before, but it did have to be very intense. She asked us to imagine that the air was also entering our genitals, and that all our energy was accumulating in the first chakra, located in the pelvic region. Soon I again notice a tingling in my fingertips, and a certain altered state of consciousness. I don't yet have any reaction in my genitals, but suddenly, after a few minutes there is a female exhalation heard in the room that approaches a moan. As if that had given us permission, more moans are heard. They start out subtle, but gradually intensify.

Barbara warns us to prepare to pass the energy to the next chakra. I lose my concentration: I don't have anything to pass on. Barbara gives the order and some sighs are heard. I continue at my own pace, trying to focus on the exercise and accumulate some energy; it's okay if I'm going slower. I moan a little bit, out loud, to see if that helps.

Soon there is moaning all over the room, the energy has already moved from one chakra to the other, and I've given up. I covertly decide to just observe. The 30-something woman next to me is crying inconsolably. She moves her pelvis, shuts her eyes tightly, her contorted mouth is half moaning and half sobbing, tears fall from her eyes and her nose is running. There is more crying around the room. Dana is in front of me with her back arched, lifting and lowering her pelvis as if she were simulating a sexual encounter. The moans from one of the lesbians are much, much louder than the entire rest of the group put together. One guy doesn't seem to be reacting at all, and I see another breathe in and out as if he were in the middle of coitus.

Barbara is encouraging us to breathe more and more quickly and concentrate all of our sexual energy. We should feel the energy built up in all of our chakras. The tension and screaming in the room was reaching a fever pitch. Then, at a certain moment, as she had done earlier, Barbara asked us to press our arms hard against the floor, hold our breath, try to push our bodies as much as possible against the floor, press as hard as we can downward and expel all our energy upward at once, letting ourselves get carried away by ecstasy. It was amazing. A collective explosion. The entire room started to howl as if everyone was having the most intense orgasm of their lives. And simultaneously. "It's usually hard to just get two people in sync…" I thought. I'll admit I was feeling pretty astonished. The situation seemed inconceivable, too exaggerated, but at the same time I realized that something was going on in that room that I hadn't been able to reach with my powers of suggestion. I knew that there were scientific studies that analyzed physiological reactions derived from Eastern practices. I was going to check them out. But first, once the workshop was over, I decided to delicately ask the attendees if they had really had orgasms.

Only the loud lesbian and another young woman told me that they'd had a genital orgasm. "I had to change my underwear," said the former to convince me that her orgasm was real. I didn't doubt her. The woman who had been crying inconsolably explained that she had felt the same as when climaxing, but not in her genitals. Another woman and a guy said something similar, suggesting that there are many different types of orgasms. I asked the guy if he'd had an erection and he said no. Another confessed that he hadn't felt anything and thought that "this was more for women, especially if they have some problem." Six or seven women—including Dana—told me that they hadn't felt any kind of sexual excitement, but an enormous pleasurable release of energy. The Japanese woman who had come after watching a documentary featuring Barbara Carrellas on the Discovery Channel was exultant despite her eyes being swollen from tears. She hadn't had an orgasm, but was feeling very sensitive. A woman in her 60s spoke of ecstasy, and a Peruvian woman acknowledged that she now felt more prepared to enjoy sexual pleasure. I spoke with Barbara and I told her guiltily: "Barbara, I didn't feel anything…" "I saw that you were distracted. Have you done yoga or meditation before?" she asked me. When I told her that I hadn't, she said it was normal not to feel anything at first, that I should take it slowly. I left the room a bit disappointed and thinking that science wouldn't be able to bring anything relevant to this experience. But then I discovered that it was quite the opposite: it seemed to reinforce it.

Hyperventilation to Activate the Sympathetic Nervous System

What Barbara Carrellas was doing with us when she asked us to breathe deeply at an abnormally fast rate is exactly the same thing scientists do in the laboratory when they want excess oxygen in the blood to activate the sympathetic nervous system. And as we've seen in earlier chapters, this activation can facilitate the sexual response.

In 2002, Professor Boris Gorzalka, from the University of British Columbia in Canada, carried out a study with 61 pre- and post-menopausal women to test what effects activating their sympathetic nervous systems had on their sexual arousal. (The sympathetic is the system that activates in the face of a sudden stressful situation when the body needs to prepare itself quickly for action.) The researchers excited the sympathetic nerves by rapid and frequent breathing with a technique called LIH (Laboratory Induced Hyperventilation). After this anomalous hyperventilation, the body increases its heart rate for several minutes, muscular tension rises and fibers of the sympathetic nervous system are activated.

Previous studies had suggested that this activation of the sympathetic system through stress could suppress the subjective sensation of sexual desire in women, and yet facilitate their genital response. The effect was less marked in men. In some pioneering experiments that had been replicated on several occasions, the psychophysiologist Cindy Meston, from the University of Texas, empirically established that higher anxiety usually inhibits sexual desire, but that moderate anxiety could increase it. Gorzalka's group wanted to check whether that was true, and compare the reactions between young women and older ones. In order to do so he would show erotic images to various groups of pre- and post-menopausal women with hyperventilation and without, measure their vaginal arousal with a photoplethysmograph, and ask them about their subjective perception with a questionnaire.

The results indicated that the activation of the sympathetic nervous system did not increase the subjective sensation of being aroused in either group, but that in the young pre-menopausal women—not in the older ones—it did produce a significant increase in genital excitement. In 2009 Boris Gorzalka and psychologist Lori Brotto repeated this study comparing women who complained of sexual arousal disorder with a control group of women without sexual difficulties, and they proved that the activation of the sympathetic nervous system through hyperventilation did increase genital arousal in the control group, but had no effect on the women with problems. It could be a coincidence, but it is very similar to what happened during Barbara Carrellas's workshop. The young women with no sexual problems were those who most recognized having gotten physically aroused, while the effect was null on men and insignificant among older women. What Gorzalka and Brotto's experiment did make clear was that the hyperventilation induced by Barbara could perfectly trigger a physical sexual response, especially after an entire day talking about eroticism and sensuality.

Women Have More Control Over Their Bodies

One relevant aspect of the tantric workshop, and demonstrated by various laboratory studies, is that women seem to be able to better control their physical arousal than men can. In some experiments they've asked women to try to facilitate or inhibit as much as possible their sexual excitement in the face of erotic stimulation. It is a very standard procedure used to check the degree of control that women have with arousal disorders, diseases, under the influence of alcohol, or when they want to analyze risky behaviors. Every study is different, but in general what they do is show erotic stimuli to three large groups of women, ask some to simply observe, others to encourage their excitement and others to repress it, and measure the genital response in all of them. The process is more sophisticated than that, but it is confirmed that most women can significantly increase or diminish their sexual response using mind control. In fact, the differences are greater when they ask them to control their excitement than when they ask them to control their inhibition. In men, on the other hand, the effects are more muted, which could be another explanation for my skepticism at the fact that the women in the workshop could get so excited at will, while I didn't notice anything no matter how much breathing, concentration and effort I devoted to it.

Meditation and Yoga Increase Sexual Pleasure

Sometimes it's surprising how Western medicine can ignore certain practices of ancient cultures that can be a wonderful complement to our wellbeing.

It is true that science sees chakras and energy flows as something esoteric and difficult to measure, something that is more part of the symbolic world than the real one. But it doesn't underestimate in the slightest the power of suggestion nor deny that different mental states can generate different physiological responses in our organism. Scientists feel confident when they experimentally confirm that cortisol increases in states of anxiety and lessens during relaxation, because they can objectively measure that and propose a mechanism of action that is coherent with everything we know about human physiology. But when they face more implausible phenomena, like healing with pin pricks or spiritual therapies, they usually display one of two attitudes: either they turn up their noses saying that they are anecdotal and have no empirical evidence to back them up, or they explore them believing that they may hide a mystery that would be interesting to solve. Critical thought is very advisable, and we shouldn't be gullible or trust every hack with a wise man's beard who tells us what we want to hear. There is a lot of hot air, exaggeration and fraud in some of these Eastern practices and alternative therapies, but there is also some truth. Our minds and our bodies are intimately connected, the key is having the tools to discern the fact from the fiction.

Lori Brotto is a Canadian psychologist who has been studying the effects of meditation on female sexuality for years. She is not outside of the scientific system, she is an elected member of the International Academy of Sex Research, the editor of several indexed scientific journals, and all of her works are published in peer-reviewed articles. The first time I spoke with her she was emphatic: "The basic idea is that meditation and other Eastern practices can improve certain cases of female sexual dysfunction and be incorporated as part of the therapy." The mechanism, ignored according to her by contemporary sexual medicine, is—among other aspects—reinforcing the link between body and mind, and increasing sensitivity.

Brotto insists that there are many women whose genitals work perfectly, but who are incapable of mental abstraction and feeling physical excitement. She also spoke to me about others who, even though they reach orgasm, don't feel satisfied by sex. According to Brotto, it's as if they were lacking a wisdom and control over the reactions of their own bodies, and she has data that supports that, in such cases, Eastern wisdom has much to offer.

In a first experiment she recruited 26 women between 24 and 55 years of age who came to a center in search of treatment for lack of sexual desire or arousal. The women did several 90-minute meditation sessions, and a few weeks later Lori gave them a series of questionnaires and made several physiological measurements of arousal. In most of the cases she noticed significant improvements in lubrication and subjective perception of sexual desire. Lori Brotto herself recognizes a limitation of the study's lack of a control group, but the positive results encouraged her to continue.

Brotto carried out another investigation with 31 cervical cancer survivors who were suffering sexual difficulties. After surgery, patients can have pain, nerve damage and various physiological problems that involve a loss of sexual response. Obviously, if the physical problems are serious and sexual function completely disappears, meditation isn't going to help them, but Brotto's hypothesis was that in those cases where sexual response was diminished but not eliminated, working on concentration with meditation could amplify it. She again recommended meditation sessions for women in the rehabilitation phase, and this time she compared them with an equivalent group on the waiting list for a treatment. She didn't observe differences in the physiological response, but there were significant improvements in the indexes of satisfaction. She identified something similar in a study published in 2012, in which she confirmed that meditation improved different indexes of sexuality in 20 women traumatized by sexual abuse in the past.

It is a very new subject of analysis for science. Meditation is based on concentrating all your attention on one point—it could be an object, image or part of the body—and letting your thoughts flow and wander. In functional magnetic resonance studies greater activity has been observed in areas related to attention and emotional response. And this, along with the undeniable relaxation generated by meditation, can help you concentrate on an attitude more conducive to pleasure.

Brotto is not against modern sexual medicine, in fact quite the opposite. She recognizes that there has been much progress since the dubious treatment of sex in early psychoanalysis, and with the evolution to behavioral therapies and the big advances brought by drugs like Viagra. But she insists that Western medicine has tended to ignore practices that historically have been very effective against the satisfaction problems that are so frequent today. She is looking for empirical evidence to get practices such as meditation considered therapies, and she is well on her way.

As for yoga, whose practice includes physical exercise in addition to meditation, evidence seems even more solid. In 2010, *The Journal of Sexual Medicine* published two studies carried out in India with a group of 65 men and another with 40 women who were starting to practice yoga. The first study established that the practice of yoga increased male sexual desire, improved erections and ejaculatory control, and boosted confidence and satisfaction after coitus. In the case of women, the authors claim that it improved everything, from desire, arousal, lubrication, pain diminishment, ability to reach orgasm and satisfaction. It is obvious that a large part of these improvements can also be achieved with any other type of exercise, and surely if you have a sedentary lifestyle and you start practicing dance or Pilates you will also see an improvement in your physical condition and sexual response. But it seems that the mental effort and the component of psychic wellbeing involved with yoga contributes to reinforcing that effect even further. The study also showed that the long-term improvements were especially notable in women over 45 years old.

Although more studies need to be done, these results suggest that in the case of people with sexual problems, the continued practice of yoga or meditation could surely enrich different aspects of their sexuality, but as for general satisfaction the most significant improvements would be seen in people with injuries, traumas or difficulties.

In a review published in 2012, Brotto insisted that the lack of synchronicity between women's genitals and their subjective feeling of sexual arousal is very well documented, and that Eastern therapies reinforce a unity of mind and body that is fundamental to sexuality and that pharmaceutical efforts and many contemporary sexual therapies are ignoring. She also explains that there are studies that indicate that acupuncture can lessen vaginal pain and increase desire in women with hypoactive sexual desire disorder. This again demonstrates that we mustn't underestimate the effects of suggestion. In cases of sexual response they may be more powerful than other interventions.

The Placebo Effect Also Improves Sexual Response

In a study published in 2011, Cindy Meston gave a placebo drug for 12 weeks to 50 women who complained of female sexual arousal disorder (FSAD). She measured the changes in their sexual function parameters at 4, 8 and 12 weeks, and found that in a third of the women, the improvements were clinically significant. In scientific terms, "clinically significant" implies an effect that is actually noted and is not just "statistically significant." The indexes of lubrication, pain, satisfaction, orgasm and desire improved in similar magnitude, and curiously they were correlated with the number of sexual encounters the women had with their partners over that time period. In her discussion of the results, Meston placed a very high value on this effect of increased intimate relations, saying that it could distort the conclusions: suggestion could make a person more predisposed to enjoy sex, and when engaging in it, enter into a dynamic of constant improvement that is reinforced by the frequency of sexual encounters. Meston warns that this factor should be taken into account when evaluating pharmacological interventions, but that it also reinforces the idea that behavioral changes are very effective, and that as we mentioned when discussing the circular model of sexual response published by Rosemary Basson, having sex despite not initially feeling desire can, in some cases, facilitate improvement. Mind and body are intimately connected, in both directions, as I saw in the tantric sex and mental orgasm workshop I attended.

11

Pornography: From Distortion to Education

It's one in the morning and I'm having a coffee with the porn actress Sophie Evans in a bar near Barcelona's Sala Bagdad, where I've seen her naked on stage just minutes earlier. I also witnessed one of her coworkers putting a condom on a member of the audience to try to fellate his shrunken penis, and watched how a couple of actors had sex with all the coldness, overacting and male dominance typical of pornography.

Sophie is charming, smiling and has an unexpectedly candid gaze. She assures me that her life away from the cameras is totally conventional and she even blushes when I ask her if she gets physically and/or mentally excited when filming scenes. "Oh!… that depends on a lot of things… there are girls who love it and others who always fake it," she replies, trying to avoid the first person. We continue talking about how the pornographic industry has changed with the Internet, about whether her fans prefer some scenes over others, and even about science. Suddenly Sophie tells me that she is reading a popular science book entitled *The Brain Snatcher*, and I give her a flustered look and say, "I wrote that!" Sophie's eyes grow wide, she covers her mouth and nose with both hands in a sign of surprise and we laugh at the coincidence. She hadn't recognized my name when I'd set up the interview.

After a little while I feel comfortable enough to ask her if she considers conventional porn to be denigrating to women and give the wrong idea about sex. Sophie responds, "Maybe it seems violent sometimes, but sometimes women like it that way. In any case, there are a lot of different kinds of porn, all kinds of opinions, and everyone has different values." Sophie is right, the debate over whether porn degrades women isn't scientific and it absolutely shouldn't be.

More interesting—if not more important—is analyzing how porn consumption affects later sexual attitudes, or if some scenes can generate mental rejection in us even as they excite us. This can be measured and, therefore, scientifically analyzed. "But we should go back to the Bagdad so you can talk to more people," Sophie says.

There, an actor tries to convince me without any empirical data beyond his biased opinion that most women really like hard-core abusive porn. Another tells me countless strange, morbidly fascinating and hilarious anecdotes about his experiences in the industry, and when I ask him if he takes anything to maintain erections he answers, "Few will admit to it, but yeah, it happens." A very beautiful transsexual who confuses my brain claims that straight men are obsessed with looking at big penises, and a couple from the audience explains that they come there to add a little spice to their relationship but also because they always discover new ideas. This last point is the one most often brought up by those who defend the educational value of pornography.

The atmosphere is incredibly friendly. Super welcoming Juani, the owner of the Bagdad, is pleased that people come just to have a good time and see something different from their day-to-day, and an actress assures me that the contents have changed a lot in the last 10 years and that now people want more violence. I tell her that a recent scientific study concluded that they really didn't, and she hesitates and says, "Oh, well, I don't know."

I found it very interesting to meet—not only there at the Sala Bagdad—people from the pornographic industry. Away from the cameras—both on sets and amateur—and the media where they continue to be in character, they are more normal folk than we think. And that's not just my opinion, science confirms it. A study published in November of 2012 entitled "Characteristics of Pornography Film Actors: Self-Report versus Perceptions" asked 399 psychology students what they imagined about the personalities of porn actors and actresses, their sex lives away from the cameras, relationships, rates of abuse, etcetera, and compared their responses to the data contributed by 105 porn actors and 177 porn actresses in the Los Angeles area. The results showed that the stereotypes revealed by the students were not at all indicative of the reality. The actors and actresses had much higher self-esteem, romantic ideals and concern over sexual diseases than the students thought. The prevalence of sexual abuses in childhood was identical to the general population, and they'd had their first sexual experiences much older than those surveyed believed. There were very significant differences. For example, the students thought that porn actors and actresses earned an average of $224,000 and $250,000 respectively, when actually the average salary is $74,000 (men, average age 35) and $74,000 (women, average age 26). Those surveyed were close when guessing the number of different partners the actresses had filmed erotic scenes with (72 real versus 78 guessed), but they were way off with the men (97 guessed versus 312 real). Both actors and actresses also had more sexual partners off camera and said they enjoyed sex a lot, contradicting the idea that they found their work "boring." In general the actresses said they enjoyed themselves a lot during their scenes and were much more satisfied with their profession than the students thought they would be. In fact, on this last point, investigators compared various personality indexes of the actresses with a comparable sample of women, and while they did find a higher rate of drug consumption, number of partners and concern over diseases, their indexes of self-esteem, sexual satisfaction, personal wellbeing and social support were higher than average.

That reminded me of when I attended a small gathering of former porn legends like Candida Royalle, Veronica Hart and Gloria Leonard at the Museum of Sex in New York. Now retired, they displayed enormous satisfaction and pride about their porn careers, although they did recognize that times are much, much harder now: there is more competition, less money, more abuses, it's more "industrialized," and it is hard for young actresses to turn down certain scenes. Candida Royalle is now a director, like many other former stars, and she called for a better porn that also took the preferences of the growing female audience into account. But here, like whether condoms should be used on set, there began to be differences of opinion.

In my superficial contact with the porn world I've heard almost as many different opinions as people. I'll admit that after talking with the warm folks at the Sala Bagdad I left with a more positive vision than I'd had as just as a spectator, but I was also disconcerted by so many varying perspectives, some of them even contradictory. I realized that I should return to science to get a more objective viewpoint, especially if I wanted to find out whether porn consumption has individual or social risks, and even just to get clearer on what sort of pornography women liked more.

Women Prefer Lesbian Porn Over Gay

Taken out of context, some scientific investigations can be very comical. In earlier chapters we saw that erotic images are used in sexual psychophysiological laboratories to measure genital and subjective arousal in certain circumstances or sexual dysfunctions. Researchers must be convinced that said erotic images perform their function as a scientific tool well, and that gives them motives to conduct a rigorous investigation into what type of porn women prefer.

It's simpler with men, since various experiments have shown that videos are much more stimulating than photographs, erotic literature or audio recordings, that the more explicit the better, and that "the regular" is plenty to increase arousal. But with women, investigators have some doubts. First they confirmed that, even though the difference was less than with men, women are also more excited by videos than by reading or photographs. But later, high methodological principles demanded that the contents of the erotic videos used by different labs be standardized and meticulously analyzed to find out what stimuli are most appropriate for optimizing the female sexual response, as the authors set out to do in the study "What kind of erotic film clips should we use in female sex research?"

While they could ask some experts in the field, again, opinions are so varied. One of the most interesting conversations I had while researching this book was with the well-known porn director and activist Tristan Taormino. When we met in the New York Financial District, she told me that women prefer scenes without force, ones where they perceive a connection between the couple and where you can see that the woman is enjoying herself. Women seem to watch the actress and empathize with her more than they pay attention to the attributes of the actor. But half-joking she admitted that "there are a lot of opinions on the subject; the truth is not even I know for sure, and I wish they'd do a scientific study on it."

They have: researchers at Wayne University in Detroit showed 90 different porn clips to a wide group of women who had watched pornography in the past, but who weren't regular consumers. This is an important point, and implies that the results of the study won't necessarily coincide with online sales or searches. Porn consumption among women has risen dramatically due, among other things, to the anonymity of the Internet, but scientists aren't attempting to find out what most stimulates habitual consumers, but rather to find the images that work best in the most standard possible population; which is to say, any woman who they put an erotic film on the screen in front of. This is also why they don't fully rely on the conclusions of the pornographic industry aimed at female audiences. There is a certain agreement that women prefer explicit but consensual heterosexual scenes that are more romantic, and that they don't like close-ups, ejaculations onto people, or lesbian and anal sex. But they wanted to investigate it for themselves, and to especially find out whether there was a correlation between the subjective psychological response and the physiological one. Scientists don't like preconceived ideas.

The 90 clips were each a minute long, and 88% showed sex between men and women, and the other 12%, homosexual relations. There were different positions, group sex, ménage à trois, interracial, fellatio, cunnilingus, anal sex, masturbation, sadomasochism and bondage, among other practices. The women studied had a minimum of a high school education, ranged in age from 18 to 57 with an average of 31, and every race was represented. After each video they were asked to respond what degree of physical and mental stimulation they experienced.

The statistical analysis of the results didn't offer big surprises. The heterosexual scenes of vaginal sex were by far what they most liked, they preferred positions in which the man took the initiative, and there was a very good response to scenes filmed indoors. Sex between a man and woman was preferred to homosexual, although there was a clear inclination—an interesting difference from male response—for scenes involving two women instead of two men. This last point is coherent with the hypothesis we discussed in Chap. 3, on the impact that erotic videos could have on activating mirror neurons, involved in imitation, making us feel part of the action. The lowest scoring for most women was anal sex, fellatio, and behaviors considered abusive, while a considerable number of them described these scenes as not mentally stimulating but physically so. This is an interesting bit of data, which is also consistent with other studies that show that many women have sexual fantasies about acts they don't really want to do.

The investigators are aware of the limitations of their study, especially because there is such a wide range of tastes, but they insist they looked at factors like the attractiveness of the actors, the habituation the participants reached by the last clips, and the laboratory conditions, and that they did find some general trends. I doubt that the industry will pay them much mind, but their data and that of similar studies is available in Pubmed.

Porn Can Exacerbate Some Problems, but Doesn't Cause Them

I confess that, in general, I find porn to be like the McDonald's of sex, and I don't like the role it usually gives to women. But in the United States they have gone much further in the discussion of its possible negative effects and there are books and conservative movements that consider pornography something apocalyptic that is destroying the values of respect and morality. An example is *The Porn Trap*, whose authors argue that pornography has become more extreme, that the Internet has made things worse because it makes it more individualistic and provokes social isolation, that porn erodes couples because the regular consumer loses his attraction for his wife and makes her feel she has to compete, that its continued consumption causes mood disorders, that it affects the brain like a drug, that it perceives women as objects and increases both risky behaviors and the number of sexual aggressions. I should point out one small detail: all their arguments are based on partial opinions and not on proven evidence. And some aspects, such as whether porn is addictive or not could, to a certain extent, be studied scientifically.

The basic idea is that watching porn is a pleasurable act that activates the brain's reward system, making us secrete high amounts of dopamine from the ventral tegmental area to the nucleus accumbens, and like other pleasurable activities, if we do it to excess in isolation from other pleasures and in an obsessive way, it could become a behavioral addiction. Although many neurobiologists reject the use of the word "addiction" and prefer "dependence" or "compulsive behavior," if they define gambling as an addiction they should do the same with online porn. They are tremendously similar. In fact, some understand porn addiction within "Internet addiction," which is to say, addiction to the constant novelty and excitement the online world offers us. We are compulsive novelty seekers, and that is why we can spend hours and hours surfing the web and following links or interacting on social networks. The screen absorbs us. We can go onto a website looking for information or an erotic video and end up reading or watching twenty. Every new tempting image is a release of dopamine that motivates us to click. In this first stage, the "routine" of watching porn online isn't very different from the habituation that becomes a necessity, of checking Facebook or certain websites constantly. We get into a cyber-routine from which it is hard to escape. But with porn the situation can get worse, for two reasons. First, our brain is programmed to prefer sex to news about politics or science, and it is normal that it is more of a trap for our limbic systems. But also, if it is accompanied by masturbation, the release of dopamine during sexual arousal is even greater, the brain begins to associate that activity with intense pleasure, and with repetition this conditioning is reinforced. The dopamine levels will never be as drastic as during cocaine or heroin consumption, but they will be forging a behavioral dependence, which in some cases of isolation and lack of other motivations could become a serious obsession that interferes in our personal and work lives. This only happens when there is a loss of control, which in addictions is manifested by less activity in the prefrontal lobes of the brain.

This activity is exactly what neuroscientists Donald Hilton and Clark Watts, at the University of Texas, are analyzing when they compare fMRI studies of people with different obsessions and dependencies. Their results concluded that compulsive consumers of pornography do indeed show alterations in the frontal lobes of their brains similar to those found in any other dependence, and therefore the activity could lead to loss of control and addiction. However, in the same journal, other authors criticize that interpretation, saying that what was actually observed was not that pornography causes problems, but that people with certain problems abuse pornography. The latter authors believe that the indiscriminate use of pornography depends on the context and is induced as a response to psychosocial stress. Which is to say, the association with problems exists, but pornography is not the cause but rather the consequence.

Notice that here we are not talking about sex addiction or hypersexuality, which we'll discuss later in the book, but about the routine of looking at porn on the Internet (a therapist told me that many of her patients watch videos for hours and hours but only masturbate sometimes), that will only really develop into an obsessive disorder in someone who has a strong predisposition and preexisting psychosociological problems.

Another curious effect is being studied to find out if porn often causes demotivation and erection problems. There are a fair amount of reports from sexologists and urologists of patients who, after a period of frequent online pornography consumption, say they feel less arousal, have weaker erections and more difficulties reaching orgasm when having sex with a partner. Here the idea is that these people are using a certain type of women and activities as a reference point, making real encounters less stimulating. While this hypothesis may seem logical, and it evidently happens in some cases, I have not found any scientific study on the subject. Italian researchers led by Carlo Foresta say they have done a massive survey and found an association between the abusive consumption of pornography and erectile dysfunction, but at the time of this writing the data doesn't seem to have been published anywhere.

Yet, beyond the consumer him or herself, does pornography negatively affect attitudes towards women? Psychologist Neil Malamuth, from the University of California at Los Angeles, is a world authority on the effects of pornography. In our conversation he is unequivocal from the very start: "In most people pornography does not have positive or negative effects, but when there are other risk factors like having suffered abuse, a family history of aggressions or a narcissistic personality… then we do see that extreme pornography increases the risk of aggression towards women." This statement must be interpreted carefully. Malamuth recognizes that "books like *The Porn Trap* have no scientific basis," but he doesn't say that pornography is completely harmless.

In one of his studies, Malamuth proved that porn doesn't affect conventional men's behavior towards women, but that it does slightly in those who have psychological problems and consume extremely violent porn. Similar results were replicated in Denmark by Gert Martin Hald in one of the most frequently cited studies on the subject. Analyzing 688 heterosexual men and women between 18 and 30 years of age, he found a slight but significant association between the consumption of extreme pornography and a higher tolerance to violence against women. The study also revealed that men use pornography fundamentally to masturbate and women out of curiosity or as part of partner play. One out of every five women said they had never seen pornography, as opposed to only 2% of men. Of the women who said they had, 7% watched it three or more times a week, while the percentage increased to 39% in men. Analyzing the impact that consumers felt porn had in their lives, the Danish generally evaluated it as much more positive than negative.

In fact, there are studies that even give a net benefit to pornography consumption. On a social level it is argued that porn decreases the rates of sexual crimes, because it acts as an escape valve, a substitute. Studies realized in Japan, the United States, Denmark and the Czech Republic have confirmed positive associations between greater access to porn and less sexual aggressions. On an individual level, supporters defend that for single people and those with privation, moderate use can be a great help, and that within a couple it has a certain educational value in adding possibilities to their lovemaking repertoire. The dilemma is actually when someone morally conservative complains that "then they want to do what they see on the screen" and someone more liberal responds: "What's the problem in that, if it's respectful and consensual?" This is the main argument of directors such as Tristan Taormino, who tells me straight out that "part of what I do with my filmmaking is educational porn." Tristan has been giving talks on sexuality to both adults and teens for many years, and she says that she is constantly surprised at the limited and repressed sexual perspective and behavior that so many people and couples have. She admits that most porn always follows the same script, but says that with her films she tries to "expand the definition of sex" and "show many more of the erotic possibilities that we have." She insists that there are many types of porn, even while allowing me my comparison to McDonald's, but she seems convinced of the educational and even therapeutic potential of good erotic films for couples with low libido.

It is a delicate subject because there are countless aspects to analyze, and I don't want to overlook those cases of people for whom exposure to porn has had damaging consequences. But everything seems to indicate that if we open ourselves up in a critical and selective way to better erotic cinema like what Tristan and many other filmmakers defend, that undoubtedly the benefits for those who want and know how to enjoy porn far outweigh the risks.

12

Let's Do It Tonight, Dear, I Have a Headache

Oh, Lucía! Not tonight, I've got a headache…

Don't be like that, Manolo, really I'm doing it for your own good. I read in a book that sex increases the levels of some neurotransmitters and it can cure your migraine…

Come on… really, let it slide, Lucía… I've had a rough day at work and I'm really stressed out.

Even better! Sex is great for stress relief! It's very relaxing. Yeah, it's true… it's scientifically proven. It's not that I'm really horny or anything… but I'm worried about your physical and emotional wellbeing…

Yeah, right! Please, don't insist, I already told you that I'm not in the mood.

Come on, Manolito… I know you like it when I stroke you like this… And in the book I read that, even if you aren't in the mood, physical contact can start to arouse us like a reflex… A lot of therapists recommend trying to just go with the flow…[1]

Oh, Lucía, Lucía… Ummmmmm… You're so bad… Oh, oh, ooohh… You dirty girl! Alright, if it's good for my health…!

Another thing that the book Lucía read says, in case you are tempted to use the same strategy, is that it's not like taking a pill that you just have to swallow and rapidly starts to take effect. A quickie isn't going to do it. You should

---

[1] Even though this last argument sounds awful, it's not as bad as it seems. It is true that the body interprets headache as an illness and considers that it's not the best moment to reproduce. But, as we've seen earlier in this book, sexual desire doesn't stem solely from a mental predisposition. The right kind of caresses can awaken our libido suddenly and, with obvious limitations of course, many therapists recommend to those who want to improve their sexual function to just let themselves go with the flow and start intimate contact despite not feeling initially aroused.

know if you are planning to use the analgesic excuse on your partner, that you have to at least help them have an orgasm.

Obviously no one has sex as if they were going to the gym, motivated by the health benefits. But due to the growing awareness of the very positive effects—both physical and emotional—of a satisfying sex life, there are increasingly more adults who take their sexual health very seriously, establishing a series of routines, and make the space and time to enjoy their intimacy. Of course sex is practiced for pleasure, fun and the desire to express love towards our partner. But many studies confirm that the benefits of sex on the organism can be a small added motivation.

If we start with something as common as the headache, scientists propose two ways that sex can alleviate migraines. On one hand, the sexual stimulation itself lessens the pain threshold. If in the laboratory they gradually pinch the finger of a woman who is being vaginally stimulated and another who isn't, the one receiving stimulation will be slower to feel the pain. As we will explain in other chapters, the overexcitement of some nerve fibers seems to cause relaxation and an analgesic effect. This is a temporary and non-specific effect that perhaps doesn't really apply to the case of migraines, but the second way is more plausible: the release of endorphins and oxytocin after orgasm contributes to reducing pain and discomfort, especially when followed by relaxation. In fact, according to several surveys, about 10% of women masturbate to reduce menstrual cramps, and many recognize that sexual arousal does diminish headaches. One could respond to Lucía's therapeutic initiation of sex by saying that no one has proven that her method of eliminating migraines is more effective than a regular analgesic pill, but she could insist by saying that it also helps in relaxation, getting off to sleep and reducing stress.

However, it should be pointed out that in some people orgasm has exactly the opposite effect and causes a sudden, very bad headache. That is what's called orgasmic thunderclap headache. It's rare but there are many cases in medical literature of people who go to the emergency room, frightened by these unexpected pains. The physiological causes of the orgasmic thunderclap headache are not well known, it usually appears in people who already have a predisposition to migraines, and it could have a similar origin to those headaches that sometimes occur after intense physical exercise as a result of the rise in blood pressure (at the height of orgasm it can suddenly reach a maximum of 200). For a long time these headaches were considered harmless, but recently neurologists began to suspect that they could be involved in problems of cerebrovascular regulation, or be a sign of reversible cerebral vasoconstriction syndrome, something that should be studied because it is easily treated. In fact, the International Headache Society classifies headaches that stem from sexual activity into three types: the first is short-lived, happens at the start of intercourse and is more related to the muscular contraction of the neck and head. The second is the severe kind that appears abruptly right in the middle of orgasm and lasts 15 or 20 minutes. And there is a third that shows up a little later and can last hours or days.

The idea that sex calms the brain is well documented. The word hysteria comes from the Greek *hyster*—uterus—and since Antiquity it was believed to possibly originate in a woman's lack of sexual pleasure. As I already mentioned, it was from this sexist perspective that the first vibrators were designed in the early twentieth century as tools to generate therapeutic orgasms to reduce hysteria without the need for physical contact. Obviously, that is not the cause of hysteria, but it is true that vaginal stimulation calmed it.

Stress is another story altogether. It is widely shown that the oxytocin secreted by the hypothalamus after orgasm—or when a mother nurses her child—has a profoundly calming effect, and we don't need many statistics to show that both men and women sometimes resort to masturbation to relax. Some time ago it was thought that this strategy was much more common among men, and perhaps it still is; but, again, surveys of women confirm that the relaxing properties of orgasm are one of the primary motivations for self-stimulation. And not only to help them fall asleep, although it helps with that as well.

But beyond these momentary benefits, is sex really a healthy activity for the organism or is that mostly a myth? You can respond with opinions or gut feelings, but the best way to be sure is to ask science.

Sexus Sanus in Corpore Sano

If we want to defend the idea that having sex is marvelous for our bodies, we can surely do a study to show that people with more active sex lives enjoy better general health. But we would be overlooking the fact that perhaps health favors a full sex life and not the other way around.

If, on the other hand, we want to argue that sex takes a physical toll and has risks of infection we could also explain that nuns and priests usually live longer, but again we would have to take into account that their lifestyles are usually healthier in other aspects. It isn't easy to be rigorous in epidemiological studies.

Focusing only on the cardiovascular aspect, one could state that sex is a both super-healthy exercise and that there are documented cases of heart attacks due to rising blood pressure during intercourse. Both things are valid. That is why scientists are investigating to objectively figure out how sex affects our health.

But before I continue, allow me to point out the subtle but enormous difference between the various types of epidemiological studies. With a retrospective study we can, for example, gather up 50-year-olds who eat fruit every day and others who never eat any and compare their levels of cholesterol. This, if the sample is large, can give us important information. But no matter how much we look at variables such as smoking or exercising, it is very likely that there are other ignored factors that influence the results. We can also take a sample of people of similar age and socioeconomic circumstances, some of who've suffered coronary disease and others who are healthy, and ask them all how much fruit they've eaten in the preceding years. This second type of retrospective study has a better design and the result is more informative, but it still isn't ideal. The most reliable studies—and also the most expensive—are the prospective ones: they select a group of people with similar characteristics, they measure their initial cholesterol levels, and they divide them into two groups. They ask one group to eat fruit every day for a month and the others to eat the exact same diet but without fruit. A month later they measure their cholesterol levels, and the conclusion will undoubtedly be much more scientifically valuable. When in the press we see an eye-catching headline about the effect of a certain substance or activity on our health we have to take into account the size of the sample and the kind of study done.

Going back to the relationship between sex and health, one of the most peculiar experiments was published by American scientists in 2007, in the *American Journal of Cardiology*. Researchers measured the variations of heart rate and increase in blood pressure of 19 men between 47 and 63 years of age and 13 women between 44 and 58 in two different situations: at home having sex and in the gym doing a training program on a treadmill. After correcting variables, the conclusion they came to was that sex can be considered physical exercise, but that it is actually a very light one, equivalent at most to strolling quickly. Which is to say, that it is no substitute for sports.

Other studies analyzing the metabolic consumption of sexual activity have had similar results, showing that we don't burn many more calories during intercourse than walking up several flights of stairs. At least if we do it with our partner, since other studies suggest that in the case of an infidelity or a new relationship we exert twice the energy.

Obviously there are a wide variety of sexual practices, but the conclusion that all the studies come to is that in terms of physical exercise, sex is much better than being a couch potato, but in and of itself it won't help us lose weight[2] or get us out of having to hit the gym.

And this also applies to the supposed risk of heart attacks during intercourse. While there have been cases recorded, a recent review of the medical literature on the cardiovascular effects of sexual activity concludes that they are anecdotal and, except in combination with drugs and anomalous stressful situations, the risk is negligible. In fact, statistically, the number of orgasms over our lives is positively correlated with longevity and with less cerebrovascular accidents, and in 2012 the American Heart Association wrote up a very thorough report, addressed to health professionals and published in the journal *Circulation*, in which they concluded that, except in those with very serious cardiovascular problems, it is better to recommend having sex than abstaining. So it definitely has more benefits than risks, but the benefits are not very substantial.

There is more data that links sex to the prevention of illness. An extensive study by the American National Cancer Institute published in 2004 found a slight association between frequency of orgasms and less incidence of prostate cancer in men. They don't know why, but the hypotheses are that it would alleviate tension in the area and that ejaculation could eliminate toxins. A British study confirmed in 2008 this slight but significant protection against prostate cancer with a case-control study in men under 60. The results are confusing because they seem to indicate that in young men the risk increased. The literal conclusions were that "alone, frequent masturbation activity was a marker for increased risk in the 20s and 30s but appeared to be associated with a decreased risk in the 50s." Which is to say, it could be positive, but there's no need to consider it an obligation for maintaining our health.

It has also been argued that frequent sexual activity reinforces the immune system. This is relatively easy to measure. In a study analyzing two homogenous groups of students, some who had sex at least twice a week and others who rarely did, they found higher levels of the antibody immunoglobulin A in the blood of those who had frequent sex. The investigation was done

---

[2] There are those who say that sex could make us gain weight. The Indian professor Ritesh Menezes published a letter—not an article—in the journal *Medical Hypotheses* arguing that since sexual activity and orgasm increase the levels of prolactin in the blood (four times more in coitus than masturbation), and that prolactin levels are associated with weight gain (observed in laboratory animals but also in humans with hyperprolactinemia), sex could contribute to weight gain this way. Hmm. It would be fun to see if the energetic expense makes up for that.

by psychologists with 44 men and 67 women but, to be honest, those same authors also demonstrated that petting a dog for 15 minutes increases levels of immunoglobulin A, which raises some doubts.

I have no need to inflate results. Sex is a healthy activity, but much more so emotionally than physically. It improves self-esteem, reinforces partner relationships, contributes to our wellbeing, makes us feel more alive and balances our psychic health. And all of this has a physical repercussion, strengthening both body and mind. But, really, our motivation for having sex shouldn't be good health, but the other way around: being healthy in order to have good sex.

Corpus Sanum in Sexu Sano

It's not just about going to the gym to be more attractive, but also to be better in bed. Being healthy and in shape translates into a more satisfying sex life, for both men and women.

Countless studies have associated sedentarism, obesity and tobacco use with higher risk of sexual dysfunction. We have known that for some time, but investigators from Atlanta took it a bit further and analyzed if regular exercise improved sexual response and erectile capacity in a sample of young, healthy men without sexual problems.

In layman's terms, the conclusion was that men between 18 and 40 who practiced sport and were in better physical shape had "harder" erections and expressed more satisfaction with their sexual encounters. It isn't that the sports increased their libido, since their level of sexual desire was the same as in those who didn't practice them. What it significantly improved was their sexual response. We already know that we shouldn't apply general tendencies individually, and here we aren't insinuating that an athletic guy is going to be better in bed than a guy who goes to the bar ever night and wouldn't even recognize the inside of a gym. Every person is his or her own control, and the information you should take from this is that if you exercise, you can improve your own sexual "performance." In fact, another investigation concluded that middle-aged men who begin to exercise regularly lessen their risk of sexual dysfunction by 70%, and that is certainly a figure high enough to get your attention.

All of the above also holds true for women. There are myriad studies relating sexual satisfaction with one's perception of their own body image, but there are also purely physiological reasons: exercise improves blood flow in the genitals and increases hormone levels related to arousal.[3] Additionally, it facilitates the activation of the sympathetic nervous system, which is involved in orgasmic response. There are numerous studies that have linked physical exercise in women with greater sexual desire and response, so the conclusion is clear: sex may be good for your health, but having good health is undoubtedly beneficial for your sex life.

Sex in Old Age

A film featuring transsexuals, lesbian sex or fetishes is no longer shocking. That's all been done. If some director really wants to be provocative and reflect a much larger hidden reality, they should show an erotic scene with people over 80. Sex in old age is still a real societal taboo. It's not shown, but it's also not talked about or studied, despite the fact that half of married people over 70 have had vaginal sex at least once in the last year and 20% at least once a month, according to the survey of sexual behavior published in 2010 by Indiana University and the Kinsey Institute. The survey reflects that 42% of men and 21% of women between 60 and 69, and 30% and 11.5% of those over 70, have masturbated during the last month. And, 23% of men 70–79 years old and 19% of those over 80 have taken erection drugs in their last sexual encounter.

In more general terms, an extensive study published in 2007 in the *New England Journal of Medicine (NEJM)* with more than 3000 adults between 57 and 85 established that, in the United States, 26% of people between 75 and 85 are sexually active, 53% of those between 65 and 74, and 73% of those between 57 and 64 years of age. In this survey, "sexually active" was understood as having had sexual relations at least once in the last 12 months. It is lax criterion, but even in the group of those over 75, 14% responded that they have sex with a partner two or three times a month, and 6% at least once a

---

[3] There is an important distinction to be made when mentioning this hormonal aspect. It is often said that exercise increases testosterone levels and that is what boosts sexual desire. It's not clear that that's the case. In 2010, Cindy Meston published an article about a study where a group of women had to fill out a questionnaire for twenty minutes while others ran on a treadmill. Then they showed them all erotic images and measured their genital excitement. They also measured their saliva for levels of testosterone and alpha-amylase, a molecule that indicates activity in the sympathetic nervous system (SNS). The analyses revealed that their testosterone levels hadn't changed but their alpha-amylase levels had, indicating greater activity in the SNS during exercise and, indeed, the women who had been physically active had greater genital response. The conclusions reinforce the very widespread idea that the activity of the sympathetic system facilitates arousal, but they offer doubts about the relationship between exercise, testosterone and arousal.

week. It was also calculated that one out of every seven American men over 57 take Viagra or a similar drug.

Sex in old age is a subject of growing importance. In March of 2012, John DeLamater, a sociologist at the University of Wisconsin, published a review of all the scientific bibliography on sexuality and aging, to date, and concluded that many healthy people between 70 and 80 continue to be sexually active, and that the frequency of sexual expression is related to higher indexes of physical and mental health. In his detailed work, DeLamater cites surveys according to which 59% of men and 35% of women over 45 agree with the statement "sexual activity is important to my overall quality of life," and only 3% of men and 20% of women state that "I would be quite happy never having sex again." Thirty-nine percent of men and 37% of women agree with the statement "sex becomes less important to people as they age," and only 2% and 5%, respectively, with "sex is only for younger people."

It is difficult to know how all this data has evolved over the last few decades. Paradoxically, the largest surveys on sexual behavior conducted in the United States and the United Kingdom only began to include those over 60 very recently. Sex in old age is a taboo even for science. We do have statistical data from recent decades in Sweden in the periodical sexual surveys they began to conduct in 1971. In 2008 an article was published in the *British Medical Journal (BMJ)* based on these surveys, and it established that between 1971 and 2001, the percentage of married men over 70 who had had intercourse in the past year had risen from 52% in 1971 to 68% in 2001, and in women from 38 to 56%. In unmarried men over 70 it rose from 30 to 54%, and—pay attention here—in unmarried women from .8 to 12%. This last figure is perhaps the most significant. Among those sexually active, in 1971, 10% of Swedish men over 70 and 9% of women had sex weekly, compared to 31% of men and 26% of women in the year 2001.

All those figures may be overwhelming, but even taking into account that illnesses and disabilities are the main hindrances to sex, the numbers clearly reflect that there are increasingly more healthy people who can and want to continue enjoying their sexuality throughout their lives. This, along with the benefits but also possible risks that sexual practice entails—especially injuries—makes it outrageous that medical research isn't more closely studying sexual behavior and function in old age.

For two years I worked at the American National Institutes of Health (NIH), in Bethesda, a little town on the outskirts of Washington DC that has perhaps the largest biomedical research complex in the world. With 6000 researchers on its campus, the NIH is divided into 27 different institutes, including the Cancer Institute, Mental Health, Allergies and Infections, the Eye,

Human Genetics, and, of course, Aging. My conversation in late 2011 with Dr. Richard Hodes, director of the National Institute of Aging (NIA), went as follows:

PERE	At the NIA, you study all aspects of health and aging, right?
DR. HODES	Yes, yes, of course. And not only illnesses but also how to maintain quality of life over adulthood, prevention, health recommendations…
P	Such as types of physical exercise that have less risk of injury, for example?
DR. H	For example.
P	And what about sexual function and behavior?
DR. H	… (four seconds pass)
P	I mean if you study subjects such as renal function, muscular loss, cardiovascular health, cognitive deterioration, vision problems… if you study all the organs and vital functions, you must also study the decline in sexual function and how to adapt it throughout old age, right?
DR. H	Well… there are studies that say it contributes to good health… but we… send us an email and we'll get you some information.

The head of public relations for the NIA was with Dr. Hodes at the time. A few months later I sent her a message asking for information and a recommendation on whom I should interview. Her response was: "We have no experts on sexuality on our staff," and she gave me the contact of people at the University of Chicago with projects financed by the NIA. I insisted by asking if there really wasn't anyone in the entire Institute of Aging at the NIH who was researching sexuality and she told me: "As far as I know, there is no researcher at the NIA focused on that subject." I couldn't believe it, and I searched through scientific publications until I discovered the document *Living Long and Well in the 21st Century: Strategic Directions for Research on Aging*, developed in 2007 by the NIA to mark the guidelines on aging and quality of life, and I confirmed that there was not a single reference to sexual medicine. It was paradoxical. The fear of sex that the American Government seems to have, even in its scientific and medical aspects, is remarkable. I also contacted the institute of the NIH responsible for spinal injuries to ask if they took the sexual response of the disabled into account and they didn't even answer me, and the director of the National Mental Health Institute, Thomas Insel, when I asked him in person whether they investigated pederasty, gave me the excuse that "we don't receive scientific applications on that subject." Human sexuality is censored in the American public agency for biomedical research, and to a lesser degree in science in general. And it's a shame.

Getting back to aging, something special about scientific research is that it usually shows us hidden aspects of society and anticipates tendencies that will be relevant in the future, but in the case of sexuality it seems to buy into the false stereotype that sex disappears in old age and it doesn't seem to see—or want to see—the growing importance of sexual relations in later adulthood.

We know that a satisfying and safe sexual life offers physical and psychological benefits in old age, that our population is aging in increasingly better health conditions, that the desire to remain sexually active is on the rise and there are obvious limitations both physical and psychological. But few doctors have enough tools to deal with the subject with their patients based on empirical data.

In some men and women, sexual desire diminishes with age, but sexual function doesn't necessarily have to. It's obvious that cardiovascular diseases and diabetes generate erectile problems in men, and that the radical lessening of estrogens in women after menopause brings with it lubrication problems and that, over time, muscular atrophy can increase pain during intercourse. The article in the *NEJM* deals with the large number of medications that have effects on sexual response that are not discussed by doctors with their patients, and it is clear that diseases such as cancer affect sexuality in many different ways. But there are not yet enough studies and recommendations on the subject.

This is the key point: lessening desire doesn't have to be necessarily considered a problem, but when desire is there and physical response fails it is problematic, and that should be investigated and treated by medicine. Especially because while hormonal, psychological and social factors can diminish libido with age, this doesn't affect everyone equally, nor does it make sex less pleasurable or emotionally satisfying. In fact, several studies indicate that older women know their bodies better and some reach orgasm more easily than when they were younger.

One of the most detailed studies on sexuality in older women was published in January of 2012 by researchers at the University of California, based on specific surveys about sexuality given to more than 1300 women between 40 and 97 years of age who lived in the same community outside of San Diego. These women took part in the Rancho Bernardo Study, and their life conditions and health parameters have been monitored for years for use in different epidemiological studies. In this case, researchers divided the women into four groups: 40–55 years old, 55–68, 68–79 and over 79, and they compared different aspects of their sexuality. The results were significant: while sexual desire and number of relations lessened over time, and difficulties such as lack of lubrication increased, the frequency of orgasms during the sexual act was practically identical. And the most striking: the indexes of satisfaction seemed to increase with age. Which is to say, there was less sex and with more difficulties, but it was a fuller and more satisfying experience. That reinforced the idea that actually it is more important to ask about satisfaction and well-being than frequency or dysfunction. The study's general conclusion is fundamental: the quantity of sexual encounters decreases but not the quality, and desire is lost but physical and emotional pleasure remain. It is obvious that sexual relations change over time, but less than we think.

Also from a more sociological viewpoint it is curious to note how practices have changed with time. For example, with regard to oral sex, DeLamater's review mentions that 51% of women ages 50–59 have had intercourse during the last year and 36% have had oral sex, while in those over 70, 21% had sex with penetration and only 7% oral sex. Proportionally oral sex is less frequent in old age, when in theory it should be simpler. And we haven't mentioned anal sex or the differences between heterosexuals and homosexuals because of the flagrant lack of data available.

As we are constantly repeating in this book, none of these statistics should be a source of pressure. If there is no desire for sex, then there's no problem. What is a delicate issue is when there is a desire but it is accompanied by arousal problems due to physical reasons having to do with the aging of the body, or psychological reasons due to changes in the couple or loss of confidence in body image, etc. Here is where experienced therapists equipped with medical and scientific data have much to offer. Sexual satisfaction in old age should cease to be a taboo subject, both for society and for science and medicine.

13

Sex in a Wheelchair, for Love and Pleasure

Quico is a young, robust, smiling, lively, handsome, chatty young man, who has a lot of success with women despite having been in a wheelchair for the last 14 years.

On December 25th, 1998, after Christmas dinner with his family, Quico was riding his motorcycle home. Driving calmly along a street in his hometown of Sant Andreu, Spain, suddenly a car failed to yield and steamrolled over him. Quico went flying and fell on his back with a thud against the curb. His T12 vertebra, in the dorsal region, located slightly above the navel, sunk several centimeters, pinching and breaking the nerve fibers of his spinal cord. In the hospital they reduced the inflammation and put the vertebra back in place, but the damage was irreversible. The connection between his brain and the nerves in the lumbar and sacral regions of his spine was lost forever. Quico would never have sensitivity in the lower part of his body or voluntarily control any muscle below his waist. That included his leg muscles and the group of pubic muscles that control excretory function and the ejaculatory reflex.

They were very difficult moments for Quico and his family, who had to face the challenges of this new life. But 14 years later, Quico Tur doesn't feel disabled in the least. In fact, when we met on the Passeig del Born in Barcelona in August of 2012, Quico told me that the "wheelchair has given me freedom and a lot of opportunities," and that "thanks to it I've been able to fulfill big dreams." Perhaps the most striking is being Spain's number one player of wheelchair tennis for the last 9 years, and taking part in the Athens and Beijing Paralympic Games. He was super-excited about being about to travel to London for the next ones.

Quico assures me unequivocally that he doesn't mind not being able to walk. He has full autonomy. He lives alone, travels alone, he arrived at our meeting driving his own adapted car, and he has a completely independent life. He doesn't care about the walking. He says that the lack of control of his excretory function is somewhat more uncomfortable, but that by recognizing his body's signals, he's learned to anticipate his needs and adjust with no real problems. What he—and most people with spinal injuries—would really like is to regain sexual function. That is much more important to him than walking.

When I say most people with spinal injuries, I'm not just speculating. In 2004, Kim Anderson, of the University of California, published an oft-cited study in which she asked 681 people with spinal injuries what would they choose first if they were able to recover one specific physiological function. In the case of quadriplegics, almost half answered the ability to move their arms and hands, with the second most common reply being sexual function. But among paraplegics like Quico, 27% said sexual function, 18% excretory control, 16.5% stability and strength in the upper part of their body, 16% walking, and 12% being free from chronic pain, followed by other functions in lower percentages. The results are revealing in many ways and force us to reflect on the importance of sex in our lives. When I mentioned this study to Quico, he wasn't the least bit surprised.

Quico loves women. In fact, he dates a lot and has an active sex life that is far from negligible. But it's not the same thing. When he is with a woman, he takes Viagra and can have sex with no problems except for the fact that he has no sensitivity whatsoever in his penis. Often he doesn't mention that and pretends to be feeling pleasure so that the women are relaxed and not inhibited. He does enjoy himself despite not having genital sensation. He says that he gets very excited by giving pleasure and that over time parts of his body have gained sensitivity. Now he spends a lot more time on foreplay, caresses and sexy games, and he's learned new techniques that can leave a woman incredibly satisfied. Quico enjoys sex as much as the next guy, but he admits to some frustration: "It's like a volcano that never quite explodes," he says to me about his lack of orgasm, and he again says, convinced, "Sexual sensations are the only thing that I really, really would like to be able to experience again."

As strange as it may seem, if Quico's injury was further up he would have better sexual response. His parasympathetic nerves in the lumbar and sacral region would be intact, and while they wouldn't receive direct orders from the brain, they would be able to react and have an erection without Viagra as a reflexive response to genital stimulation. Also, if the fibers of the sympathetic nervous system that come out of the middle of the spine reached his pubococcygeus muscles, he would be able to ejaculate and feel something close to an orgasm.

In the hospital of the University of Quebec, in Montreal, specialist Frédérique Courtois sums it up perfectly for me: in injuries below the T11 vertebra, the nerves in the lumbar region are completely disconnected from the brain and there is no sensitivity, no erection from physical contact and no climax, even though in certain cases there can be the slight stirrings of an erection through mental stimulation via nerves that reach the penis from higher areas of the spine. In injuries located in the middle of the torso, between the T11 and the T7, there is no sensitivity nor any reaction through mental stimulation; however, the nerves in the lumbar region are intact and there can be erection and ejaculation as a reflexive act from direct physical contact. The problem is that here the fibers of the sympathetic system are damaged and the climax is not felt. On the other hand, in injuries above the T6 verterbra there is erection from physical contact, ejaculation and a sort of "ghost orgasm" or non-genital orgasm. Blood pressure rises, the sympathetic system activates, there are spasms, tachycardia, flushing, hyperventilation and an orgasmic sensation that can sometimes be pleasurable. Frédérique Courtois shows me a sort of extremely strong vibrator that he uses to generate erections and ejaculation in the physically disabled. They don't feel anything in the penis, but they see it become erect, that arouses them, allows them to masturbate or have sex with their partners, and sometimes ejaculate. Courtois explains that that very morning he had provoked the first ever orgasm for a 27-year-old man, who actually left a bit dismayed. The first few times, when they feel that strange, sudden reaction of their bodies that they don't recognize as an orgasm, they usually define it as off-putting, and it is with practice that it is experienced as pleasurable. Frédérique has published many scientific articles on the sexual function of men and women with spinal injuries, and the sensations they describe during ejaculation. And while he explains that there is much individual diversity, he assures us that many men and women with spinal injuries experience sexual pleasure (and even orgasms if their injury is high) without having any sensitivity in their genitals.[1]

---

[1] In Chap. 3 we saw that the sensitivity of the penis and clitoris depends on the pudendal nerve and the sensitivity of the vagina depends on the pelvic nerve. These nerves of the parasympathetic system begin in the sacral and lumbar regions in the lower spine, and that is why spinal injuries make genital sensitivity impossible. However, when the injury is not very high, some women can still react to deep penetration in the most profound part of the vagina, because hypogastric nerve fibers reach there from the middle part of the spine. But most surprising was when Barry Komisaruk was experimenting on women with higher injuries, and in theory absolute insensitivity, and he observed that some of them noted very deep stimulation in the cervical region, even reaching orgasm. This had never been described before, and after more experiments Barry proved that the vagus nerve (a nerve that instead of going through the spine travels from the brain inside the body, informing on the state of the vital organs) reached the uterus and could transmit sexual stimuli.

In any case, reaching orgasm or not is actually the least of it. At Barcelona's Guttmann Institute of Neurorehabilitation, the man in charge of the spinal injury department—Joan Vidal—agrees with Frédérique Courtois about the enormous importance of studying sexual function in the physically disabled. Joan explains that, especially among the younger ones, sex is their main cause for concern. But not just the search for conventional physical pleasure, which, in fact, is impossible. There is still a drive to maintain sexual desire, in some cases it is a question of self-esteem, or wanting to please a partner, and especially for the emotional wellbeing that comes from sharing intimacy, caresses, arousal and pleasure with someone you desire.

Men and women with spinal injuries are a living testament to the fact that sexual expression goes far beyond individual genital pleasure, and that giving pleasure can be as desirable as receiving it. Psychologist Anna Gilabert knows this full well, both from the patients she treats at the Guttmann Institute and in her own personal experience.

Like Quico, Anna is also in a wheelchair since an accident she had when she was 19. In her case, the first question she asked the doctors was if she would ever be able to be a mother, but soon she also started to worry about sex. Now, at 32, she has been married since September of 2011 and assures me she has a completely normalized sex life. "I don't even remember what it was like before, but I'm sure it's much better now," she says convinced. Anna explains that obviously sex is different, and that at first it seemed strange and not exactly pleasurable. But that over time her breasts and other parts of her body gradually gained much more sensitivity, that many sensations like movements, fantasies and visual stimuli became eroticized, and while orgasms are not like before, she sometimes does feel some sort of high point of pleasure followed by relaxation. Anna says that she hasn't lost a bit of her sensuality, that areas of her skin are now much more sensitive, and even that the genital component has reappeared as part of a more subjective, but not necessarily worse, arousal. In fact, she disagrees that satisfying your partner is the main motivation for those with spinal injuries to have sex, and she states that most of her patients at the institute have sex for their own physical pleasure, especially the women, who are more able to eroticize other parts of their body and enjoy the mental play and stimulation. Anna's example proves the maxim that sex can be fabulous if we expand it beyond genitals and penetration.

Quico and Anna's words left a big impression on me, personally and above all because they made me realize the enormous importance in our lives of expressing our sexual intimacy, and how taboo it is in our society. When I mention it to friends, few of them have ever thought about sexual relations among the physically disabled. And, it seems, moralistic science has only begun to consider it very recently.

Neurosurgery to Regain Genital Sensitivity

Science and medicine continue to talk about "walking again," despite the fact that many quadriplegics are well-adapted to their lives in wheelchairs and what they are really want is to have sensitivity in their genitals again. There is one way to do it: sensitive nerve bypass.

When Joan Vidal, of the Guttmann Institute, told me about this possibility, he was referring to trying to improve control of both sexual and excretory function. As for the latter, the muscles we use to hold in our urine and to defecate when we have to bear down are activated by nerves that come out of the sacral and lumber regions of our spine. When there is a spinal injury in that area, the orders from the brain don't arrive and voluntary control of the pubic muscles is lost. But what would happen if we could connect nerves from another part of the body? Imagine them joined to the nerves responsible for, say, strength in the upper abdominals. If this were possible, when a disabled person noticed that he was about to urinate he would flex his upper abdominal region and perhaps also contract the muscles in his bladder, closing off the release of urine. Doctors have been trying to do this for some time, but the results are not entirely satisfactory because of the atrophy and progressive loss of muscle tone produced after the injury.

Which is to say, that over time the muscle disappears and restoring motor function is very complicated, but the nerves remain intact and getting back sensitivity through a nerve bypass should be much easier. In fact, Dutch surgeon Max Overgoor has already achieved men with spinal injuries regaining feeling on contact with the penis and he is still researching with the clitoris. My interview with him was among the most awe-inspiring of all the ones I did for this book.

In 2006, Max Overgoor published initial results with three patients affected by spina bifida. They were three boys, aged 17, 18 and 21, who maintained an interest in sex and fantasized, had normal testosterone levels, got mentally aroused and could achieve an erection with Viagra, but whose lesions impeded any penile sensitivity. They could feel touch and caresses on their inner thighs, because the sensory nerves reached there from a part of the spine that was above their injuries. The procedure was breathtakingly simple: surgeons made an excision on one side of the pubis, identified the end of the ilioinguinal nerve, which goes from the spine to the inner thigh, and connected it with microneurosurgery to the dorsal nerve on the base of the penis. A few months after the operation and with a lot of rehabilitation exercises, the patients began to feel a strange sensitivity in their penises. At first, when they massaged their glans it seemed like they were being touched on their inner thigh, but over time and with training their brains gradually reconfigured its corporeal image and the three boys ended up feeling the contact on one side of the glans (not the rest of the penis). One of them had a partner, and in the article he describes that 12 months after the intervention, not only did he have more sensitivity in the glans, but he also experienced it as erogenous. Sexual activity with his partner had increased considerably (from one encounter a month to five), he felt much more excited, he had five times as many morning erections, and his satisfaction with his life and his sex life had improved enormously. The two other boys regained sensitivity and felt the results were positive, but since they didn't have sexual partners the feeling of improvement was much less.

It is a fascinating subject. At one point in the interview, Max talked to me about the results that he was publishing in early 2013. With a similar procedure, he had widened the sample to 30 patients with spina bifida or spinal injury and had achieved unprecedented success: after the operation and several months of rehabilitation, 24 patients (80%) gained sensitivity in the glans. Eleven of them noted touch on their penises and 13 still felt it on their inner thighs, but for all of them that improvement signified a large erotic benefit and more reflexive erections from direct contact. The psychological study revealed higher indexes of sexual satisfaction, and three of the patients who had never had an orgasm before the microneurosurgery did achieve it afterwards. According to Max Overgoor, the patients with spinal injuries who had had sexual relations before their accidents recognized their orgasms as different, less intense and with a more psychogenic sensation. But they all stated that having sensitivity in their penises again after the operation enormously improved their sexual motivation and experiences. And there were no postoperative problems or discomfort.

Max explains to me that for the moment the technique is limited to injuries that are very low (beneath the L1 vertebra) that allow them to directly connect the ilioinguinal nerve with the one in the penis, but that he is contemplating the possibility of, in higher injuries, cutting a nerve in the leg and transplanting it to connect the sensory nerves of the penis with others that come out of areas of the spine that are above the injury. He admits that such an operation is more complex and he doesn't have as much confidence in the initial results being positive, but he is convinced that it is worth attempting and continuing to work on new strategies that allow people with spinal injuries to have better sexual function. He is also beginning a project with women to try to regain sensitivity in the clitoris. He assumes that it is a more difficult surgery, but he is optimistic. Talking to him, I am surprised that something like this hasn't been studied before, and I ask him if he is using tools that were previously unavailable. He says no, that he has been working in this field for 10 years and this type of neurosurgery is relatively simple, but that at first he had a very difficult time convincing the ethics committees to approve his projects. I am astounded. How could it be unethical to try to improve the sexual function and quality of life of disabled people who want that? Science boasts of having revolutionary ideas and anticipating cultural change, but as far as sex is concerned it is as fearful as the rest of society. Or even more. Max is not the only scientist I met who is afraid of his colleagues considering him frivolous for studying sex. They will discover how outmoded their ideas are when these procedures become routine and people like Quico can fulfill their dream of regaining genital sensitivity.

This is truly important and really profound. Obviously Quico would love to feel genital pleasure again, but what most motivates him is being able to express his sexuality to the fullest when he is with a woman he likes. Sex is not only an act of physical satisfaction but also one of love and exchange that brings people together chemically and emotionally. It is something so wonderful that it is hard to understand how it remains such a taboo in our society. Of course, if anyone finds anything unseemly in Quico's desire, if anyone blushes or thinks that because of his condition he should repress his sexual feelings, there is still time for them to open their mind and get past all the destructive shame they were raised with. Quico's story is exemplary in many ways. As an example of overcoming personal challenges, with true optimism about life and a real fighting spirit. But as far as this book is concerned, it is also vital to show the importance that a healthy sex life has on our wellbeing, and how science can contribute to that.

14

Science in Sexual Orientation

What I transcribe here is a literal quote from a scientific article, published in 1968 in the prestigious British science journal *Proceedings of the Royal Society of Medicine*, about therapies to correct sexual deviations. I discovered it when I was searching for old research to analyze how science viewed homosexuality decades earlier. Imagine my surprise when I read this in the introduction:

> [Studies] have shown that it is possible to reorientate homosexuals with psychotherapy, but the treatment is time consuming and the success rate is low. [...] it is important to find out to what extent behavioural techniques such as aversion therapy answer this need.
>
> Aversion therapy aims to associate noxious stimuli with some aspect of the deviant behaviour or attitude. In earlier methods the noxious agents were chemical, e.g. apomorphine was used to produce nausea and vomiting. With this method Morgenstern *et al.* (1965) treated 13 transvestites; 7 were much improved and 5 showed some improvement. Recently electric aversion has largely supplanted chemical aversion as it is safer, easier to control, more precisely applied, and less unpleasant. [...] the present authors have so far used electric aversion in 40 male patients—16 homosexuals, 3 paedophiliacs, 14 transvestites and transsexuals, 3 fetishists and 4 sadomasochists. This paper is a preliminary report of results to date.

If this introduction shocks you—I couldn't believe what I was reading—it gets worse. In the methodological part it explains that electric shocks were applied to the patients' arms, "associated with three different aspects of the deviant behaviour: (1) *With the deviant act;* e.g. shocking the transvestite as he is cross-dressing. (2) *With the deviant fantasy;* e.g. shocking the masochist as soon as he signals the presence of a masochistic fantasy in his mind. (3) *With the erectile response* to deviant stimuli; e.g. shocking the homosexual as soon as he starts to develop an erection to a picture of an attractive male…"

I won't go on. The results indicated that "at a year's follow up, thought only 6 cases (15%) were completely successful. Transvestites, fetishists and sadomasochists improved the most; results with homosexuals were less satisfactory and transsexuals did badly."[1]

The article was entitled "Electric Aversion Therapy of Sexual Deviations," and was published by John Bancroft and Isaac Marks in one of the most serious scientific journals of the period. Incredible. Especially because it wasn't an isolated study. I continued my bibliographic search and found many titles such as "Change in Homosexual Orientation" (1973), "A Case of Homosexuality Treated by in Vivo Desensitization and Assertive Training" (1977), and "Alternative Behavioral Approaches to the Treatment of Homosexuality" (1976). Lastly I read a review entitled "Toward a New Model of Treatment of Homosexuality" (1978) whose summary showed the beginnings of a big change: "A review of the literature of outcome studies in the psychoanalytic or behavioral treatment of homosexuality reveals limited results when "heterosexual shift" is the goal. Recently, however, a growing body of empirical knowledge has accumulated that challenges the illness or maladaptive model of homosexuality. Consequently, a new model has been emerging that is designed to assist homosexuals to recognize, accept, and value their sexual identity and to help them adjust to this identity in a predominantly heterosexual society. Unfortunately, only a few studies exist that examine the results of this new approach."

It is disturbing that a text like this is so recent, but let us not forget that homosexuality was not removed from the Diagnostic and Statistical Manual of Mental Disorders (DSM) until 1973, and even the 1987 version of the DSM-III maintained the diagnosis "ego-dystonic homosexuality" to define the "persistent lack of heterosexual interest" and "distress from a sustained pattern of homosexual arousal." Until then scientists believed they had arguments for writing articles with offensive titles such as "Aversion therapy of homosexuality" (1970), also by Dr. John Bancroft.

---

[1] Atrocities aside, the results are coherent with what we now know about the relative flexibility of sexual behavior. Sexual identity (feeling like a man or a woman) is not something you "choose," but rather something that comes already defined from embryonic development and the earliest stages of life. A transsexual doesn't choose to be one, instead showing disconformity with his or her body from childhood. Sexual orientation (feeling attracted to men, women or both) is more fluid in women, but once it is defined is very stable in men. On the other hand, a large part of stimuli that condition a greater sexual desire are learned through experiences during different stages that are key to sexual development, and they are more susceptible to intensification, lessening or being substituted by other practices.

Wait a minute… John Bancroft? Imagine my absolute bewilderment when I realized that those studies had been conducted by charming John Bancroft, former director of the Kinsey Institute, one of the most prestigious sexologists in the world with more than 200 scientific articles published, whom I spoke with in July, 2012 in Estoril (Portugal), during the meeting of the International Academy of Sex Research. I called up Bancroft and talked to him again.

Before I go into what he said, it's only fair to mention that John Bancroft has written explicitly about how those experiments he conducted at the start of his career soon provoked a moral dilemma in him and that he became convinced that "there is no basis whatsoever for considering homosexual orientation a pathology instead of a variant of human sexual expression, and therefore offering treatment is professionally unethical and, according to my value system, immoral."

In fact, when I mention the subject to Bancroft he quickly defines it as "the most embarrassing aspect of my career" and admits that the criticisms made of him were fully justified. But he also argues that those were different times, both for society and for psychology. That second part is particularly interesting.

John Bancroft explains that he began his training in psychiatry analyzing methods to treat pedophilia. At that time one of the tools that was used to try to revert sexual preferences was electroshock therapy: they showed patients inappropriate images or stimuli while measuring their penile reaction, and if there was the start of an erection, they would deliver an electric charge so that the individual would associate the stimulus with a negative response. The idea behind this method was that our sexual behavior was easily conditionable. They tried to apply Pavlov's theories of conditioned reflexes, thinking that in humans it would be just as easy to induce fear or modify behavior as in the rest of animals. Bancroft tells me: "The truth is I don't even know how I got into that," but at some point they started to apply the same procedures used on pedophiles on homosexuals who felt stigmatized and wanted to modify their sexual orientation. John recognizes that "now it is ridiculous, and the therapy is based on accepting yourself, not on correcting," but that "in that period two circumstances converged: many gays wanted to change because homosexuality was not socially accepted, and scientifically we thought that sexual behavior could be easily modified. Now both aspects have made a radical shift, but at that time it wasn't seen the same way." In fact, John made several scientific studies comparing two types of therapies: one was an aversive therapy associating electroshock with homosexual stimuli, and the other systematic desensitization, which was usually used for anxiety problems or phobias. "At that time the interpretation was that some homosexuals were homosexual because they had a phobia of heterosexuality or anxiety over intimate relations with women," John explains, pointing out that desensitization therapy consisted in gradually offering different heterosexual stimuli to gays so they would, little by little, lose their fear of the female sex and "cure" their homosexuality. Again, John explains that in some cases there were slight behavioral changes with both therapies, but that, seen in perspective, they were superficial, transitory, and really didn't modify sexual orientation. Bancroft recognizes that part of our behavior can be conditioned, but that sexual orientation is much more solid than it was believed to be before the 1970s. Because of that, and because of the social changes that have come about, Bancroft was categorical that "reparative or sexual reorientation therapies" no longer have any place. In his defense, he insists that he never considered homosexuality a pathology, that he simply wanted to help people with a desire to reorient their tendencies, but he acknowledges that he soon realized that that attitude, as innocent as it seemed to his eyes, was aligned with homophobia and perpetuated the stigmatization of homosexuals. He immediately stopped practicing those "therapies" and began to denounce them.

The subject is controversial and offensive to some sensibilities, but it is also intellectually stimulating. On one hand, we know that sexual orientation is a continuum and that the labels homo-, bi- and hetero- are not airtight. Between the black and the white there is a full range of grays, both when studying behavior and when studying hormonal levels and brain structures. Alfred Kinsey reflected that in his Kinsey scale when he observed that 0 corresponded to those who defined themselves as 100% heterosexuals and 6 was those who self-defined as 100% homosexuals, many of those surveyed said they were 5s or 2s or some other intermediate number. In more recent surveys, when asking about the three orientations one finds that, for example, about 3–4% of men and 1–2% of women define themselves as homosexuals and 1–2% of men and 2–5% of women as bisexuals.[2] But if in addition to those three options, the surveys add the categories "almost always gay" or "almost always straight," a not inconsequential percentage of people—more women than men—choose those. They aren't bisexuals or homosexuals in transition, but rather people who feel they are homo- or heterosexuals but

---

[2] Data taken from Chandra A, Mosher WD and Copen C, (2011). "Sexual Behavior, Sexual Attraction, and Sexual Identity in the United States: Data from the 2006-2008 National Survey of Family Growth." *National Health Statistics Reports*; n.º 36. Hyattsville, MD: National Center for Health Statistics. Depending on how the questionnaires are carried out and what questions they ask, some surveys give slighter higher percentages. The proportion of people who define themselves as heterosexuals but who have had relations with individuals of the same gender is significantly higher, and in this group we seen the influence of socioeconomic, educational, religious and cultural values.

at some point have fantasized or even had relations with individuals of the non-preferred gender. Sexual orientation is a continuum, and exclusivity is possibly, in some cases, a result of socialization.

On the other hand, it is also obvious that in gays and lesbians as a whole there is enormous internal diversity, which is not well reflected in most scientific studies. When researchers want to compare psychosocial and biological aspects between homosexuals and heterosexuals they usually separate by sexual orientation as if they were homogenous groups, without making any distinction, for example, between people who have always felt homosexual and attracted to their same gender since childhood and those who, after varied experiences, have discovered their homosexuality as adults. They are undoubtedly among the subgroups with different characteristics that are often not contemplated in studies.

Investigating the origins of homosexuality is fairly irrelevant in practical terms, but if we are curious, the discussion with Bancroft actually offers us a very good example on the long-standing philosophical and scientific debate between the relative weights of biological determinants and socialization in the development of human behavior. In other words, if we set our prejudices to one side, the Bancroft case conceals the well-worn question of whether a homosexuality is nature or nurture, or under what conditions biology interacts with environment to define a sexual orientation that, once shaped, seems to be so set in stone.

Homosexual Fluidity: Behavior Is Not the Same as Orientation

During my visit to the Kinsey Institute I was having a beer with evolutionary biologist Justin Garcia and French sociologist Georges-Claude Guilbert, visiting professor in the Gender Studies Department at Indiana University. At one point in our animated conversation I used the expression "testosterone during pregnancy" and Georges-Claude exclaimed with a smile, "Oh! Another biologist convinced that sexual identity is determined by chromosomes!" When I said that, in large part, I was (identity, not orientation), and I asked him what he thought was behind our feeling of masculinity or femininity, he shrugged his shoulders and, still smiling, he told me, "In gender studies we assume that we aren't born biologically conditioned and that it is all the exclusive result of socialization and the roles that education, family and society imprint on us." "Exclusive?" I interjected with surprise. Justin laughed and said, "Don't bother, I've had this discussion with him a thousand times and there's no convincing him."

I mention this anecdote to explain that the greatest academic extremism I encountered while researching this book was in a subgroup of sociologists who dogmatically reject any biological factor in the development of sexuality. For them it is all the fruit of social conditioning. On the other hand, very few biologists reject the influence of sociocultural influences nor pretend to justify sexual preferences with merely hormones and genes. It is true that there are some scientists who are too reductionist and determinist, but many fewer than is believed.

Actually, this extremist concept of nature and nurture is only discussed in some university departments, and beyond that we all have no problem accepting that biology and environment are not mutually exclusive. But the dichotomy perfectly reflects the two academic approaches on the development of sexual orientation: one which gives much more weight to an individual's socialization and one which searches for determinants in genes, hormones and brain structures.

The vision that prioritizes socialization defends the idea that we are not born with a determined sexual orientation, and that it is gradually shaped by all the experiences and determinants we have over our lives, especially in childhood and puberty. In this sense, Freud was convinced that all humans are born bisexual, that both girls and boys little by little developed their heterosexuality, and that if this development was interrupted by traumatic events or alterations in their relationships with their parents, then homosexuality could flower. Freud did not consider homosexuality a problem and on numerous occasions he refused to treat it in patients who requested it.[3] But I begin by quoting him because, paradoxically, some homophobes have used his hypothesis to defend the idea that homosexuality is the result of traumas or abuse during childhood development, and as a justification to try to "repair it" using psychoanalytic therapy. Unfortunately, this practice is still used in many places, including a supposedly advanced country like the United States.

However, the more biological vision defends that natural selection has favored the presence of two genders who are mutually attracted in order to reproduce, and that that is something already established in the embryonic

---

[3] In 1935, Freud sent a famous letter to a mother worried about her son's homosexuality, telling her: "Homosexuality is assuredly no advantage, but it is nothing to be ashamed of, no vice, no degradation; it cannot be classified as an illness; we consider it to be a variation of the sexual function, produced by a certain arrest of sexual development. Many highly respectable individuals of ancient and modern times have been homosexuals, several of the greatest men among them. (Plato, Michelangelo, Leonardo da Vinci, etc). It is a great injustice to persecute homosexuality as a crime—and a cruelty, too. [...]If he is unhappy, neurotic, torn by conflicts, inhibited in his social life, analysis may bring him harmony, peace of mind, full efficiency, whether he remains a homosexual or gets changed."

stages by genes, hormones and brain structures, and that physiological factors could also determine some cases of homosexuality.

But before delving into the debate we must make a fundamental distinction between sexual orientation and behavior, and mention one of the most cited investigations of recent years in the study of human sexuality: one conducted by researcher Lisa Diamond, at the University of Utah, on sexual fluidity.

In the early 1990s, Lisa Diamond began to follow the evolution of a group of 89 women who at the start of the study defined themselves as lesbians, bisexuals or "unlabeled" women, none of whom felt heterosexual. At the time, the women were between 18 and 25 years of age and Diamond simply sought to analyze how their sexual orientation influenced various social aspects, what problems they found, how people around them behaved, what couple relationships they established and how they adapted through the transition from youth to adulthood. But, highly unexpectedly, she observed that, over the first 13 years of study, two thirds of the women changed their definition of their sexual orientation. Some of them went from bisexuals to lesbians, others to "unlabeled," some of the lesbians became bisexuals, and a few of them even switched to define themselves as heterosexuals. Diamond explains that one third of the women changed definitions twice or more during those 13 years, and that she only observed greater stability in the group that originally defined themselves as lesbians. Among the various conclusions of her work was that sexual behavior is very fluid, that bisexuality isn't a transition to homosexuality, that in some cases the difference between bisexuality and homosexuality is a question of degree, and that the terms have to be more lax, especially in the female gender. Diamond doesn't have data on men, but argues that they have less fluidity. Paradoxically, when Diamond is asked if her results imply that it is easier to modify sexual orientation, her answer is no, that what changes is the expression and behavior.

This last statement is important, and coherent with studies that show that in prisons, armies and single-sex schools there is more sexual contact between people of the same gender, without that implying a path toward homosexuality. Those people will continue to have heterosexual preferences despite being more open to encounters with partners of the same sex.[4]

The situation in prisons offers an interesting social experiment. In the context of HIV contagion, many studies have been conducted in recent years

---

[4] I've always found it peculiar that many people immediately identify as gay the guy who has the courage to experiment with another man, but never label as lesbian a heterosexual woman who occasionally has relations with other women. This may reflect a historically higher intolerance toward male homosexuality than toward lesbianism, which is in fact one of the reasons why it is argued that science has studied male homosexuality in such exaggerated disproportion to female homosexuality.

about the sexual behavior of inmates, and one of the observations is that indeed there are men and women who have their first same-sex encounters in prison and that does not "convert" them into homosexuals. In the vast majority of cases, when they leave the prison system they continue to have heterosexual preferences. For example, a study published in 2012 surveyed more than 2,000 inmates in Australian prisons and showed that 95.1% of them defined themselves as heterosexuals, even though 13.5% declared having had sexual relations with men at some point in their lives. In this last group, only one out of every five had had homosexual relations only in prison. Many of these encounters did not include anal sex. In addition to concluding that intimate experiences in prisons are much less common than is generally thought, the authors suggest that, in adulthood, sexual orientation is quite inflexible, but on the other hand sexual expression and behavior is more adaptable to special circumstances, as suggested in Diamond's work.

Is orientation more elastic in adolescence? Same-sex schools, particularly boarding schools, could represent "similar" same-sex cohabitation as found in prisons. Are there more gays and lesbians among those who study at same-sex schools as opposed to co-educational environments?

Oddly, even though it would be very easy to carry out, no one has ever financed a large-scale study that tackles this question. Only one study in British schools and several hypotheses by experts have reached the consensus that there is a lot of homosexual play and contact at same-sex schools, but that it doesn't condition future homo- or heterosexuality. There are just as many gays and straights among those who spent their childhood and teen years surrounded by classmates of the same sex as among those who were reared in mixed environments. This would also reinforce the hypothesis that sexual orientation is a psychodynamic process, but that it comes predisposed in large part from earlier phases.

As for the childhood phase, in the late 1960s Richard Greene, a sexologist at the University of California, carried out a famous study that has led to various interpretations. Greene followed the development of 66 boys from 4 to 10 years of age, over 15 years, who displayed very effeminate behavior (he called them sissy boys), some of whom even declared that they would prefer to be girls. Greene also followed 56 boys with typically masculine behavior who lived in family environments similar to those of the effeminate boys. Fifteen years later, all except one of the 56 boys in the control group were heterosexual, while approximately two-thirds of the effeminate boys defined themselves as homo- or bisexuals. None of them had wanted to change sex. As we keep repeating, the data only reflects what it reflects. Many gays never showed any effeminate behaviors during childhood, and it should not be overlooked that a third of the sissy boys finally became heterosexuals. But the study does indicate that in some cases sexual orientation is already quite well marked from before infancy.

And here is where biology, without negating the abundant cases of men and women whose homosexuality manifested itself in adulthood, maintains that in other cases sexual orientation can come biologically conditioned from the earliest stages of life, and even before birth. Under this premise, our brain structure would already be shaped for us to be more attracted to someone of the same gender, or the reverse. This would explain why sexual orientation is so tremendously stable and that trying to change it, beyond being an aberration, is a practically impossible task that only leads to worse suffering.

In fact, one of the most cited cases to illustrate that the hormonal environment in the prenatal stage is more decisive than later socialization (in this case not only as regards sexual orientation but also the entire gender identification) is that of Canadian X Reimer. X was born a boy in 1965, but at 7 months his penis was accidently destroyed in a phimosis operation. After some initial uncertainty, well-known sexologist John Money, at the John Hopkins Medical School, advised his parents to completely feminize X, change his name, dress him and raise him as a girl, and at 2 years old, give him a complete sex change. X became Brenda. They removed his testicles, constructed a vagina for him and in puberty he was treated with estrogens. Money was convinced that gender identity was something social that developed in function to how we are raised and educated. During many years, Money used the case of Brenda to defend his vision of gender's malleability, arguing that her development was like that of any conventional girl. But that all changed when it was discovered that Brenda reached puberty feeling like a man, was attracted to women, started to wear men's clothes and soon changed her name to David. David began testosterone therapy, had a breast operation and had his penis reconstructed. His brain had always been male. David married a woman but, for unknown reasons, killed himself in 2004 at the age of 38. His tragic case shows the power of biological predetermination in the face of social influence.

Yes, You Can Be Born Gay

Not just in bonobos, dolphins, goats and giraffes, actually homosexual (or at least bisexual) behaviors have been seen in more than 1,500 animal species. Although that is actually a deceptive figure, since animal homosexuality is very different from its human counterpart, especially in terms of exclusivity. The fact that male whales rub their genitals together, that some female koalas in captivity reject the rough males and prefer to get with their girlfriends to share more meticulous forms of pleasure, and even monogamous penguins have flings with others of the same sex, clearly indicates that homosexual behavior among animals is not something strange or "unnatural" as has too often been said. But it is not comparable, nor does it gives us information on human homosexuality, since in most cases it is play and not an exclusive preference.

The fact that 8% of male rams prefer to mate with other males despite having females in heat nearby, and that an investigation at the University of Oregon found differences in parts of the brain in that subgroup similar to those described in previous studies on humans, can be more informative, but the extrapolations have also been subject to much criticism. And it is true that various groups of scientists have managed to get male fruit flies to do mating dances for other males just by altering a gene responsible for transporting glutamate in the brain, or giving them a drug that exclusively affects those same brain cells; although what they are inducing is a trick, since what they've done is make the flies perceive the other males as if they were females, not turn them into "gay flies."[5]

The parallels between animal and human homosexuality are not correct and have been misinterpreted. The real lesson to learn from nature is that sexual contact between members of the same gender is not something strange, that there really is a wide range of diversity, and that within that range there are biological factors that lead some individuals toward different preferences than others. Based on this reasoning, and especially in light of how constant sexual orientation is in our species (especially in men), in recent decades science has been tracing genes, hormones and brain structures that could be implicated in human homosexuality. And even though nothing has been proven yet, there are no indications that the search and the debate are coming to an end any time soon.

The first revolution in this field happened in the early 1990s, with the publication of several controversial works. In one of them, the Dutch neuroscientist Dick Swaab carried out autopsies on hetero- and homosexual men and women until he found that the suprachiasmatic nucleus of the

---

[5] We too often interpret animal behavior with an excessively anthropomorphic gaze. Edward O. Wilson offers a fabulous example: he observed that in several ant communities, when one dies, the others pick up the corpse and take it to an ant cemetery. They are actually doing that so that it doesn't decompose in the nest, but some biologists interpreted this by suggesting that, despite seeming to us like insignificant insects, ants have a certain awareness of the difference between life and death. Wilson explains that if we inject a live ant with a chemical substance that is given off by dead ants, and place it in the middle of the nest, the other ants will pick it up and take it to the cemetery despite the fact that it is moving around as much as they are.

hypothalamus (a dimorphic part of the brain between men and women) was twice as large in homosexuals than in heterosexuals. Using this same methodology of autopsies, in 1991 Simon LeVay published that another area of the hypothalamus, the INAH3, was between two and three times smaller—and similar in size to women's—in male homosexuals than in heterosexuals. That same year, Michael Bailey carried out a study with twins, according to which when a monozygotic twin (with 100 % identical DNA) was gay, his brother was as well in 52 % of cases. However, in dizygotic twins (with 50 % identical DNA) the correlation was only 22 %, indicating that sexual orientation had a genetic component. These results generated a huge stir and were used to defend that homosexuality has a biological predisposition, but they were also subject to very solid criticism. It was argued that the differences in brain structures could be a consequence and not a cause of homosexuality and, as for the twins studies, in addition to the fact that the sample size was small, it was argued that if half of the gay twins have a genetically identical brother who isn't, that implies that genes are not so determinant.

The research has progressed, and since then they have identified other cerebral dimorphisms between gay men and women and straight ones; although, on the other hand, they have downplayed the importance of the INAH3 area. On a genetic level, one study suggested that polymorphisms in the Xq28 gene could be associated with homosexuality, but later investigations did not confirm the hypothesis. Then in 2015, a study directed by the psychiatrist Alan Sanders at Northwestern University compared the DNA of more than 900 gay brothers and heterosexual members of their families, and found a link between sexual orientations and two regions of the genome: the Xq28 and a gene in chromosome 8. It is also speculated—much more logically but still without proof—that epigenetic regulations could be implicated. One of the observations yet to be interpreted is that among homosexuals there is a 39 % higher rate of left-handed and ambidextrousness, according to the conclusions of a meta-analysis of all the scientific literature on sexual orientation and laterality that compared almost 6,987 homosexuals and more than 16,000 heterosexuals. It is speculated that it could indicate differences in brain organization from as early as embryonic development. Another interesting result was seeing that the brains of homosexuals reacted differently than those of heterosexuals when they were exposed to androstadienone, a typically masculine pheromone found in men's sweat. Although, again, we cannot establish whether this difference is innate or learned.

It is a field of study of questionable interest, which makes advances based on fairly anecdotal results. However, over time, two hypotheses on the biological conditioning of homosexuality have gained solidity: the first argues that hormonal levels during pregnancy could affect brain differentiation and therefore sexual orientation; the second, by Ray Blanchard, suggests that the mother's immune system after several male pregnancies provoke the partial feminization of the brain of her subsequent sons. Let's start with the first.

Levels of Testosterone in Pregnancy and Sexual Orientation

Let's remember that, after conception, the fertilized ovum starts to divide before defining a sex, and that it isn't until the sixth week of gestation that a fetus possessing a Y chromosome begins to develop testicles. They start to secrete testosterone from the eighth week, reaching the highest peak between the 12th and 14th weeks of pregnancy. Right then is when the brain begins to develop, which will gradually masculinize if there are high levels of androgens in the blood. Many studies have confirmed that certain brain areas are already dimorphic in men and women from this stage of fetal development. As a result, the hypotheses that defend the idea that some people are born homosexuals suggest that alterations in that spike of testosterone could make brain structures like the hypothalamus develop more or less masculinized, conditioning a future predisposition to hetero- or homosexuality.

The testosterone levels in human fetuses cannot be directly measured, but there is a very peculiar indirect clue: curiously, the difference in the length between the ring and index fingers (2D:4D ratio) is related to levels of testosterone exposure during pregnancy. The more testosterone that ran through our bodies in the fetal stage, the longer our ring finger is compared to our index. The 2D:4D ratio of men is clearly higher than women's, and although with sometimes contradictory results, they have established differences between heterosexuals and homosexuals. The last meta-analysis published suggests that the differences between lesbians and heterosexual women are clear (lesbians would have longer ring fingers than heterosexuals, which would indicate greater exposure to fetal testosterone), but the results are more incongruous in men.

Studies with laboratory animals also show that prenatal levels of androgens play a key role in adult behavior. Even though we have underscored that it is not comparable, if they modify the hormone exposure levels during gestation of male and female rat fetuses they manage to get the adult males and females to show sexual conduct typical of the other gender, like males with lordosis and females trying to mount other females. Here it is important to point out that the "homosexual" behaviors in rats could also be induced after birth. Mexican

Genaro Coria-Ávila put various male virgin rats in the same cage and injected some of them with a substance called quinpirole, which acts antagonistically against D2 dopamine receptors. These receptors are involved in the generation of sexual desire and the formation of ties with a partner. What Genaro Coria observed is that, after some time, the male rats injected with quinpirole had erections in the presence of other males, played more with them, showed female precopulatory behaviors such as darting and in occasions they tried mounting. By telephone, Genaro insisted that sexual behavior is a continuum and that in rats—and perhaps in humans—homosexuality can be learned. He believes that if while cohabitating with people of the same gender, activities that release large quantities of dopamine are performed, such as sex, play and drug use, one can develop a conditioned preference toward individuals of the same sex. Other similar studies injected oxytocin into some cohabitating female rats and over time observed "lesbian" behaviors only in those injected. According to Genaro, the stimuli associated with the first sexual experiences are very determinant, and the plasticity of the brain allows sexual orientation to be shaped by learning.

To conclude this section on prenatal hormones and sexual differentiation, as we will see in the chapter devoted to intersexuality, there are people with XY chromosomes who secrete normal levels of androgens but whose cells, due to a genetic mutation, don't have testosterone receptors and they develop with the body, mind and sexual orientation of women. Or XX women with congenital hyperplasia who secrete much more testosterone and as adults experience proportionately more attraction to their same gender. If we put together all of these facts, it does seem that androgen levels during pregnancy could explain some cases of homosexuality.

The More Older Brothers, the More Gay

The other hypothesis on biological determinants of homosexuality is somewhat outlandish, but I cite it because several independent investigators have assured me that it is solid and because it points to a highly interesting question. In 1996, psychologists Ray Blanchard and Tony Bogaert found that, on average, gay men have many more older brothers than heterosexual men do. This fraternal birth order study has been replicated in several cultures and is statistically well demonstrated. In fact, Blanchard and Bogaert calculated that the probability of being gay increased by 33% for each older brother. Blanchard is a recognized sexologist who has spent his entire career studying different aspects of human sexuality. When, on his 67th birthday, I asked him whether the most logical interpretation of this fraternal birth order finding wasn't that a homosexual preference developed just from living with older brothers, he explained that:

> From the beginning we always sensed that there had to be some biological effect. My experience made me think that in men, sexual orientation was defined in very early developmental stages, and that the interaction with brothers would be something secondary. And that was proven in a later study by Tony (Bogaert).

The study that Blanchard was referring to was published in July 2006 by Anthony Boagert in the journal *PNAS*: when comparing a total of 994 homosexuals who lived with older brothers, it was established that the link between homosexuality and a greater number of brothers only happened when they were brothers from the same mother. He even observed that the effect was also maintained in cases in which the biological older brother had died or didn't live with the family. I also interviewed Bogaert, who stressed that "the presence of biological older brothers and not the cohabitation was what increased the proportion of homosexuality."

Those results are proven, and they served to create Blanchard and Bogaert's maternal immune hypothesis, whose theoretical basis is that during pregnancy it is very common for some fetal cells to pass into the mother's blood stream. If these cells are from a male fetus, they could contain some male cell molecules and the mother's immune system could create antibodies against them. In a later male pregnancy, those antibodies could affect the development of the fetus, perhaps impeding certain brain areas from masculinizing and conditioning sexual orientation. According to this hypothesis, the more male pregnancies, the stronger the immunization and the higher the probability of homosexuality. Blanchard admits that this mechanism is still not experimentally proven, but he tells me that they are collecting blood samples from mothers who have had gay sons to analyze their antibodies and see if the hypothesis plays out. When asking about how the maternal antibodies could alter brain development, Blanchard recognizes that he isn't very sure but that he has seen that mothers develop specific antigens against proteins associated with the Y chromosome. Far from being confirmed, and with criticisms of various types, Blanchard and Bogaert's maternal immune hypothesis is still a reference in the scientific study of homosexuality.

Biological predisposition does not imply determinism, and no naturalist ever ignores the role of surroundings and socialization when it comes to shaping our behavior. We know that we have a tremendously plastic brain that allows us to learn and adapt to different circumstances in our environment, and that experiences define many aspects of our personality. But the inflexibility of sexual orientation in adults does indicate that it should be configured from a very early stage, and denying that has a lurking danger: homophobic groups defend the idea that homosexuality is exclusively a result of experiences after birth because that allows them to design reparative therapies to revert that learned tendency. These practices are damaging and condemned by psychiatric associations, and beyond the fact that we may find them barbaric, the intellectual question remains: is it possible to change sexual orientation?

The Solidity of Sexual Orientation

For decades electroshock therapy, drug treatment, conditioned masturbation, the use of prostitutes and psychoanalysis have been used to try to revert homosexuality. And even though therapists assure us that they are successful and some of their "patients" say they've changed, scientific scrutiny has never confirmed that the changes are real. Modified behavior has been observed and even certain attraction to women, but when carefully analyzing patterns of arousal, it is always seen that they continue to be attracted to men. Which is to say, the consensus was that they could revert behavior but not sexual orientation. Until the polemical study published in 2003 in the journal *Archives of Sexual Behavior* by psychiatrist Robert Spitzer of Columbia University stated that "there is evidence that change in sexual orientation following some form of reparative therapy does occur in some gay men and lesbians." The article caused a media firestorm, since Spitzer was not a defender of reparative therapies, but rather a highly prestigious psychiatrist with extensive experience in the study of sexual orientation who, in fact, in 1973, had been one of those responsible for removing homosexuality from the list of mental disorders in the *Diagnostic and Statistical Manual* of the American Psychiatric Association. Also, the sample in his study was very representative (200 individuals) and the results were significant. In his investigation, Spitzer chose 143 men and 57 women who had been voluntarily subjected to reparative therapies and who said their sexual orientation had been modified. Spitzer carried out structured interviews to measure the characteristics of the change and concluded that, while complete reorientation was rare, in most cases—and with a higher proportion in women—they did shift from "predominantly or exclusively homosexual" to "predominantly or exclusively heterosexual." Spitzer concluded that the participants weren't lying or deceiving themselves (as was reported in many prior studies) but rather that, indeed, they had modified their sexual orientation.

The study was welcomed by religious communities and homophobes and received with indignation by homosexual associations, and with severe criticism by the scientific community. In the same *Archives of Sexual Behavior* several authors rechecked the methodology, especially the fact that the sample selection had been made among men and women who had sought treatment for religious reasons and were presenting themselves as evidence that the change was possible and desirable for others (93 % declared that religion was very important to them and 78 % had given public talks defending these types of therapies). That implied a basic, huge distortion in the study. Additionally, in Spitzer's original work, the dimension of the change was established only based on the patients' declarations of their own behavior and fantasies before and after the treatment. Which is to say, there was no way to control that they weren't exaggerating to promote this practice. John Bancroft was the author of a harsh critique, arguing that while there is a basis to conclude that sexual orientation is not always fixed and immutable, Spitzer's results are not representative and in no way justify a reparative therapy that reinforces the homophobia and social stigmatization of homosexuals, and that it causes enormous psychological problems in those who decide to undertake it.

The study had an enormous impact in the United States and continued to be used by anti-gay collectives to justify reorientation therapies, until in April of 2012 the most condemning criticism came, a retraction by Robert Spitzer himself, who sent a letter to Ken Zucker, editor of the journal *Archives of Sexual Behavior*, in which he expressed his revision of his study in these terms:

> I was considering writing something that would acknowledge that I now judged the major critiques of the study as largely correct. [...] I decided that I had to make public my current thinking about the study. [...] From the beginning [the basic research question] was: "can some version of reparative therapy enable individuals to change their sexual orientation from homosexual to heterosexual?" [...] the study design made it impossible to answer this question. [...] The fatal flaw in the study [was that] there was no way to judge the credibility of subject reports of change in sexual orientation. I offered several (unconvincing) reasons why it was reasonable to assume that the subject's reports of change were credible and not self-deception or outright lying. But the simple fact is that there was no way to determine if the subject's accounts of change were valid. [...]I believe I owe the gay community an apology for my study making unproven claims of the efficacy of reparative therapy. I also apologize to any gay person who wasted time and energy undergoing some form of reparative therapy because they believed that I had proven that reparative therapy works with some "highly motivated" individuals.

Spitzer doesn't deny that there may be a certain degree of flexibility, but he recognizes that what science has established is that what is truly damaging is not accepting homosexuality.

What's Damaging Is Homophobia, Not Homosexuality

We have talked a lot about the causes and possibility of changing sexual orientation, when those are the least of the concerns of most researchers in this field. For example, Ron Stall, director of the Center for LGBT Health Research at the University of Pittsburgh assured me: "I don't care if the origin of homosexuality is older brothers, family environment or testosterone before or after pregnancy." What he wants to understand is risky behavior associated with drug use, how to minimize HIV contagion, to clarify the role of the papillomavirus in the rise in anal cancer and the concrete causes of depression in adolescence, and to study many other factors to improve quality of life and wellbeing in the LGBT community.

Ron Stall is the author of the largest study realized to date on the health disparities in the gay population. Ron meticulously analyzed medical data from 2,881 homosexuals in different age ranges and districts in Chicago, New York and San Francisco, and he found that "among the homosexual population there are higher rates of depression, suicide attempts, sexual abuse, drug consumption and HIV infections." Ron explains that those rates had already been separately documented in previous, smaller studies, but he wanted to stress his syndemic theory according to which all those risk factors are interconnected, reinforce each other and could have originated from common causes and should therefore be tackled together from a public health standpoint. In fact, Ron Stall cites various studies conducted in his center on teenagers, according to which "the initial most important cause of this rise in syndemic problems that reinforce each other is the lack of acceptance of homosexuality in the family and school environment." Ron showed that among homosexual teens who suffer bullying at school and family rejection at home, the suicide attempts, depression, substance use and risk of infections are much more frequent than among those who have internally and externally accepted their sexual orientation. Ron Stall insists on a blunt message to society and families: "If you really want the best for your son or daughter, don't try to change their sexual tendencies, because not only will it not work, but you can cause them a lot of harm. Concentrate your efforts on correcting homophobia, not homosexuality."

Does Male Bisexuality Exist?

While we were having a drink in a bar in Boston, Anne confessed that she was "one hundred percent bisexual." She told me that she had had sexual and romantic relationships with both men and women, and that "I fall in love with or feel drawn to the person, independent of their gender, and for me that's very normal." Anna, a 32-year-old German woman, believes that her bisexuality is due to growing up with her mother and her mother's lesbian partner; she assures me that her mind is free of stereotypes, and that for her a gaze or a pair of lips can be equally attractive if they are part of a man's body as part of a woman's. She says that usually it all starts with a friendship, and that when she feels attached and close to someone she may start to feel a desire for physical intimacy with them. Anne has had stable, exclusive relationships with both men and women, and when she is part of a couple with either gender she doesn't feel the need to have affairs with the other sex. She says that her sexual relationships with men were absolutely "normal" and that no guy had ever mentioned noticed anything strange. At that point in her life she was looking to meet someone special, possibly for the long-term and having a family, but she didn't have any preference whether that was a man or a woman.

Anne could define herself as 100% bisexual, but her case certainly isn't the most representative, since in many studies most of men and women who identify as bisexuals have a preference for one gender.

It's actually a matter of definitions. In surveys, approximately 1–2% of men and 2–3% of women classify themselves as bisexuals, but when we understand "bisexuals" as people who feel sexually attracted to both genders, even in a very wide range of degrees, the numbers are higher. And in a most restricted idea of bisexuality in which the degree of attraction to men and women is of similar intensity, then the rates are a fair amount lower, especially in guys. That is precisely why some academics believe that it is an ambiguous term and that, particularly in men, absolute bisexuality is pretty rare.

In 2005, Professor Michael Bailey, of Northwestern University, published a study that really made waves among bisexual collectives. Bailey recruited 38 men who defined themselves as homosexuals, 30 heterosexuals and 33 bisexuals, took them one by one to a laboratory, placed a plethysmograph on their penises to measure subtle changes in its size, and showed them various videos with erotic images either of just men or just women. According to the results, the homosexuals showed a clear pattern of arousal with the male images, the heterosexuals with the female, and in the group of bisexuals, three fourths showed a pattern identical to the group of homosexuals and one fourth similar to the heterosexuals'. Bailey told me that he did not observe any different pattern characteristic of bisexuals; he concluded that there was always a preference for one gender and that therefore, despite many men defining themselves as bisexuals, strict male bisexuality was actually very uncommon. In his article, Bailey also cited other studies, according to which there is a transition seen in young men, from bisexuality toward homosexuality.

Bailey's work received much criticism for drawing such provocative conclusions with such a small sample group, and especially for not contemplating the emotional aspect. Within the bisexual collective, there are people who feel more emotionally attracted to one sex and more physically to the other, and this wasn't reflected in the study.

In response to the criticism, Michael Bailey decided to repeat the study following stricter criteria for choosing the bisexuals. He recruited only those who, apart from defining themselves as such, had had at least two different sexual encounters with men and women, and romantic relationships with both genders that had lasted more than 3 months. To the methodology he also added erotic videos of men having sex with both guys and girls at the same time. Published in 2011, the results then indicated that in this subgroup of bisexuals the arousal patterns were right in the middle between homosexuals and heterosexuals, and that they could clearly be identified as a different sexual orientation. The conclusions that time were that, while to a lesser degree than a survey on sexual orientation would reflect, male bisexuality does indeed exist.

This study and some of the previous ones on the causes of homosexuality raise a very valid question: are they at all useful? As we said in earlier chapters, we should be guided by culture, not nature, and in a matter such as this one every social decision will always be more important than any physiological information. In that sense, knowing the biological causes of homosexuality is in many respects irrelevant. But those who scientifically investigate sexual orientation defend that the goal of their studies is simply to give us information that allows us to understand ourselves better as a species and as individuals. That alone justifies it; knowledge is always a good thing in and of itself. At the same time, they argue that at a certain point the scientific perspective contributed to reinforcing that many gays and lesbians weren't that way by choice, but because their sexual orientation formed an intrinsic part of their nature and had to be defended as such. In fact, there is another collective that is demanding scientific interest to dispel the lack of comprehension surrounding them, and they are asking to be, alongside homo-, hetero- and bi-, to be considered a new sexual orientation: asexuals.

Learning from Asexuals

Rebecca was the third asexual person I met. She must have been about 25 years old. She was super nice, and so open that I dared to confess to her how strange it was for me to try to imagine never feeling any sexual attraction for anyone. Rebecca turned and said, "You see that old guy over there in the white shirt?" "Yes." "Are you sexually attracted to him?" "No…" I answered, disconcerted. "Would you go to bed with him?" "Of course not!" "Well, that's how I feel about everybody, I just never see anyone I want to have sex with."

So simple, and so clear. I was finally beginning to understand how an asexual person might feel. But Rebecca continued, "And right now, do you feel sexual desire?" "No…" "You don't feel any inner arousal that makes you predisposed to any sexual activity?" "Not right now." "Well, that's how I always am; I never feel spontaneous sexual desire." Again, that was illuminating. Those were the common traits in the very diverse community of asexuals, people who feel no desire or sexual attraction for anyone and can be very happy that way.

This last point is important: many asexuals insist that they have no problem, that it's those who want to have sex and can't who are anxious about it, but that a life without sexual attraction—which doesn't always mean without sex—doesn't necessarily generate any stress. Actually what makes them most uncomfortable is the lack of understanding they feel from a large part of society.

Most asexuals do have romantic feelings, and they admit that their lack of desire represents a serious challenge when it comes to maintaining a relationship. But otherwise they feel completely normal, and many of them even consider asexuality a new sexual orientation. They argue that just like there are people who feel attracted to their same gender, others to the opposite gender, and a third group to both, they don't feel attracted to either. And that doesn't bother them. In fact, some asexuals do feel a certain aversion toward sex, but most simply feel an indifference that allows them to have sexual relations with their partner if that's the agreement, and masturbate periodically if they believe it's good for their health. Their genital arousal usually functions correctly.

Anthony Bogaert, of Brock University in Canada, is one of the top world experts in asexuality. His study published in 2004, based on a sample of 18,000 Brits, established that 1 % defined themselves as asexual and he began the empirical study of this characteristic that he didn't consider a disorder but a sexual orientation. Since then, Bogaert has met and analyzed hundreds of asexuals, and while during our conversation he keeps insisting on the vast diversity among them, he assures me that most asexuals have erections and get lubricated without any more problems than the rest of the population. Nor have they found differences in androgen levels, many of them masturbate often, and most don't feel any phobia or repression of sex. Simply, just like I would never sleep with that old guy in the white shirt, no matter how aroused I was, or I would prefer to masturbate than have the slightest physical contact with someone who repelled me, asexuals feel that same discomfort when imagining themselves naked or having sex with anyone.

Bogaert tells me that "it is a very diverse group and I don't rule out that in some cases asexuality is due to psychosocial factors, but we have very rarely found traumas or religious determinants, and we have found physical differences such as short stature, worse health and a large proportion of left-handed and ambidextrous people. I think that it is biologically conditioned and that's what leads me to think of it as a different sexual orientation." When Bogaert explains that many asexuals form romantic partnerships, I ask him if in those cases, having relations over time, some of them gradually become sexual. "There may be some cases but it's not usual. Like other sexual orientations, asexuality is usually very stable. We are analyzing whether in some asexuals who want it, increasing their testosterone levels could raise their degree of desire; but we aren't convinced that it can work," Bogaert explains, admitting that the term "asexual" is vague and can easily overlap with people who have an extreme case of lack of desire. According to Bogaert, the key difference is that asexuals don't have any problem with their lack of sexual desire and no wish to try to correct it. They don't feel attracted to other people, but that doesn't keep them from having a positive view of sex and curiosity for experiencing different sexual expressions.

After talking to Johanna, a young asexual woman, for 30 minutes, I couldn't help myself. "Look, excuse me for saying this, but you seem like a very sexual person to me." "Right?" she responded, laughing. My incredulity came after listening to Johanna say that she felt no physical desire whatsoever and that when she'd tried to have sex with someone she had been deeply dissatisfied, but that sometimes she masturbated to relax. She had tried sadomasochism and bondage because she found it somewhat compelling, and when she passed a sex shop she would always go in and have a look around. When I heard that last part I couldn't help but ask, "And when you see a vibrator don't you want to give it a try?" "Well… I have used them." I know that it sounds paradoxical, but I continued my line of questioning and asked her if she watches Internet porn and she replied, "No, never. I can look at it out of curiosity, but I don't feel any attraction toward other people or get aroused." That is the key point. Asexuals aren't antisexuals. They just don't ever feel desire. And understanding why is a tremendously fascinating mystery for science.

Lori Brotto, of the University of British Columbia, wanted to begin her investigations with good knowledge of the characteristics of asexuals, so she made a detailed study of a group of 54 asexual men and 133 women. Her results showed that 32% of the women had had stable relationships of more than 7 years, and 0% of the men. Seven percent of the men never masturbate but more than 50% did it at least once a week. 42.7 percent of the women never masturbated, but 23% of them did between one and four times a month, and 6.8% several times a week. However, when they were asked about their ideal frequency of intercourse, 86% of men and 94% of women responded between 0 and 2 times a year. And as for the frequency of kisses and intimate touch with their partners, 84% of men and 75% of women responded that it was none.

Among the study participants who had some type of sexual activity, she observed that the indexes of desire were clearly less than in the general population, but the erectile capability in men and the lubrication and incidence of vaginal pain in women were similar. In women, orgasmic response and satisfaction was comparable to women with hypoactive sexual desire disorder, HSDD.

According to Anthony Bogaert, the subject of masturbation is key. He explains that "Clearly there are asexuals who never masturbate, but many others who do. Some simply do so to eliminate tensions and for their health, but others have sexual fantasies. What's interesting about these fantasies is that they never imagine themselves in them... in other words, they have a pattern of sexual fantasies that are very different than most people, and that is what we are investigating now." Bogaert is very convinced that the study of asexuality will offer very unique and groundbreaking views on conventional sexuality. For example, he considers asexuals as proof that romantic love can exist and be maintained with no need of sexual desire.

This subject is one of the most trendy in sexology, and according to Lori Brotto, it has enormous implications. She is a member of the group working on defining sexual disorders for the next DSM-V, and she wants to establish whether asexuals' permanent lack of desire and arousal should be considered a very severe case of hypoactive desire disorder, whether it implies the existence of a psychic disorder or if, on the other hand, asexuality should be seen as one extreme of normality. The question is not trivial, since the first two options would mean considering it a pathology that should be treated, and not the third.

Brotto is aware of the challenges inherent in the fact that the population that defines themselves as asexual is very heterogeneous and the term is vague, but she is very interested in studying asexuals with a depth unprecedented in science. Lori tells me that in an unpublished investigation they are comparing biological and behavioral factors among a group of 403 asexuals, 131 controls and 130 people with HSDD (hypoactive sexual desire disorder). At the time of our conversation, the preliminary results clearly suggest that, while there are cases of overlap, asexuality is qualitatively different from HSDD. Perhaps the most divergent element is that among asexuals there are far lower indexes of discomfort over their condition. There are also important differences in the types of sexual fantasies they have: in controls and people with lack of desire they usually include other people, and in asexuals they are less frequent and generally revolve around objects and situations. Asexuals who masturbate do so with a frequency similar to cases of HSDD, but they have much fewer sexual encounters with a partner. Lastly, people with lack of desire do not define themselves as asexuals.

As for considering it a psychic malady, Lori Brotto is of the opinion that it is not, since asexuals' levels of depression and disorders are low and their alexithymia (inability to identify their own emotions) is only slightly above the control group and lower than the group of people with lack of desire.

In the more physiological part of her studies, Lori is analyzing three biological markers that are dimorphic among heterosexuals, homosexuals and bisexuals: number of older brothers, relationship between length of ring and index fingers and proportion of left-handed and ambidextrousness. The preliminary results have not found differences in the 2D:4D ratio. As for brothers, asexual men seem to have, just like gays, more older brothers than heterosexuals, and among women they have not found any differences. However, the most important data is in the proportion of those who are not right-handed. It is well established that among gays and lesbians there is a higher percentage of left-handed and ambidextrousness, which, according to some investigators, could be defined by hormonal levels during pregnancy and brain development. In the case of asexuals, both in men and women, the proportion of non-right-handed people is more than double what it is in heterosexuals. Lori insists that more samples must be analyzed, but it all seems coherent with asexuals' claim that this is their sexual orientation, not a problem of socialization but rather part of their nature. This is a relatively new subject, but perhaps after some time, in addition to hetero-, homo- and bi- we may consider asexuality a different sexual orientation, and viewing asexuals suspiciously and with an eye to reparation will be as frowned upon as considering homosexuality abnormal is today. Living without sex is only a cross to bear if it is accompanied by desire and frustration. Not being interested in sex, on the other hand, is not necessarily a problem.

15

Learning from S&M Clubs

There were three things that particularly intrigued me on my visits to sado-masochistic clubs in New York, and in my meetings with members of The Eulenspiegel Society[1] for BDSM (Bondage Domination Sadism Masochism) education: (1) The exquisite communication over boundaries and sexual preferences between those who practice sadomasochism, so much more fluid than in most conventional couples. (2) The very well delineated roles in dominance and submission. They take it to an extreme, but when incorporated into the erotic play of non-S&M couples it can also be exciting. (3) Physical pain as a trigger of arousal. I found some of the practices frightening but, again, within your own personal limits it can be a very interesting ingredient. And, this last point does have a lot of meat to it, scientifically speaking. That our brain interprets the same stimulus as painful sometimes and pleasurable other times has a basis in the profound distinction between physical emotion and feelings.

My meeting with members of TES was set for a Saturday at 9:30 PM in a restaurant in the Chelsea section of Manhattan. I had already participated in a couple of their weekly discussions on different aspects of BDSM, and they invited me to go with them on one of their outings organized with other fetish and S&M clubs in New York. First they gathered in a private room of the restaurant in Chelsea to have dinner, meet those who were newcomers, talk calmly with the beginners and then all go together to the hidden club Paddles.

---

[1] The mission of The Eulenspiegel Society, founded in New York in 1971, is to educate, and to support members of the BDSM culture. The term BDSM includes sadomasochists, fetishists, those who practice bondage and other types of sexual expression that involve inflicting or receiving physical pain or humiliation. TES organizes workshops and weekly discussions for its members at its headquarters on 36th St. Most of the meetings are open to anyone interested. I attended several, and found it very enlightening.

I showed up a few minutes late so I wouldn't be among the first ones there. I was feeling a little nervous and hesitant. I didn't tell anyone where I was going that Saturday night. When I arrived, there were already about 30 people. Bo—one of the most avid TES members and the person who was leading that night's gathering—greeted me with a smile, and her slave invited me to take a seat at one end of a long series of tables set up in a U shape. I introduced myself to the 50-something guy who was sitting next to me, dressed in a black tee-shirt and leather pants with tattoos visible on his neck and arms, and the smiling, heavyset, middle-aged woman of color with a plunging neckline, who was seated in front of him. They soon resumed their conversation, and I looked around at the other folks and nodded in a friendly way. The age range was from about 20-something to 50-something. Except for a few who did have a "strange" appearance by conventional standards, most of them looked absolutely run of the mill, wearing street clothes, and their behavior was extroverted and festive. Seen from a distance, the event could pass for a regular work party with people of different ages, races, styles, behaviors and socioeconomic conditions. And even though later, once we were in the club, their clothing changed and some of them seemed to transform, it seemed to me that outside of the club the atmosphere was a very natural one.

I have no intention of drawing conclusions about personality based on my few encounters with sadomasochists. Not from what I saw with my own eyes, nor from what I was told by members of the collective, who could well be affected by cognitive bias and unconsciously distort the reality of their own subculture. That's what we have science for: that supposed neutrality and objectivity.

A huge Australian study surveyed almost 20,000 people on different aspects of their sexual conduct. 2.2% of men and 1.3% of women acknowledged having had sadomasochistic practices in the last year. It was more frequent among homosexuals and bisexuals than among heteros. Researchers analyzed the BDSM practitioners in detail and confirmed a higher prevalence of oral and anal sex, more use of pornography and a greater number of sexual partners, but they didn't find significant differences in their perception of infidelity, anxiety, problems associated with sexuality nor, curiously, interest in sex. One significant piece of data was that, on average, sadomasochistic men felt slightly less psychological stress than the rest. The researchers interpreted BDSM as just another cultural practice and it is not associated with any pathological symptom, past abuse or sexual difficulties. This is controversial because other studies did indicate that among masochists there are usually more arousal problems, personality disorders and difficulties establishing conventional relationships. From my experience and deep conversations with sadomasochistic men and women, those problems do exist, but I wouldn't feel confident saying that they are more common than in the general population, nor that the practice of sadomasochism increases or diminishes them. My feeling is that some people with dissatisfaction or anxiousness resort to BDSM, but not that BDSM necessarily generates dissatisfaction or anxiousness.

There are a couple of important nuances here. Firstly: sadistic lack of control and sexual aggression is definitely described as a paraphilia in the DSM. But that is not what dominants and submissives practice with full consent and pre-established boundaries on both sides. A sadistic disorder such as an illness associated with psychopathy, antisocial behavior and personality disorders is not the same as consensual sexual sadism practiced in the BDSM context that does not cause any discomfort in its practitioners. In these circumstances it should not be considered a disorder to treat. Secondly: various studies have observed that when a person has psychological problems, some practices such as sadomasochism can become obsessive and exacerbate those problems. I have seen cases of submission that really did seem—to me—to be covering for obsessions and serious emotional imbalances. But the inverse does not occur: BDSM is highly unlikely to generate problems in healthy people who've simply chosen that as an option to expand their sexual repertoire. Like Sioux.

Sioux and Anais arrived together and sat down in front of me. They both had totally conventional appearances and I asked them if they were new. Anais told me that she'd been involved in BDSM for 5 years and that she defined herself 24/7 (meaning 24 h a day, 7 days a week, something that was an integral part of her personality). Sioux had only practiced it four times before and very sporadically. She had a boyfriend but he wasn't into it, so she came occasionally with girlfriends. "It's like a hobby?" I asked her. Anais interjected, with a resentful expression, "For me and many others it is a way of life." Sioux smiled at my tactlessness and answered, "Well… in my case I guess you could call it that."

Sioux was tall, svelte and attractive according to conventional sociocultural standards. To be honest, she was one of the few attractive people I saw at the dinner and later in the club. There were plenty of guys who obviously went to the gym regularly and women dressed in provocative outfits, but in general the atmosphere wasn't exactly glamorous. It is acknowledged among practitioners that there is a higher rate of overweight people in sadomasochistic circles, especially among women. It is obvious that beyond the leather clothing—which is very typical—it is impossible to establish stereotypes, but of course I didn't see anyone like Christian Grey, the protagonist of the bestselling *Fifty Shades of Grey*, which tells the story of his dominant relationship with submissive young Anastasia Steele. But there was someone who reminded me of the character of Anastasia: her name was Haddass.

Haddass arrived smiling and waving nervously. She was 21 years old, a bit plump, and she looked like a college student, with big eyes and sweet chubby cheeks, and her expression was somewhere between tense and eager. She sat down by my side and soon told me that it was her first time. I think my "mine too" reassured her. She explained that she had been very curious about it every since she was little, that she'd tried a few "things" with boyfriends, that she'd read a lot about it on the Internet and that she'd finally decided to try to join in on some real sessions. I asked her if she was dominant or submissive, and she replied, very convinced, "submissive!" I asked her again if she remembered anything that could have awakened her interest in BDSM, and after thinking for a few moments she said no. She said that from her first sexual experiences she had always liked feeling dominated but she had never taken it very far, and that now she wanted to try something more intense. Right away I thought about following her evolution closely.

Before continuing, I want to explain that, beyond pain, what defines sadomasochism are the roles of dominance and submission. Dominant is the one who has absolute control and submissive is the one who completely relinquishes it. Generally the dominant one is the "top" who inflicts pain, whips and humiliates the "bottom" (submissive), but a dominant can also demand that the submissive step on them, hit them or do what they ask. There are people who always have fixed roles as dominant or submissive, but some—generally couples who practice BDSM as just another of their sexual activities—are called "switch" and exchange roles. In the consent lists the tops and bottoms use to agree on their limits and preferences there are dozens and dozens of practices, from genital torture to verbal insults and the demand to obey any whim. When someone only practices the lightest form of BDSM and sporadically, they are called "vanilla" (this isn't derogatory). Those are the basic concepts and there are many more variations, like owners and slaves, depending on whether there has been a commitment agreed to, sisters, or relationships of protection when someone is just starting out. But then it gets complicated. You can find more information and people on fetlife.com, the largest online BDSM community in the world.

When dinner is over, the one-to-one introductions are made and Bo gives her tips about what to expect and how to behave at the club, and then we all go up Seventh Avenue as a group to 26th Street and make a left to get to a door that looks like the typical emergency exit of any commercial establishment. There was no sign. Bo knocked, they opened what turned out to actually be an emergency exit door, we walked down a long hallway and a narrow staircase, we paid 25 dollars at the booth and we went in to Paddles. Some of us were dressed normally (dark clothes are a must), and others wore leather clothes with varying degrees of flesh visible. The group immediately dispersed. Several went to change their outfits. The place was a tangle of hallways with a lot of walls that created hidden corners, a common area with a bar where they served no alcohol and several rooms with various devices, like chairs where you could immobilize someone, benches where you could lean while someone whipped your ass, structures to tie people to and whip them and cots for all sorts of activities.

The first thing I stopped to look at was a naked girl stretched out on a cot with her hands tied behind her and a man inflicting small burns all over her body. She looked like she was afraid, terrified that he would get too close and really burn her badly. When the man brought the flame to her armpits she writhed in anguish, but when he brushed it over her nipples she seemed aroused. And the little smacks to her hairless genitals seemed to generate a tension that, as she explained to me half an hour later, after the session when she was untied left her with a sensation of deep excitement.

In one of those hidden corners there were two women shooting darts with a peashooter that stuck into the rear end of a guy on all fours, girls sitting with men licking their shoes, and many people smacking each other with flat paddles, thin semi-flexible bars, whips with leather fringe, or just spanking. While I watched a scene of a stocky man running sharp knives over a woman's tied-up body, I heard a scream behind me.

It was Sioux with her back up against the wall, her arms tied over her head, wearing only a G-string and being whipped by her friend Anais. To be honest, she didn't seem to be having such a great time.

I talked a lot with Bo, who is a real BDSM encyclopedia. There was a moment when we were both in front of a large woman, on all fours on a small bench while a guy whipped her with a switch so hard that it sent shivers down my spine. The woman was sobbing and after an incredibly hard blow she gave off a scream that made a pained expression cross my face. "That must hurt!" I said to Bo. "Well, that's the idea," she responded, insisting that nobody there was doing anything they didn't want to. The limits and signals were agreed to ahead of time, and they could make a gesture at any point that would immediately halt the action. I couldn't help but ask her if that really was sexually arousing. She answered that sadomasochism isn't always about sex. That many times it is, but other times it is just about trying to reach that altered state of consciousness brought on by the pain, accompanied by a later feeling of catharsis, profound wellbeing and a strong bond with your partner.

Strange as it may sound, science corroborates Bo's statement. In one of the few studies made with sadomasochists, American researchers measured the levels of testosterone and cortisol (the stress hormone) of 58 dominants and submissives before starting a session, immediately after and 40 minutes later. The sessions lasted an average of 57 minutes and included dominance and humiliation, inflicting pain and sensory deprivation. Both men and women participated, and in addition to measuring all of their hormone levels they gave them a questionnaire when the session was over. The results were quite significant. Testosterone is a hormone related to aggressiveness, struggle and sexual desire, and as was to be expected the levels in their saliva rose slightly after the session, both in dominant men and women, and lowered a little in submissive men. But the most unexpected result was a marked increase in submissive women. In the article, the authors don't dare to suggest an explanation for this fact, but they say that they could reflect an improvement in mental state during the session and an increase in sexual desire. However, the most revealing results were in the levels of the stress hormone, cortisol. In dominants there was practically no alteration during the sadomasochistic activity, but in the submissive who received pain or humiliation, the cortisol levels increased significantly during the session, and dropped radically 40 minutes later among those who said the experience had been satisfying. In this subgroup, cortisol levels decreased even past the level they were before beginning, and corresponded with the sensation of feeling greater attachment to their partners.

An important observation of the study is that cortisol lessened more in those who had higher levels before the session. So we can interpret that as meaning that if we are physically or psychologically stressed, masochism can alleviate that, and if our stress levels are normal, the effect will be negligible; but in any case it is coherent with the affirmation by masochists that they feel physically relaxed (in a few paragraphs we will talk about the emotional aspect) after having suffered pain and humiliation. In fact, when the couple that Bo and I were observing finished their whipping session, the woman spent a few minutes crying in the arms of her owner, and after a little while I saw them hugging and half-asleep on a couch in the club.

We will look at other studies on the connection between pain and pleasure, but first I should mention that I was disappointed to not find more serious investigations that analyzed the case of sadomasochists. I remember a fabulous conversation on the physiology of pain with Fernando Cervero, a Spanish neuroscientist who has been living in Canada for more than 30 years and is one of the top experts in the world on the study of pain. When I visited him in the summer of 2012 at his laboratory at McGill University, he talked to me about the causes of hypersensitivity, about lobotomies on people with chronic pain done in the early twentieth century in which their pain didn't disappear but did stop bothering them, about phantom pain, about fibromyalgia, about people who are born without pain and usually die very young, about nerve receptors specific to pain, about the difference between the peripheral pain of a sunburn and the internal pain of an irritable colon, about the different levels of pain tolerance in people, and above all about the importance of the psychological factor and how pain can increase or diminish depending on how it is contextualized. I asked him if someone had studied sadomasochists, and he told me not that he knew of. That's surprising, and another example of the enormous shame with which science regards sex. Despite how interesting it would be to study the physiological and psychological aspects of pain in sadomasochists, and analyze those differences—if they exist—with respect to those who don't practice S&M, few scientists are willing to step out of their academic bubble and ask for financing to do so.

When Pain Produces Pleasure and Takes Away Another Pain

It is startling to realize that scientific research has been investigating pain and pleasure separately, when they have so much in common. The signal from a gentle caress on the nape travels along the same nerves when it suddenly turns into a scratch. However, that signal does not activate the same parts of the brain. The insula, the thalamus, specific areas of the cerebral cortex, and levels of transmitters respond differently depending on the sensory information they receive, but also based on how we interpret the stimulus. It is fascinating. That same scratch can cause pain if it is accidental and unexpected, or pleasure if it takes place in the middle of lovemaking. But the relationship between pleasure and pain is much more profound and complex than that. Some chapters back we looked at a study by the Croatian Sasha Stulhofer that described women who practiced anal sex despite feeling pain because they eroticized it. What does it mean to eroticize pain? Where is the sensory threshold that makes a pinch on the nipple go from sensual to painful? And why does that threshold increase with sexual excitement?

Siri Leknes is a Norwegian scientist who studies exactly which contexts condition an experience to make it more or less painful or pleasurable. One of those determinants is as simple as our expectations. Imagine you are taking part in a television game show and you are watching a roulette wheel spin that will decide if you are going to win a thousand, 2,000 or 5,000 euros. If it lands on a thousand you will be sadder than if you won the same amount when the probabilities were 100, 500 or a thousand. In the same way, Leknes measures the skin conductance and brain activity of people who are about to receive a slight burn from a laser. They tell them that the laser's intensity will be x and that they will show them the value on a screen. While the participants in the study see that the laser's intensity is less than they'd been told, their body physiologically experiences less pain than when it is more than they were expecting, and just by changing the value of x, the pain threshold that each person tolerates also changes.

Siri Leknes is one of the best-known researchers on the study of the relationship between pain and pleasure. I ask her if she has worked with sadomasochists and she says no. I confess my surprise and she admits that a couple of times she's had her financing rejected. I insist that it's not about understanding sadomasochists per se, and that science could use those cases to comprehend mechanisms and then apply them to other disorders, and she responds, "I completely agree, but there are certain taboo subjects, even for science." Despite that, Siri explains that there are various hypotheses about the relationship between pleasure and pain—both physical and psychic—both in masochists and non-masochists. A first lesson: the former don't seem to have any difference in sensitivity from the latter.

That was one of the points that most intrigued me: do masochists have less sensitivity to pain? Does their tolerance get higher? Leknes cites other studies and answers no. Firstly, because pain doesn't diminish with exposure, no one ever gets habituated to back pain, for example. Perhaps we could get used to feeling it, but it never disappears, since it is an essential signal, evolutionarily speaking. And secondly, based on statements by masochists themselves, who do not believe they have a higher tolerance to pain. I remember asking Niki at Paddles, whose response was, "If I stub my toe against the table, it hurts me just like anyone else." Everything seems to be a question of context.

There is a small study published in 2010 that compares pain perception in sadomasochists, controls and fibromyalgia sufferers, concluding that the somatosensory cortex of those with fibromyalgia reacts much more intensely to the first painful stimuli, that the masochists respond equally to the controls, but that if the stimuli are repeated there is a certain mitigation in the masochists. In any case, they have yet to study the physiological differences between regular practitioners of BDSM and conventional people, so let's move on to the eroticization of pain in moments of arousal. It is something that happens to everyone: during sexual encounters, pain becomes pleasure. Why?

A first step toward understanding this is assuming that pleasure and pain are a continuum. To explain it in a very crude way, if we are hungry seeing high-calorie food could make us feel desire, and if we are full, it could make us feel nauseous. To put it another way, the signals of attraction and repulsion are one way to force the restoration of the body's homeostatic balance.

If we arrive home freezing cold in the winter and jump into the shower, the hot water will feel more pleasurable than the exact same temperature in summer. The body seeks to restore its internal balance as quickly as possible, and in that sense, eliminating pain or discomfort generates pleasure. Which is to say, the mere fact of the pain lessening increases pleasure, but how is that interpreted in the brain? One aspect is explained by the neurochemical relationship between the opioid system involved in feeling pain and the dopaminergic system of pleasure. There is one very clear effect: sexual pleasure increases dopamine, which in turn induces the release of opioids generating an analgesic effect that induces relief and increases the pain threshold.

As I've explained earlier, Barry Komisaruk pointed a laser at the tails of several rats and gradually increased its intensity until they moved their tails away. When he stimulated their clitorises at the same time, the rats took much longer to feel the pain and shift their tails. And Komisaruk proved something similar in women, by incrementally increasing pinching pressure on their fingers while they masturbated and seeing that they were much slower to complain. But what I'm saying is nothing new, we all know that pain diminishes in the midst of the sex act. The real enigma is why pain can sometimes generate pleasure.

There are several theories. On one hand, very intense pain can release attenuating endorphins that, once in the bloodstream, generate a feeling of wellbeing immediately after the pain stops. This could explain the state of mental relaxation some sadomasochists experience after a session. On the other hand, Siri Leknes speculates that since they share circuits, the same pain could be activating parts of the dopaminergic system (dopamine is a neurotransmitter related to anxiety and motivation) and increase sexual desire. It wouldn't exactly be an increase in pleasure, but according to Siri it could justify why controlled painful stimulations are exciting. It actually makes a lot of sense if the body knows that sexual pleasure is going to relieve its pain through the mechanisms described earlier, and in fact, in experiments with rats, the ones that were pinched were quicker to seek out mating. They say that the opposite of love isn't hate, but indifference. In much the same way, perhaps the opposite of pleasure isn't pain but insensitivity.

Siri Leknes mentions other possibilities for why pain could increase pleasure, and uses the analogy of spicy food. The spice injures the inside of the mouth and in theory we shouldn't like it: in fact, the plants synthesize that hotness in order to avoid being eaten. But it has the secondary effect of leaving the mouth, lips and body more sensitive to other stimuli and stimulates other sensitive endings, and that is pleasurable for us. In that same way, pain could make us more sensitive to sexual contact. It is well documented that people with more difficulties attaining sexual arousal—like some of the masochists I've spoken with—use pain to increase their sensitivity threshold. Following that same logic, Leknes points out that, in extreme situations, fear and pain can also excite the sympathetic nervous system and favor sexual response, even activating the orgasmic reflex. In my limited experience with the sadomasochistic community, they did mention that it is frequent for some people to only be able to reach orgasm if it is accompanied by physical or psychological pain.

But, speaking of psychological effects, Siri suggests I look at the work of the Australian Brock Bastian, and mentions a very powerful concept: one pain can eliminate another pain. Both physical and emotional. On a physical level, a very simple example is itching. In cases of eczema or intense itch, some therapies use pain to suffocate the itching. The pain is a relief, and in fact that is what we all do when we rabidly scratch at an annoying bug bite. But what is important is that it is not only about localized desensitivity, but to a lesser extent it would also happen if we were to scratch really hard on some other part of our skin. The basic principle is that the feeling of pain is so important that it blocks all the other secondary functions and discomforts. The pain demands concentration from the body; it is saying that something very serious is happening here and requires the body to focus on it. Those who suffer chronic pain that keeps them from concentrating on certain tasks know this very well. Extrapolated to other realms, Brock Bastian has proven that self-inflicting pain can also reduce the discomfort generated by feelings of guilt. That is not new; Brock recognizes that prior studies have proven that self-inflicted pain is a way to alleviate psychological pain in people tormented by guilt, and that physical pain and suffering are used in many religious rites to redeem affliction from guilt. The concept is once again that one pain diminishes another, not only in the moment but also for some time afterward. In a study published in 2003 it was seen that when someone feels sad, the release of μ-opioids is reduced in several brain areas and that physical pain increases the activation of μ-opioid circuits. We aren't going to reduce everything to neurotransmitters, but it has been well proven that physical pain provokes relief of mental pain, and the more intense the pain, the greater the feeling of wellbeing when it disappears. It is highly controversial, but Brock Bastian believes that it is possible that some people start a masochistic practice in order to liberate inner tension. I don't like to use individual cases, because they are often not a reflection of the norm and can lead to mistaken interpretations, but I can't help but cite the case of a dominatrix (a woman who charges for services of domination) who explained to me that she had a client with a micropenis who asked her to humiliate him, mock his penis, hit him and insult him. She herself made the exact same analogy of it being like scratching hard at an itch in order to alleviate it.

But the most interesting thing about Brock's work is that he is analyzing said relationship between pain and guilt in everyday situations of people without discomfort. The two hypotheses he is trying to prove experimentally are as follows: (a) when we feel guilty we are more predisposed to self-inflict some sort of suffering, and (b) feeling pain reduces our sensation of guilt.

In an article entitled "Cleansing the soul by hurting the flesh," published in 2011, Brock Bastian separated 62 students into three groups, asking one group to write for 15 minutes about situations in which their behavior in the past had been immoral, and another group to write about moments when they had behaved well and the third, about whatever they had done the day before. Then he asked each of them separately to stick one hand into a bucket of water with ice at 0 °C for as long as they could. After drying off, he asked them to indicate, on a scale of 1–5, how painful the experience had been. As he expected, those who wrote about an immoral action had their hand beneath the ice for longer and stated that the experience had been more painful. Extrapolating from a study such as this one is tricky, but Brock interprets that in general those who feel guilty are more predisposed to suffer and accept pain. Brock Bastian wanted to prove whether the relationship was also true in the opposite direction, or in other words, if experiencing pain makes us feel less guilty. In another study done in 2012, entitled "Physical pain and guilty pleasures," Brock Bastian carried out a similar experiment with another 58 students: he provoked a feeling of guilt in some of them and not in others, and asked half of them to put a hand in ice water, and the rest in water at 37 °C. Then he told them he had to go look for some papers, and he offered them some chocolates. Those who had put their hands in the ice water without feeling guilty took more chocolates than the rest. From this methodologically questionable study, Bastian infers that suffering a painful experience redeems us of guilt over later pleasures.

Perhaps the unconscious use of pain and suffering as a strategy to calm other physical and emotional discomforts is much more widespread than we think. And maybe also in the sexual context, physical pain can excite because it releases opioids and dopamine, causes wellbeing from endorphins afterward, reduces stress levels produced by cortisol, activates sympathetic nerves responsible for orgasmic response, leaves our body more sensitive to physical contact, or simply because it stimulates us sexually if a scene or activity seems perverse. In each of those senses, except in pathological cases, the label "sadomasochistic" could be interpreted as just a question of degree as compared to conventional people who also have their roles and fetishes, except that those are perhaps more hidden and less acknowledged even by them.

Almost at the end of my night at Paddles I found Haddass in her underwear being whipped very lightly by another member of TES. She had decided to give it a try. After her session she fervidly explained that she'd discovered that she loved the lashes on her rear end and, much to her surprise, she was very excited by other people watching her. Although there were other things that she had thought she would like and they'd left her cold, or even irritated. Overall she was thrilled and tremendously satisfied with the big step she had taken toward getting more in touch with her sexuality.

I'll admit that at some points inside Paddles I was confused and puzzled, and there was no point where I was tempted to join in. I wasn't at all attracted to hitting someone, no matter how pleasurable they found it, and much less to being humiliated or having pain inflicted on me. But I confess that I saw some kinky practices and more vanilla role-playing that I did think could be stimulating if incorporated into another context. It really is an odd world, and in general unfairly characterized by society. I went a few more times—out of scientific interest, of course—to some other BDSM clubs and parties in New York and spoke with many more members of these tribes and subcultures, always as a spectator. I remember the unforgettable moment when Yearofye (a consultant by day and fetishist by night) stood in front of me in her ultra-tight bustier and said, "So, in the end, what are you: dominant or submissive?" My jaw dropped like an idiot and I said earnestly, "Well, to tell the truth, I hadn't thought about it… if anything, dominant… but, can I be neither?" "Not until you try it!" exclaimed Yearofye with a naughty smile. Honestly, don't venture to imagine how you would react in the face of an emotionally new situation, because you don't know. And even though I did see cases that seemed sick to me, I don't think we should make too many assumptions or draw hasty conclusions about people who use pain and games of dominance and submission.

Sexual expression is incredibly diverse, and perhaps what best reflects that is found not in the relationship between pain and pleasure but in fetishisms, those objects, behaviors or situations that have a Pavlovian effect on our arousal.

Fetishists from Head to Toe

John is a married lawyer with two daughters who loves licking women's shoes. I met him at a party organized through the website fetlife.com. He explained that his wife usually wore high heels when they made love, and that they had agreed that he could go to fetishist events every once in a while to pay to play with feet and all sorts of women's shoes. According to John, he has always been fascinated by shoes, and he remembers as a little boy feeling very attracted to a teacher who often wore short skirts and high heels. He doesn't perceive his fetish as an obsession, merely a preference. It is true that feet and shoes help arouse him, but they aren't essential; he can have sex without them, he tells me.

John is a fetishist, but he doesn't have the characteristics of a paraphilic fetishist with a psychiatric disorder. According to the DSM, the diagnostic criterion for fetishistic disorder must include an exaggerated and continued sexual desire for at least 6 months for inert objects or non-genital areas, but, above all, a lack of control despite it causing problems or interfering in the wellbeing of the fetishist or other people. In other words: if it doesn't hurt or bother anyone, fetishism is not a disorder that should be treated. Actually it is comparable to any other hobby that only becomes problematic when it becomes an obsessive addiction.

Tamara has a less patronizing viewpoint. She came to New York from Eastern Europe, started working in a shoe store and, one day, a regular client suggested she be a model at a fetish party. Why not? she thought. As we drank a beer in a bar on the Lower East Side, Tamara tells me that she would sit comfortably in the club in shoes and a miniskirt as men approached her, and if she gave her permission they paid her 20 dollars for every 10 minutes of licking her shoes, stroking her bare feet and—within limits—running them over their faces or other parts of their bodies. Despite the tempting offers, she refused to go any further, both in that club and the one where she worked for a while as a stripper.

The activity in the fetish club was a very easy way for her to earn some extra cash but despite that, she stopped doing it and admitted that she ended up feeling pretty repulsed by it. She recognizes that for some clients it could be just something they have a fondness for, but says there are many others who display pathological behaviors and would spend a ton of money on their uncontrolled obsession. The line between normal behavior and paraphilia is very fuzzy. Fascination for and symbolism with feet is one of the most common fetishes, and it has been present in numerous cultures throughout history. In the *Kama Sutra* and many ancient artistic representations we already see sexual acts involving feet; the Greeks considered that having the second toe longer than the big one was a sign of masculinity; in China they bound little girls' feet to keep them small and deformed, and female shoes have been a clear sign of eroticism since the Middle Ages. Academics interpret that for some the curves of the feet could be sensual; since they are such an overworked part of the body, keeping them soft and in good shape is a sign of youth and high class, and that some people with insecurities about their body or their genitals could unconsciously shift erotic attention to their feet, or simply that the sensitivity of the soles of the feet can be a highly erogenous zone. In the article "Erotic and sadomasochistic foot and shoe," Frenchman P.H. Benamou describes the case of a New York woman who would rub her feet in food so that her two little dogs would carefully lick them and heighten her sexual arousal, and he explains that since Ancient Egyptian times several queens and courtesans have had servants who stroked and tickled their feet. This all suggests that looking at the feet as an erogenous zone is not at all "abnormal."[2] The larger question is why, in some people, this becomes an obsession.

John says that he remembers masturbating while thinking about his teacher in high heels, and he thinks that perhaps that could have contributed to his fetish, but he knows fellow foot-fetishists whose obsessions appeared in adulthood. Fetishists are a community that is constantly experimenting and trying new things, and looking merely at their childhood and adolescent experiences to explain their preferences would be absurd. But there is much scientific literature that suggests that the earliest references for sexual pleasure can have an important impact on sexual development, and that classic Pavlovian conditioning can contribute to reinforcing what we consider erotic. And that holds true for all of us, not just for fetishists.

We Are All a Bit Fetishistic

In the early twentieth century, the Russian Ivan Petrovich Pavlov would ring a bell right before feeding his dogs. He repeated this until the dogs associated the two stimuli, and they unconsciously learned that after the bell came the food. After some time, the dogs would start to salivate just at the sound of the bell. Those were the first studies that prove classic conditioning in animal behavior.

Jim Pfaus's group carried out something similar at Concordia University in Canada. As we explained in Chap. 1, researchers separated virgin male rats into two groups. They had one group have their first sexual encounters with females who wore leather jackets and the other with "naked" females. After several sessions they allowed them all to mate a few times with naked ones.

---

[2] The concept of normality is different if we look at it from a sociocultural or Darwinian perspective. For example, several authors interpret female exhibitionism as a natural instinct to show one's genitals for reproductive ends when in the fertile phase, or to foster relationships of competition or comradeship between females. While it shouldn't be viewed as something abnormal from a Darwinian point of view, excessive exhibitionism is a paraphilia when it is problematic for third parties and the individual themselves. Again, the nuance lies in the lack of control, in compulsive desire and its negative effects.

They let some time pass and they then exposed the same males to females with jackets and without, and they observed that those who had had their first copulations with naked females showed no preference for either group, while the "fetishistic" males were clearly more drawn to the rats in leather jackets. In fact, when some individuals were only given the option to mate with rats without jackets, they were much slower to begin coitus and needed more thrusts in order to ejaculate. Some couldn't even achieve it. According to Jim Pfaus, the males had associated the jacket with sexual reward, and without it they had arousal deficits and failures in the activation of their sympathetic nervous system needed to ejaculate.

Comparable experiments have been repeated with neutral stimuli, like lighting or almond essence, but Jim wanted to find out if even repulsive agents such as cadaverine—a smelly substance produced by rotting flesh—could turn into exciting elements. His experiments showed that the virgin male rats whose only option was mating with females stinking of cadaverine, as adults they lost their natural repulsion for said substance. Which is to say, that even a noxious stimulus could become eroticized. Rats don't always react like humans, and less so when we are dealing with something as culturally influenced as sexual expression, but it isn't far-fetched to think that sexual experiences that leave a strong impression—because they are the first ones, or particularly intense—can unconsciously condition some preferences in adulthood. It is something that is extremely difficult to study empirically in humans, and it can only be done through indirect studies, like the one published in 2010 which suggests a relationship between a father's body hair pattern and his daughters' attraction toward more or less hairy men.

Without a shadow of a doubt, initial conditioning is important in sexual aspects, but the Pavlovian conditioning through repetition has even more of an impact. There are many studies that show that learning to masturbate in a certain way can mark us for the rest of our lives, and that our first sexual experiences leave a mark. They are not necessarily determining factors, but they do have an influence. However, what's most interesting is seeing how pleasure conditioning factors are always present.

In a series of classic experiments carried out in the 1970s, Kantorowith asked several volunteers to masturbate while he showed them three neutral photographs at different moments of the act: the first image when they began masturbating, the second in the phase before orgasm and the third in the refractory period. After repeating the experiment several times, the neutral image shown right before orgasm already would provoke a partial erection, the one shown at the start had no effect and the one seen during the refractory period reduced arousal.

In a very similar experiment some researchers in California asked 18 college students between 18 and 23 years of age to masturbate (they paid them 25 bucks) eight times in various spaced-out sessions. In each session they would show them photographs of one type of female body for various seconds before their arousal began, another type further into the masturbation and a third in the refractory phase, after ejaculation. When, 3 months later, and in the same experimental conditions, they again showed them comparable images but this time measuring their penile response, they were still more aroused by the body type shown during masturbation, and less so with the final type.

It is interesting to point out that this conditioning doesn't only influence what we consider more or less attractive, but it also modifies certain behavior patterns. In 1974, Silberg and Alder published, in *Science*, another very curious experiment: they allowed a group of control rats to copulate freely with receptive females for 30 minutes. They let another group start to copulate, but after the seventh intromission they took the female out of the cage (I should mention here that in the case of rats the male chases after the female, manages to penetrate her, but she escapes, and it goes on like that several times before, in one of his intromissions, he ejaculates). With a third group they also took away the female at the same point, but without counting the number of intromissions. After several repetitions, the second group learned to ejaculate prematurely before the seventh thrust, and the third to copulate more quickly and eagerly. So if you masturbate as fast as you can while hiding in the bathroom, be careful.

Since Pavlov and the radical behaviorist B.F. Skinner, conditioned reinforcement and the association of neutral stimuli with varying emotional states of pleasure or aversion are well documented in animals and humans, and without a doubt, something so basic as sexual development must follow those same parameters. They have even explored neurochemical mechanisms: if we inject morphine or oxytocin into a male rat after mating him in the presence of a specific stimulus, an association is produced, which indicates that opioid receptors and oxytocin are involved. And if we repeat the experiment but blocking those same receptors with Naloxone, the conditioning disappears.

Our behavior in non-pathological conditions is infinitely less flexible than that of a rat. But it has been proven repeatedly that there is a critical period in the development of sexual conduct, associated to our first desires, masturbations, orgasms and coupled acts, that can condition our preferences for habits or physical characteristics with our future partners. The degree to which this is more or less flexible and whether it is behind some fetishes is something that, to date, no scientific study has established beyond speculations and some anecdotal cases.

In any case, the lesson here is that, in our species, innate determinants perhaps play a role in our most fundamental sexual preferences, but after that they are completely sculpted by experiences, and that the concept of fetishism is very broad and to a certain point a question of degree. There are those who defend the idea that something as common as a male preference for huge breasts is actually a fetish created by Western culture, socially accepted, and in some individuals exacerbated by masturbatory practices. The underlying logic is that if, for whatever motive, exposure to something or someone provokes greater desire or pleasure, that can generate an unconscious preference that we want to repeat and if we reinforce it with more pleasure at each new exposure, it can remain a mere preference, become a healthy fetish, or an obsessive one.

Sexual Fantasies: Inhibiting Sins in Our Thoughts Leads to More Sins of Word and Deed

Imagine you have a sexual fantasy you consider inappropriate, that distresses you and you want to eliminate from your indomitable thoughts. It could be some homosexual desire, attraction to a family member, or extreme masochistic images that your value system interprets as incorrect. If you struggle intensely to suppress your fantasy, will it remain reverberating in your head in a some kind of rebound effect? That is the profound and highly interesting question that Laura Sánchez, a sexologist and researcher at Spain's University of Almería, is analyzing experimentally.

Sexual fantasies are a fascinating field of study. Sometimes controlled and other times spontaneous, some allowed and others unconfessable, sexual thoughts emerge in our minds through mysterious inner processes, usually awakening sudden arousal and, in some cases, bewilderment and discomfort.

Sexologists view them as a natural and healthy process associated with greater sexual satisfaction. There are very common ones, like an urgent desire to make love with our partner, visualizing the acts we'd most like to do, or, 2 minutes later, imagine ourselves having an affair with a coworker. But in the infinite diversity of sexual fantasies we also find having relations with someone of the same gender, being dominated, taking part in an orgy, imagining what the person sitting across from us on the subway would be like in the sack, watching or being watched by strangers, having sex dressed up as a nurse, clown, teenager or any other stimulating fetish, and having sexual adventures of all sorts.

Scientifically analyzing sexual fantasies is terrifically complex, but experts believe that they are a key factor in sexual development and that understanding them is vital to understanding ourselves, both on an individual level and in order to discern biopsychological and cultural influences. One of the most interesting aspects is that, despite the fact that men continue to have more fantasies than women, the gap between genders could be narrowing. In 1984, researchers at Columbia University carried out an extensive study analyzing the frequency and type of fantasies among college students, which was repeated in 1994 with the objective of bringing it up to date and comparing data. Two of the main conclusions were that among young women the frequency of sexual fantasies had increased, and that they were usually more in keeping with their experiences. This second finding was in agreement with other studies that showed that, within the enormous individual diversity that exists, women's fantasies generally have more private content and men's are more exploratory. The study in the mid-1990s showed that there were no important differences between men and women in the most common fantasies, such as touching and caressing nude bodies, sex with strangers, or trying new positions. However, the men had more visual fantasies like watching their partner undress or watching themselves having sex, being with more than one woman at the same time, practicing anal sex, having relations with a much older or much younger person. As for the women, they more often preferred masturbating their partner than being masturbated. The largest difference between men and women was the fantasy of having sex with a virgin, found in almost half of men and less than 10% of women. The only fantasies that were more common in women were homosexual ones, having sex in front of other people and being forced sexually.

General surveys are useful for finding patterns, comparing cultures or analyzing tendencies, but once those have been established, what's really interesting is analyzing specific aspects. For example, a study published in 2012 followed the sexual fantasies of 27 single women in their 20s, for 30 days, observing that they had significantly more fantasies when ovulating, and that during their fertile period they had proportionately more fantasies involving men.

Another study suggested that the types of fantasies that gays and straights have are not very different, and another that they weren't that different between single people and married ones. In fact, this last study realized by Hicks and Leitenberg was specific to sexual fantasies and couples, and 98% of surveyed men and 80% of women between the ages of 18 and 70 (average age being 33) admitted that they'd fantasized about someone outside of their relationship in the last 2 months. Perhaps the most curious finding of the study was that the number of sexual partners in the past and having been unfaithful were not factors that increased the number of extramarital fantasies in men, but in women they clearly were.

Specific studies with sexual delinquents have shown that their acts and aggressions are very clearly related to their fantasies, but there are also an enormous number of "paraphilic" fantasies that are never carried out. With pedophilia, for example, approximately 15–20% of college-age men admit to having had fantasies about girls younger than 17 and 3–5% with those younger than 12, but most of them would not act them out even if they knew with absolute certainty that they wouldn't be caught. The figures differ among studies depending on how the sample group is selected, since many people lie in these types of surveys, but they do seem to agree in that having a fantasy does not necessarily imply the intention or willingness to carry it out.

In the largest review of scientific literature to date, published in 1995 by Laitenberg and Henning, it was concluded that 95% of men and women have had sexual fantasies, that men have more than women during masturbation but not during intercourse, that with age the number of fantasies decreases, that unlike what Freud said they are not a consequence of the lack of sexual activity since people with the most fantasies are those who have the most sex, that scenes of dominance and submission are really very common, that there is a clear reinforced conditioning effect of actions that one has done or seen while masturbating, and that about 25% of people are bothered by some of their fantasies.

In that sense, researchers Nieves Moyano and Juan Carlos Sierra, at the Psychology Department at Spain's University of Granada, are carrying out very interesting work on the effects of negative fantasies. Juan Carlos explains that generally sexual fantasies are considered positive and their presence is associated with good indexes of health and sexual satisfaction, and that overall this continues to be true, but that not enough attention has been paid to the negative effects of fantasies that are experienced as emotionally disturbing. Nieves and Juan Carlos have been studying sexual fantasies for years in order to evaluate the Spanish population against international scales, and they realized that a certain type of fantasies seem to reduce sexual desire. They searched for scientific bibliography on the adverse effects of fantasies and only found one Canadian study done in the 1990s, so they began a large study with more than 2,000 people between the ages of 18 and 80 on the nature and effects of negative sexual fantasies. Nieves Moyano showed me the preliminary results in a meeting of the International Academy of Sex Research in August 2012 and, months later, when I spoke with Juan Carlos Sierra, the work had yet to be published but they gave me advance results showing that homosexual fantasies are those that men most frequently experience as negative, while women are more tormented and guilty about those involving submission and having sex against their will.

Obviously, classifying a fantasy as "negative" is something that depends on individual and cultural criteria, but the most interesting of Nieves and Juan Carlos's conclusions is that certain "undesirable" fantasies do indeed have an adverse impact on sexual functioning, generating blockage and loss of desire. In their large-scale study, Nieves and Juan Carlos are also analyzing what sociocultural and personality variables influence the perception of fantasies, and above all which would be the best therapy for reducing the related anxiety. The most logical option seems to be trying to substitute or restructure the fantasy, but here is where the research done by Laura Sánchez at the University of Almería dovetails perfectly, warning us that that could prove counterproductive.

Try Not to Have Obscene Thoughts Involving a White Bear

Laura's work on thought suppression is a continuation of Daniel Wegner's. In his seminal studies, Wegner asked a group of volunteers to express out loud every thought that passed through their heads for 5 minutes, but to try with all their might not to think about a white bear. Then he asked them to again express every thought for 5 minutes, but this time without suppressing the white bear. In both situations they were to ring a bell every time they thought or spoke about the white bear. Wegner repeated the same process with an identical group of students, but changing the order of his instructions: in the first 5 minutes they were to express every thought without having to suppress the white bear, and in the next 5 minutes doing so.

When comparing the results, Wegner observed something very notable: in the students who had first made an effort to suppress their thoughts about the white bear, when they could think freely about whatever they wanted, the white bear appeared with much more frequency than in the second group before the suppression. The conclusion was obvious: trying to suppress a thought makes it reappear more often. Which is to say, if we want to stop thinking about something that bothers us, disgusts us or makes us uncomfortable, struggling to suppress it has a snowball effect that only makes it more insistent. Wegner's experiment has been replicated several times in different contexts and some think that it could be one of the mechanisms involved in creating obsessions.

Would the same be true of sexual fantasies? In a first study with 80 volunteers, Laura asked half of them to choose a word associated with a sexual practice that they considered "appropriate" (that varies from person to person, but examples could be "intercourse" or "kiss") and the other half "inappropriate" (like "orgy" or "exhibitionism"). Then she asked them all to create a sexual fantasy based on that word, and to think about it for a couple of minutes. Then she had them carry out a distracting task, like associating figures in pairs, such as lion/lion tamer, and then half of each group received the instruction to suppress any thoughts related to the sexual fantasy while the other half were asked to focus on it. Later, Laura gave them a word association test to detect how present the fantasy was in their minds. When in September of 2012 I met Laura Sánchez at the European sexology conference in Madrid she insisted on this word association methodology. Laura explains that people usually lie in surveys about sexuality—and even more when the subject is inappropriate sexual fantasies—so she doesn't consider it rigorous enough to trust the truth of the volunteers' responses. According to her, tests that analyze response times, attitudes and word associations are much more revealing.

That said, Laura's results did indicate that in every condition there was this "rebound effect" that showed that trying to inhibit a fantasy made it appear more frequently later. But, in addition, Laura compared the effect of sexual words and non-sexual words (like "travel," "fight," "laugh," and "win") and saw that the rebound effect was significantly greater in the sexual ones. Which is to say, that trying to suppress a sexual desire generates more of an obsession than trying to inhibit other pleasures, like for example a certain food. In this first study there were no important differences observed between fantasies that someone felt were "appropriate" and "inappropriate," but Laura Sánchez assures me that in the forthcoming wider results they have observed that the effort to suppress inappropriate fantasies generates more of a rebound effect than appropriate ones. She clarifies that the terms are always relative to each person in particular, and for example "anal," "masochism" and "voyeurism" can be inappropriate for some and appropriate for others. Laura even used fMRI scanners to analyze the activation of brain regions involved in the suppression of fantasies, again confirming their behavioral results: even though it seems paradoxical, if we spend 5 minutes trying to suppress a sexual fantasy it will show up more often in our minds than if we spend 5 minutes concentrating on it. And according to Laura, "even more if we are dealing with arousal by thoughts that are frowned upon by the society we live in."

Laura's results could have important repercussions on many levels. In the more everyday aspect, it suggests that the best strategy for getting something out of our heads is letting it appear and disappear without trying to suppress it, but in a more academic environment it could be enormously useful for understanding the development of sexual fetishes, preferences, habits and obsessions. It also reinforces the idea that, in our unconscious, the sexual has more power than the neutral and the "incorrect" weighs more heavily than the "correct," but above all it has implications in the therapies for paraphiliacs and sexual aggressors that is usually based on trying to reduce the frequency of fantasies, and in some cases could be counter-productive.

It is undeniably a riveting subject, one that forces us to assume that trying to control our minds is much more difficult than controlling our actions, and that we should be more acquiescent with our Judeo-Christian idea of sinning in "thought, word or deed." We can just accept defeat with the first one, and if we don't, then we run the risk of increasing the next two.

Fantasies of Being Taken Sexually by Force

I can't end this chapter without mentioning the example that is most emblematic of the difference between thought and inclination: as surprising as it may seem, according to a review of the 20 main studies carried out on rape fantasies, published in 2008 by Joseph Critelli and Jenny Bivona, of the University of Texas, 31–57% of women have, at some point, had fantasies in which they are forced to have sex without their consent, and it is the primary fantasy of 9–10% of them. In 2009 the same authors published a second study with 355 female college students concluding that 62% had had this type of fantasy, at an average of four times a year, and that 45% of those with rape fantasies defined the experience as erotic, 46% as erotic and repulsive, and 9% as totally repulsive. Even though, on the face of it, it makes no sense that a woman would feel desire to be forced to do something she doesn't want to, and that in real life is traumatic, the authors don't find it paradoxical.

The interpretations are speculative, but in their bibliographical review, Critelli and Bivona describe empirical data and explanations published by psychologists and researchers on rape fantasies.

The first hypothesis was put forth by Deutsch in 1944 and referred to masochistic desire being more frequent in women than in men due to their greater physical weakness and unconscious search for strength in the male. According to Critelli and Bivona, the empirical data rules out this hypothesis as a generalized explanation. Masochism could play an important role in only few cases.

The most frequently cited explanation is that some women want to have sex but feel inhibited by religious, cultural, marital reasons, or simple don't want to be seen as promiscuous, and they fantasize about being forced because it allows them to experience sexual relations without feeling guilty or responsible for them. According to data from Critelli and Bivona's review, women who have been raised in more sexually repressive environments have more rape fantasies, but they are careful to point out that the theory of avoiding sexual guilt can be valid in certain groups of women, but doesn't explain the common behavior of most women who say they have these fantasies in addition to their consensual relations.

Another hypothesis says that it is simply a consequence of having a more open attitude toward sex. Women who have more diversity of experience and partners usually have a wider range of fantasies, including ones where they are forced. The authors of the study consider that while that is logical, it doesn't explain why force is introduced as an element of pleasure.

Some investigators also suggest that the desire to be forced can represent a wish to be intensely desired, to be sexually attractive to the point of making a man lose control. This hypothesis is not yet tested, but it is consistent with the fact that "feeling desired" is one of the factors that women most deem exciting. There are also anthropologists who consider it a consequence of our sexist society, and evolutionary biologists who think that in nature rape exists and is favored by natural selection, and we could have inherited a certain instinct of permissiveness. The last hypothesis put forth by the review is that the anxiety, fear and anger provoked by rape can activate the sympathetic nervous system and increase sexual desire, and that the women who have experienced more pleasure in moments of rougher sex can fantasize with even more extreme situations.

Evidently, none of these theories will give a complete answer to the intriguing phenomenon of rape fantasies, but it again shows the vast and fascinating complexity of our mind, and how difficult it is to understand it based merely on an analysis of our behavior.

16

Disorders of Obsession, Impulsivity, and Lack of Control

Melissa's downtime is all about sex. At 45, with no kids and several relationships in her past, she says she has given up for the moment on having a stable partner. She prefers to satisfy her intense sexual desire with a list of lovers that she manages in a very organized way. She meets most of them on the Internet. She asks for photos and always demands real names and links to work websites, Linkedin or Facebook. Yet she never gives her personal information straight away, and only sends photographs where you can barely see her face. If a guy doesn't accept her conditions, no problem, she's got plenty of other candidates. When a woman in New York wants sex, she can have as much as she likes.

Generally Melissa has the first date with a potential lover at a bar near her house. They have already discussed the conditions and explicit details via email, and if nothing unexpected happens, they will soon head up to her apartment to have sexual relations for approximately an hour and a half. When the man leaves, Melissa pulls out her notebook and jots down his name, telephone number and her impressions of the encounter. That will allow her to choose which lover to call on which day depending on what she's in the mood for. Some are more affectionate, others more perverse, some are young, some are older, they have different body types, some are good conversationalists, and others go straight to the job at hand and leave without being a bother. Obviously many of them do not pass the test, but if she enjoyed being with them, Melissa will usually repeat periodically until one of the two grow tired of it. Some guys only last for three or four encounters, others several months, and she's been getting together with her longest lasting lover for more than 3 years. Since he is married, that works just fine.

Melissa doesn't allow them to be the ones to contact her, and demands they respond immediately to her messages. If they take longer than 15 minutes, she'll already be setting up another date. She usually meets up with someone every other day, but admits that there've been crazy weeks where she's slept with up to ten different people. When she's left wanting more after a date, she calls up one of her trusted lovers, who she knows will hop in a cab and be at her house in 20 minutes. Sometimes she goes to private parties in clubs where she practices group sex. She loves it. She also has sex with other women and ménages à trois, as well as being game for trying any new sexual practice that catches her fancy. It is a mix of curiosity and excitement. When she travels, she really gets off on sleeping with local people and having hidden sex in public places. She has had several lovers at work, but she tries to avoid it because of the obvious downsides. She does acknowledge that she needs sex almost daily, and it's only been when she's sick that she's gone several days with just masturbation.

Some of her girlfriends tell her she has a problem. She denies it. They tell her she should slow down, for her own good; they tell her she's a sex addict. Is she? Psychiatrist Richard Krueger and therapist Meg Kaplan, both experts in hypersexuality and paraphilias and researchers at Columbia University, tell me she isn't. If Melissa doesn't lose control, if her sexual hyperactivity doesn't cause her emotional, work or economic problems, and she doesn't feel any desire to change her lifestyle, her situation is not pathological. Unlike addictions to drugs, food or gambling, with sex addiction it's not the quantity that's determinant. Consuming a lot of alcohol leads to obvious physical and social problems, but having a lot of sex doesn't, necessarily. The key factor is not what society establishes as "normal," but what results in a loss of control and causes serious problems in everyday life. Melissa admits that sex obviously interferes in her free time, but not in her work or in her friendships. For her it is more of a hobby, one that might just disappear some day on its own. She is not obsessed.

Hypersexuality Is Not an Addiction

According to the criteria that Dr. Richard Krueger was elaborating in order to, possibly, specifically include "hypersexuality disorder" in the next edition of the manual of mental illnesses, the DSM-V, someone could be diagnosed as hypersexual if they meet the following conditions: 6 months of intense and recurrent sexual fantasies and impulses or behaviors associated with three of more of these five criteria: (*a*) interference with other activities, goals

16 Disorders of Obsession, Impulsivity, and Lack of Control

or important obligations, (*b*) the appearance of sexual need in response to emotional states such as anxiety, depression or irritability, (*c*) appearance of sexual need as a reaction to stressing situations, (*d*) lack of impulse control despite trying, and (*e*) maintaining sexual behaviors without considering the physical or emotional risks to oneself and others.

Doctor Krueger tells me that he's uninterested in the pseudoacademic debate about what name to give to media-popular "sex addiction." He prefers the terms "hypersexuality" or "compulsive sexual behavior" because they are broader and have less connotations, and he doesn't like the term "addiction" because it doesn't include some of his patients whose uncontrolled desire doesn't stem from learned conditioning, but rather from illnesses or brain injuries. He assumes that the concept of addiction is easier for the general public to understand, but he is bothered by mass media's banalization of the subject, and its labeling of many people as addicts and hypersexuals who really aren't, since, according to him, that creates social pressure and excessive concern in patients who then come to him with unfounded anxiety. Another psychiatrist who prefers to remain anonymous told me that in the United States treatment for sex addiction has become a lucrative business.

Joe does meet the criteria to be considered a sex addict. He arrived at the offices of Dr. Kaplan and Dr. Krueger almost bankrupt from prostitution and with a wrist injury from so much masturbation. When I show my surprise, Meg insists that it's not that strange to find men who masturbate compulsively more than ten times a day. Joe is never satisfied. Half an hour after being with a prostitute he is looking for sex again. He spends hours and hours watching porn on the Internet and trying to find women online to meet up with for sex, even though that's almost impossible, because Joe is not attractive or socially adept. Would he have these problems if it were as easy for him to find sex as it is for Melissa? "That is an unanswered question," Dr. Krueger weighs in, avoiding controversy, but citing some studies that indicate that, indeed, frustrated desire can generate anxiety and become an obsession.

Doctor Krueger adds that "apparently the number of sex addicts is on the rise, but there aren't really reliable statistics on the current situation and much less on the situations decades ago," and he clarifies that there are different subtypes of hypersexual behavior according to the predominance of practices such as compulsive masturbation, cybersex, erotic phone lines, pornography consumption, regular visits to strip clubs, prostitution or consensual sex with adults; this last type is most common among homosexuals and the very, very few women who have pathological hypersexuality.

During her 30 years of research and therapy, Megan Kaplan has treated many hypersexual women, but she assures me that they are clearly an exception. "There are women with exacerbated, constant desire linked to a loss of control, but most cases are the consequence of neurological problems such as bipolar disorder, psychological disorders or the consumption of medications," insists Dr. Kaplan, adding that "when the root of the dependence is more behavioral, we usually find it within the context of a specific relationship and not in a wild search for promiscuity, like in men." Megan tells me of women who consume excessive pornography, others who like to show themselves on the webcam, or those who have very high sexual desire and could fit the classic concept of "nymphomaniacs," but she insists that they are rarely extreme cases, that most shouldn't be seen as problematic, that they are usually transitory situations and that the cases of "female sex addicts" are anecdotal.

Here I should explain that some cases of hypersexuality are due to purely neurophysiological disorders that distort the balance between desire and inhibition. For example, in some cases of epilepsy or accidents that injure areas in the prefrontal cortex, or with past lobotomies, they have observed damage in the brain's inhibition mechanisms that lead to hypersexuality. In those cases, sexual impulsiveness is not brought on by an exaggerated increase in desire, but because there is no "braking" mechanism. On the other hand, some prescription drugs for Parkinson's that increase dopamine levels, or in some patients with Tourette's syndrome who have excessive endogenous levels of dopamine, or in methamphetamine addicts whose consumption is linked to sex, hypersexuality is directly produced by a neurochemical increase in desire.

Obviously sex addiction can also originate in repetitive use and habitualization to a pleasurable behavior, like any other obsessive-compulsive disorder or behavioral dependence, but it is important to stress that for that to happen there have to be aggravating factors.

Sex in and of Itself Does Not Cause Addiction

The basic model of the neuroscience of addiction is the following: consuming alcohol, cocaine or tobacco increases, through various paths, the amount of dopamine in the nucleus accumbens, the pleasure-reward center located right in the brain's limbic system. This dopamine is what generates the sensation of wellbeing, euphoria and motivation. But if the practice is repeatedly frequently, the neurons of the nucleus accumbens "get used to" those unusually high levels of dopamine and synthesize more dopaminergic receptors in the synaptic areas. That leads to normal levels of dopamine, such as those offered by food, exercise or sex, being less satisfying, and that generates a state of anxiety in which the only thing that brings back our neurochemical wellbeing is the substance we have become dependent on. Additionally, neuroscientists have realized that the continued consumption of drugs provokes less activity in the areas of the prefrontal cortex related to inhibition. This is important and implies that addicts are addicts because the changes in their limbic systems make their desire to consume increase, but also because deteriorations in the prefrontal cortex lessen their capacity for self-control.

This refers to substance addictions, but for some years now neuroscientists have also seen that behavioral addicts suffer similar changes in their brains. Although there is an important nuance: sex and gambling will never release the same amount of dopamine as, say, cocaine or heroin, and are unlikely to create lesions in the prefrontal lobes. That means that having sex, shopping or exercising at the gym are not going to lead to an addiction in and of themselves. They could create a habit that evolves into a custom and even an obsession, but rarely will they turn into a physical dependence or addiction. This only happens in people who already have prior damage in the inhibitory areas of the prefrontal cortex, obsessive-compulsive disorders, or when sex is accompanied by drugs that increase dopamine levels and the brain associates the two stimuli. But, in general, if there are no prior physical, mental or psychosociological problems, having a lot of sex does not lead to addiction. And when those previous problems that provoke an obsessive, compulsive search for rewards exist, sex is usually a consequence and not the cause. In fact, in the scientific bibliography we find cases of sex addicts who are also compulsive shoppers, gamblers or have obsessions like compulsively washing their hands.

Here we should make special mention of the consumption of pornography online. According to experts, it is an especially delicate problem because facility of access and growing social isolation make it very easy to enter into a downward spiral. Several studies indicate that the cases in which watching porn online goes from a routine to an obsession are drastically on the rise. However, experts also assure us that in general it is not that hard to break this cycle with cognitive and behavioral therapy and find activities to substitute this dependency, and that in only very few cases is it necessary to resort to pharmacological treatments that reduce desire. Again, the really serious problems arise when there is a pre-existing addictive personality or psychological disorder.

In addition, many experts believe that there aren't really that many men addicted to sex as some statistics reflect, and they criticize the mass media for labeling cases like those of Michael Douglas or Tiger Woods as addiction. For those experts, on occasions labeling oneself as an "addict" is a way to justify a lack of control over a behavior that is considered socially inappropriate, and of showing the resolve to rectify it by going to a rehabilitation clinic. Well-known sexologist Stephen Levine published a very revealing study on this in 2010. Levine analyzed 30 cases of married men who had gone to rehab for sex addiction. Of those 30 patients, two were men who had done nothing more than masturbate with pornography, breaking the rules imposed by their strict wives. Five ended up at the clinic when their wives discovered sexual secrets in their past, like having had lovers or frequenting prostitutes. Four were tormented because they visited erotic chat rooms and felt they had been unfaithful to their wives. Seven simply had unconventional sexual fantasies and were closer to the definition of paraphiliacs than addicts, and five were accused of hypersexuality because they often went to strip clubs with their buddies. Over the phone, Levine tells me that "none of those categories should be classified as sex addiction, and only in seven of the 30 patients was there a downward spiral and dependence on pornography and illegal sex observed."

By no means do I wish to downplay the importance of the problem of sex obsession, because it undoubtedly exists and is radically on the rise in the case of dependence on online pornography, but I do want to stress that hypersexuality should not be evaluated in function of the amount of the desire but rather the lack of control—the big difference between Melissa and Joe—and especially if it has negative consequences on the individual or those around him (or her). Hypersexuality can be very serious when linked to paraphilias. I will talk about them and the types of therapies used to treat sexual obsessions later in this chapter, but first I would like to go a little further into the neuroscience of the balance between desire and control, between arousal and inhibition, because when it comes down to it, that's what defines our impulsiveness and decision-making, and not only in a sexual context.

The Neuroscience of Impulsiveness and Lack of Control

Impulsiveness and self-control are studied in diverse scientific disciplines that try to explain, for example, why Marcos is on a diet and can't stop eating, why Silvia ends up kissing someone who isn't her husband after a few glasses of wine, or why a football player loses control in a moment of tension and harms an opponent. Those examples are taken from everyday settings, but actually the underlying behavioral mechanisms aren't that different from those in much more delicate situations like an addict who can't stop using cocaine, a father who abuses his son or a violent person who commits a crime in a fit of rage.

Self-control is fundamental to our lives, and science is very interested in what leads us to not be able to control ourselves in the face of an impulse or immediate gratification, despite knowing the negative consequences or that it will rule out greater rewards in the future. Science assumes that our behavior is the result of the balance between desire and inhibition, and disrupting any of those two factors can lead to serious, undesirable actions. We have already seen causes that increase desire, but in our daily lives, what does our ability to restrain ourselves—when faced with a dessert, a provocation or a kiss—depend on?

In recent years, several studies have confirmed that the dorsolateral prefrontal cortex (or DLPC) is clearly implicated in impulse control. A study of ex-smokers who hadn't smoked in 3 weeks observed that those who had greater activity in their DLPC had less relapses, and another study published in *Nature Neuroscience* in 2010 used transcranial magnetic stimulation to block activity in the dorsolateral prefrontal cortex of several healthy volunteers and observed that they made much more risky and impulsive decisions. An interesting investigation in 2012 also compared activity in the anterior cingulate cortex (AAC), an area that connects the limbic system and the prefrontal cortex and is involved in recognizing whether something is incorrect, and observed that when control was lost, activity in the AAC did not change. This indicates that the action continued to be perceived as negative, and what was decreased was the braking mechanism offered by the DLPC. It is clear that the capacity for control is in our brains, but what blocks it?

On one hand we have already seen that mental illnesses, brain injuries and drug use can cause direct damage to the structures related to self-control, and we know that when we're drunk activity in the prefrontal cortex is also considerably affected. But it has also been seen that stress, physical arousal, depression, lack of sleep and even tiredness can also clearly inhibit this activity. Being worn out or in a stressful situation lessens our inhibition.

On the other hand, there are determinants of a more psychological nature, for example discomfort. It has been seen that the mere fact of being angry or having negative feelings make us more permissive with ourselves and more apt to overstep boundaries in things that have nothing to do with the cause of our anxiousness or unease. We are worried about one thing and we uncontrollably blow off steam with another. That can be interpreted in two ways: one is that the negative feeling requires a lot of mental effort and competes with our requirements for inhibition, and the other is that when we are uneasy there is a greater tendency to seek out quick alternatives that restore a sense of wellbeing. Another effect is the abstinence violation effect, according to which breaking an inhibition is liberating and we are more likely to relapse. A study with people on diets observed that those who were made to eat an ice cream as proof of abstinence violation reported that when they went back home they ate much more.

Again, control can be lost through an increase in desire or a letting up on the brakes, and that happens to us in our daily lives but is also very relevant in the case of paraphilias. For example, in Chap. 3 we saw that many men are pedophiles (they feel physical attraction for prepubescent boys or girls), but few of them are pederasts (commit sexual aggressions against minors). Here there is clearly a mechanism of inhibition involved, which many adults are capable of controlling without the slightest problem. But, as I mentioned earlier, it is believed that pederasts who regularly abuse children, apart from an abnormal paraphiliac orientation, have lesions that cause less activity in the prefrontal cortex and therefore a limited capacity for self-control. On the other hand, the rest of pedophiles who do not transgress must be careful not to expose themselves to situations of stress, drunkenness or any other circumstance that could alter their brain and make their desire supersede their inhibition. This is a subject that is being investigated and about which little is known, as is the case generally with the strange world of paraphilias.

Paraphilias: When Science Articles Are Stranger than Fiction

Walking through the poster session of the International Academy of Sex Research conference, I came across one presented by the Brazilian Janaina Reis: "Formicophilia: A case report and literature review." The poster described a single 53-year-old man from a rural area of Brazil who at the age of 14 saw an ice cream covered in ants on the ground and was curious how it would feel on his penis. He tried it, he liked it, and he repeated it several times, growing aroused and even having some orgasms. If that had been the end of it, it would have been anecdotal, and according to the definition of paraphilia from the manual of mental disorders as "sexual practices that are desired or realized compulsively and which respond to unusual and socially unacceptable stimuli [...] and cause discomfort to the individual or harm to others," perhaps the man wouldn't have earned the label of zoophiliac. But he took it a little further.

According to the poster, the individual in question is retired, Catholic, volunteers as the secretary of his town's parish and has good social and family relationships, but for a long time had sex with dogs and goats, admits having been licked and penetrated by a dog, and still often puts ants on the tip of his penis. Now he uses fire ants (Solenopsis), whose bite is more painful.

Researchers explain that they treated him with psychological therapy and 20 mg daily of Paroxetine, and that his zoophilic symptoms lessened.

Paroxetine is one of several pharmacological treatments used to diminish the sexual impulse in pederasts, rapists and extreme cases of hypersexuality. We will discuss it further on.

The DSM-IV classifies the following as paraphilias: exhibitionism, fetishism, frotteurism (touching or rubbing up against someone without their consent), pedophilia, sexual masochism, sexual sadism, transvestism, voyeurism and the category of NOS (not otherwise specified) paraphilias, which includes telephone scatalogia, necrophilia, zoophilia, autoerotic asphyxiation, partialism (sexual obsession with one specific part of the body) and all the subtypes you can imagine. There are online lists of hundreds of the most hard-to-believe paraphilias, and in medical literature we find descriptions like the formicophiliac Brazilian, the X-ray of a pelvis with a bull's horn stuck in it and a scientific article with a full-color photograph of a German man naked in a bathtub with animal intestines and guts hanging from his neck. The authors of the study explain that this necrophiliac (who feels sexual pleasure from corpses) is a 40-year-old civil engineer, married with two kids, good social relationships and an apparently normal personality, who in addition to the bathtub photo went so far as to dig up the cadaver of a recently deceased woman in her 20s, cut off her breasts and take them home with him, and I won't even tell you what he did with the stolen body of a newly dead 14-year-old girl. I don't want to shock you with these stories of extreme sexuality, but believe me when I say that scientific literature has described many cases—almost exclusively male—of sexual obsessions we could never even imagine, much less understand.

Among the most common paraphilias, pedophilia, exhibitionism and frotteurism are always classified as mental problems, while sadism, fetishism and transvestism are only classified as illnesses if they are compulsive and cause problems. Notice that the criteria is not how "normal" they may seem from a biological point of view (there are academics who have long discussions over whether feeling desire for adolescents of 14 or 15 should be considered a paraphilia), but whether they go against the laws and norms of society or not. So there are sexual activities that are considered paraphilias in one culture and not in another. One example would be that, in 1973, the DSM removed homosexuality from its list of paraphiliac activities, but in some countries it is still banned.

Pathological paraphilias affect men much more so than women, and usually begin to manifest themselves at a very young age. For example, according to Megan Kaplan's data, 59% of pederasts say that their fantasies with prepubescent children started in adolescence. Megan runs a center at Columbia University just for teenagers, and among them are many who have committed incest with or sexual aggressions towards adults and children. Many paraphiliacs have more than one paraphilia, few social aptitudes, problems of dissatisfaction, depression and psychological disorders that make them believe that their victims enjoy what is being done to them. A therapist told me that once an exhibitionist patient revealed himself to her and started to masturbate, convinced that she wanted him to.

There are occasions in which paraphilia is a substitute or an exploration, for example sexual contact with animals, that happens more frequently during adolescence and then disappears, or in the case of some adults, who practice it when they have no other options. But in many cases, a zoophiliac declares that he truly prefers animals to humans, just like a rapist desires nonconsensual sex, or a pedophile is more aroused by children than adults. This is an important point when dealing with treatment, since paraphilia is considered a specific sexual orientation, and just as homosexual or heterosexual preferences are almost impossible to change it is not a simple task to alter the orientation of someone who is only aroused by children or objects with psychological therapy. In those patients what must be treated or controlled is the behavior, not the desire.

The causes of paraphilia are still a mystery. On a biological level there is talk of brain damage, developmental problems and even genetic predisposition. They have also observed that sexual abuses suffered in childhood increase the risk of paraphilias, especially pedophilia and sexual aggressions. But according to the most accepted general mechanism, the triggering factor is the association of a specific stimulus with arousal and sexual pleasure during key stages of development, and a repetition that reinforces the link over time. It is very complicated to conduct rigorous studies, but many paraphiliacs remember striking experiences during their childhood and adolescence, and that suggests that they are a result of learned processes. We have already discussed how classic Pavlovian determinants underlie some fetishes and sexual preferences. It is possible that if one enters into a spiral of always associating pleasure with a concrete object or behavior, the dependence can grow into an obsession that, along with psychosocial problems, obsessive-compulsive disorder, hypersexuality, drug use or impulsivity, can become a paraphilia.

Megan Kaplan confesses to me that they are a bit lost because there are very few paraphiliacs who seek out treatment, and because there has been scant scientific investigation of hypersexuality and paraphilia (Megan and Richard Krueger are part of the group working on defining that section of paraphilias for the upcoming DSM-V). Megan says that they usually begin therapy with a new patient by dealing with the psychological component and that, within an acceptable level of risk, they try to have patience with behavioral change. According to her bibliographical review, aversive or reconditioning therapies have not shown good results, and she considers that sometimes drug treatment is necessary. The Paroxetine we mentioned earlier is a selective serotonin reuptake inhibitor with a strong side effect of libido diminishing that can be very useful in conjunction with psychological therapy. But Richard Krueger is of the opinion that in serious cases the most effective treatment is "chemical castration," which drastically reduces testosterone blood levels. This is achieved through injections of synthetic progesterone, with drugs like cyproterone acetate that interfere with cellular androgen receptors, or with gonadotropin-releasing hormone analogues that inhibit the function of the pituitary gland, drastically reducing the secretion of testosterone. Several studies have confirmed that this reduces sexual desire and the frequency of fantasies, and that allows for better control. Doctor Krueger cites scientific bibliography to confirm that in truly ill people the results of exclusively psychological therapy are very questionable, and that in general it is not effective enough for dangerous paraphiliacs. There are cases of pederasts, like the Californian Brian DeVries, who have agreed to be surgically castrated in order to avoid the impulse to commit crimes that they recognize as reprehensible but feel incapable of preventing themselves from doing. Because paraphilias can reach truly aberrant extremes, even to the point of bringing about the paraphiliac's own death.

Autoerotic Asphyxiation: Dying in the Search for Pleasure

The figures are confusing, since they depend on estimations and on whether the death was registered as a suicide or as an accident, but it is calculated that between 500 and a thousand people die each year in the United States from autoerotic practices. Medical literature has documented cases of death by electrocution, by hemorrhaging after insertion of foreign objects into the genitals or anus, by suffocation after wrapping the entire body in plastic, and detailed descriptions such as that of a 29-year-old man who choked to death on a cucumber while masturbating.

Once again, the scientific literature is very explicit and it is not unusual to find, in the middle of an article, a photograph of the corpse of a man who died of a heart attack while masturbating with a vacuum cleaner still stuck to his penis, or one taken by the police of a naked inert body hanging in a room, surrounded by pornographic material on the floor and on the computer screen. This last example is the most common type of autoerotic death: strangulation and asphyxiation with plastic bags during masturbation due to an obsession with generating anoxia (absence of oxygen) in the body to increase sexual pleasure.

In addition to some well-known cases, such as that of the actor David Carradine, this dangerous practice is secretly pursued by an indeterminate number—but higher than you might think—of people who forcefully create a condition of anoxia in order to augment their physical excitement. As Canadian forensic pathologist Anny Sauvageau explains in several of her reviews and articles, the victim is usually found on the floor of their bedrooms or bathrooms with his head in a bag, ropes or belts around his necks, erotic material around him, high levels of alcohol and drugs in his blood, signs of having inhaled asphyxiating substances such as butane gas or nitrous oxide, and sometimes dressed in women's clothing, with sadomasochistic elements and clear signs that this was no suicide, but rather an accident during autoerotic asphyxiation. Ninety percent of these victims are men and only 10% women. In the cases of autoerotic deaths by atypical means, the proportion of women rises to 22%.

The first question that may come to mind is why does asphyxia heighten sexual pleasure. There are several reasons. On one hand, it can increase excitement as part of a bondage ritual and extreme masochism; on the other, the lack of oxygen in the cerebral cortex can also bring about a loss of lucidity and hallucinations that could emulate the effects of certain drugs and intensify feelings of pleasure. Practitioners declare that the increase in physical arousal is very significant, which fits with the apparently illogical principle that fear heightens desire and sexual intensity.

The body doesn't know that its owner is doing it on purpose, but when it senses a serious drop in blood oxygen it experiences a reaction of absolute stress. The sympathetic nervous system activates, the entire organism goes into emergency mode and the brain can only think about escaping. The reticular formation in the brainstem sends signals to all parts of the brain, the amygdala activates and the five senses sharpen. In situations of fear, stress and tension we can hear the slightest noise or feel the most modest touch on our skin, and this extreme sensitivity, together with sexual excitement, can intensify pleasure.

As we've mentioned in previous chapters, overactivation of the sympathetic nervous system and cerebral stress can heighten arousal and bring about a more powerful orgasm. The body's reaction to anxiety and psychological and physical stress is overwhelming and is the reason why some paraphilias, sexual violence, sex in public and forbidden practices can increase genital excitement. In fact, some people can only reach orgasm when someone generates in them this stress that triggers the activation of the sympathetic nervous system, which is related to one of the most paradoxical and traumatic situations in—sometimes disconcerting—human sexuality: woman who have spontaneous, involuntary orgasms during rape.

Involuntary Orgasms During Rape

In the late 1990s, Roy Levin got a call from the police. As one of the top experts in the physiology of sexuality in the United Kingdom, they were calling him in to testify in a trial over the sexual abuses of a 15-year-old boy. The peculiarity of the case is that, during the abuse, the boy had an erection and ejaculated, and the defense counsel was arguing that that implied consent. The young man categorically denied it, stating that he lost control of his body, that he did not feel any desire or pleasure, and that the entire time he fought against the rape. Roy Levin had published several scientific articles on the lack of consistency between physical and mental excitement, and on the different mechanisms that regulated peripheral and central control (genital as opposed to cerebral). His scientific position was emphatic: it is possible to experience physical excitement and orgasm as a reflexive act without that implying desire or pleasure.

When we had lunch together in Estoril, in July of 2012 during the world conference of the International Academy of Sex Research (IASR), Roy told me that he didn't know how the case ended because "these are complex situations in which I assume many other factors and evidence have to be considered, but what I wanted to make clear was that a rape can lead to spontaneous orgasm without that signifying approval."

In fact, a few months later, the judge asked Roy Levin to testify again in the case of a woman who was raped by a coworker, who argued in his defense that woman had lubricated and reached orgasm. Roy made the same statement, even more categorically, since "in women the lack of agreement between genital and mental excitement is much bigger than in men, and for example we know that lubrication as an automatic reflex evolved in females precisely to avoid injuries and greater risk of infection in forced sexual encounters." Roy was called to testify in several more trials, until he decided, in 2004, to publish with Willy Van Berlo the largest review of the scientific literature on sexual arousal and orgasm in cases of forced and non-consensual sexual stimulation.

The first thing that Levin and Van Berlo did was discern whether the stories about physical arousal during rape were frequent or whether it was more of a question of myth, unfounded rumors, exaggerations or anecdotal cases. After a review of medical examinations of raped women they proved that indeed many lubricated, had increased blood flow to their genitals, admitted to having experienced physical pleasure against their will, moaned with pleasure, and between 4 and 5% of cases reached orgasm. When this happens, the vast majority of victims feel traumatized by a feeling of acceptance or guilt, but they shouldn't at all since it is an absolutely involuntary reaction by their body, and even though it seems paradoxical, there are various physiological mechanisms that justify the physical arousal and orgasm as a reflex with no conscious permission being granted.

As a starting point, we should return to a couple of concepts detailed earlier in this book. Firstly, that in normal conditions physical and mental arousal are intimately related, but on some occasions someone can feel mentally excited and their genitals can fail to react, and in others there can be a genital response without a subjective sensation of arousal. That is the disconnection investigated by scientists such as Meredith Chivers.

The second concept is that sexual arousal can be triggered both by mental stimuli and by a purely physical process. It is obvious that an erotic image, suggestive words or recalling a fantasy can immediately generate a genital response, but we have also seen that an unexpected caress or even just a light brush in a non-erotically charged context can lead to arousal as an automatic act. This first spontaneous arousal responds to an autonomous mechanism on a subcortical level, which later our sophisticated cerebral cortex can decide to inhibit or foster.

At the same time, there are situations in which the inhibition system can be completely blocked, like during ethylic intoxication, drug use or states of profound shock like that produced in a rape. This is the most widely accepted explanation for the feeling of pleasure during sexual abuse: the physical contact can stimulate the genitals reflexively and the extreme stress can block the ability to react or control our own body. We must remember that spontaneous orgasms can also happen while we are sleeping, that there are published cases of orgasms during certain gymnastic exercises and that, depending on the height of the injury, some paralytics can have erections and orgasms through direct genital stimulation without any information coming to or from the brain. An orgasm does not require consciousness.

In fact, in a study published in 1999 with interviews of 58 rape victims, 12 of them (21%) admitted experiencing a pleasurable physical response, and all of them without exception said that the event was, at every moment, mentally devastating. The median age of these 12 victims was 32 years old, 10 were vaginally penetrated, 9 by someone they knew, 8 rapists tried to sexually stimulate them before the abuse and 6 of the women said that before the rape they felt attracted to the criminal. Independent investigations have corroborated those unthinkably high percentages of women who become aroused during a rape, and numerous clinical reports establish that in one out of every 20–25 rapes the woman suffers an involuntary orgasm.

Yet even accepting that a state of shock in rape can impede voluntary inhibition, how can arousal be generated in the first place? Here is where the treacherous sympathetic nervous system (SNS) comes into play. Laboratory studies have shown that stress, fear, pain and repulsion usually diminish sexual response, but that on rare occasions they interfere with the SNS and generate the opposite effect: they increase blood flow to the genitals and, therefore, lubrication. In those cases, the stress of the rape would not only not be an impediment to genital response, but could even help trigger a totally involuntary automatic reaction.

In the many fewer cases of men being raped that have been described, the involuntary arousal is significantly less frequent, but Levin and Van Berlo's review also gathers data of men who have experienced pleasure during forced abuse. Again, the first thing we should remember is that, especially in young men, an erection can be caused by a wide variety of stimuli that aren't necessarily sexual, sometimes even ones involving fear or anger. In a laboratory study they induced anxiety in volunteers by threatening them with electric shocks and saw how that anxiety increased their sexual response when they were later shown erotic images.

It is abundantly clear that a woman can lubricate as a totally automatic response, that stress can block all the central mechanisms of inhibition, that there may not be agreement between subjective and genital arousal, that the activation of the sympathetic nervous system can facilitate orgasm and that, without a doubt, there can be a physical sexual response during rape, but that in no way should be interpreted as consent nor provoke feelings of guilt. Roy Levin tells me "no study has analyzed whether our work has had an impact on legal trials. I have the feeling that it has, from the reactions I've received, but where we have seen a tremendous improvement is in trauma reduction of many women who felt tormented by that unexpected involuntary reaction of their bodies."

17

Sexual Identities Beyond XX and XY

Martin is a tall, slim guy about 35 years old, with very pale skin, a wide forehead, thinning hair, and a strong resemblance to the singer from R.E.M. "I get that a lot," he says with a smile during a meeting organized to bring together people of different sexual tendencies and expressions in the New York area.

The truth is that I'm having trouble guessing Martin's role at that meet-up. At the Lower East Side meeting spot there are BDSM practitioners, the polyamorous, transsexuals, activists, representatives of gay and lesbian associations, and a few exhibitionists. Some of their appearances and attitudes clearly reveal which community they belong to, and at the start of any conversation, people immediately and overtly express their own ideas and experiences about sexuality. Yet Martin seems to avoid talking about himself and calmly behaves as if he were in a regular bar. I begin to suspect that he is an infiltrator like me, and I tell him that I'm there looking for testimonials for a book about science and sex that I'm working on. Martin shows a lot of interest, tells me that he loves science and starts knowledgeably quoting theoretical physics concepts but he discreetly continues to keep mum about what he is doing at the event. Until suddenly I just ask him, all casually, "What about you, Martin, what are you doing here?" His expression turns neutral, I notice a certain tension in his face, I think I see him swallow hard, and then he answers, as if he felt obliged to: "Well… I like to dress as a woman."

Martin seems to feel relieved, and is more relaxed as he explains that he only dresses as a woman in private, when no one can see him. It is an impulse he began to notice at a very young age, his wife knows about it and accepts it, and he has no problems filling an absolutely conventional male social role.

Martin feels like a man in every sense and defines his orientation as bisexual. In the past he had relations with men and women, but now he is monogamous with his partner. "Of course I sometimes feel attracted to a guy, just like you could be to various women, but I don't miss it. The only thing is that sometimes I feel the need to take on a female role," he says.

Martin believes that it is time to reveal his cross-dressing more naturally and is considering going out in public dressed as a woman. "But it makes me feel panicked, and at the same time I don't see why I have to," he explains. When I ask him how he wants to be treated when he is dressed in women's clothes—as a man or as a woman—he hesitates before responding, "Well… honestly I never thought about it… since I always do it in private… I guess like a woman, but you know that transvestism is very different from transsexuality, right?" "Yes, of course…" I reply.

Transvestism is a very wide term used to define people who periodically dress and act like the opposite gender, but without necessarily feeling different from their biological gender. In general they are men who feel like men, or women who feel like women, but who enjoy and feel relaxed or sexually excited by dressing and assuming social roles associated with the other gender. It could be for fun, a fetish or a practice that can offer wellbeing and inner peace—as in Martin's case—and there is no conflict between sexual identity and genital sex as there is with transsexuals. Transsexuals permanently feel that their minds and their bodies are out of sync, and they resort to hormone treatments and surgery to correct that.

Martin practices transvestism and insists that he is not transsexual, but as our conversation grows deeper he confesses something less frequent: "Well… I'll admit that in some moments of my life I have felt like a woman. It's strange, because I fully identify as a man, but there are days when emotionally I notice that I'm a woman and I have female reactions. And it doesn't necessarily happen when I'm cross-dressing, it's something internal and spontaneous. I don't know, it's perplexing." I ask him if he might have some sort of bi-identity like bisexuality and he answers, "Possibly, but that's not normal among crossdressers." A little further on we will talk a bit about little-known bigenderism and the people who alternate cycles of male and female identity, but first we should reflect—especially as we begin this chapter—on the meaning of the word "normal." I answered that "from a naturalist perspective, what wouldn't seem normal to me is that there was no one like you," and I stand by that.

If by "normal" we mean "usual," then yes, Martin is as not normal as, say, a race walker. But if by "normal" we mean something that could be coherent with nature and society, Martin would be as normal as a redhead or the aforementioned race walker. In fact, and still just keeping biology in mind, what is really strange would be not finding any ambiguity between the different levels with which we can define male or female sex.

It is true that the usual thing is for a person's gender to be defined by the XY or XX chromosomes (genetic sex), for them to be born with a penis and testicles or a vagina and ovaries (genital/gonadal sex), for a greater concentration of androgens or estrogens to circulate through their body (hormonal sex), that their brain is predisposed from childhood to mentally feel male or female (sexual identity), that the social environment reinforces that they behave, define, express and assume male or female roles (sexual role), and in parallel that they feel physically and emotionally attracted to one gender or the other, or both or neither (sexual orientation). But now we know that these categories are not always correlated (the clearest example is transsexuality, which we'll discuss further on), and that within them all there can be ambiguous or intermediary situations (bisexuality or intersexuality, which we will also talk about shortly).

However, curiously, in all these categories, the one we consider less fluid is precisely the one we ourselves have created: the social role. Martin's case is still not considered "normal" because what's "normal" is to be a man or a woman. And I insist: Martin's case isn't "usual" in the sense that sexual identity is usually very well defined, and even transsexuals feel fully identified with their mental sex and rarely develop doubts after childhood or adolescence. But, both on a biological and developmental level, it makes sense that within sexual identity itself there can exist certain ambiguity.

We all come from a little group of cells that began to divide according to genetic programming and from which a brain developed, with connections and structures that, among many other things, predispose us to have male and female behaviors. But it is biologically coherent that this cerebral formation is not absolutely static and that some of these structures can have mixed dispositions. This same continuum also takes place with prenatal exposure to androgens, since there can be intermediate values between the high and low levels, and of course varied experiences and environmental factors influence the development of sexual identity.

So, despite it being atypical, it enters into normality that Martin's mind is almost always masculine but sometimes feels feminine. From a naturalist perspective, his case may be infrequent, yet it forms part of a foreseeable and assumable diversity. What would be truly unusual would be that no one like him existed.

Intersexuality: When Chromosomes and Genitals Don't Match Up

Jen is a svelte, pretty woman who has a Y chromosome in every cell in her body.

Jen was born with female genitals and developed like any other girl until at 7 years old she began to notice discomfort in the lower part of her abdomen and a medical examination revealed that instead of ovaries she had two small internal testicles that had not yet descended. The diagnosis, following a more complete analysis, was clear: Jen was chromosomally XY but suffered from complete androgen insensitivity syndrome (CAIS). A mutation in a gene associated with cellular androgen receptors made her entire body insensitive to testosterone. Which is to say, testosterone ran through Jen's bloodstream, but none of her cells recognized it.

When Jen was a 6-week-old embryo, the SRY gene on her Y chromosome directed the formation of testicles, which began to secrete testosterone and Anti-Müllerian hormone (AMH). The AMH did its job of keeping her uterus and ovaries from forming, but the testosterone had no effect and the rest of her genitals, body and mind continued developing as if she were a woman. Jen's tiny testicles never descended and they kept secreting ineffective androgens until they were removed. The treatment continued with a hormonal supplement so that Jen could complete her development as a woman, psychological support particularly in her teen years, and the use of vaginal dilators so she could have satisfactory sexual relations. Jen always was and will be a conventional woman in every sense—except for biological motherhood—and a living example that it is hormones and not genes that ultimately lead sexual development.

One in every 10,000 people is born with androgen insensitivity syndrome (AIS), and while in cases of partial insensitivity the diagnosis is more confusing and the sex assignment must take into consideration each individual's evolution, when there is complete insensitivity those women are indistinguishable from any other woman. In fact, there are models and athletes with AIS and women who discover they have a Y chromosome well into puberty.

AIS is not without problems, but within the range of intersex conditions it is perhaps the least traumatic and most easily dealt with. Most of the other syndromes that involve an incongruence between chromosomal and phenotypic sex are more ambiguous.

The second most frequent intersex condition (1 in every 16,000 births) is congenital adrenal hyperplasia (CAH). Unlike AIS, the patients are chromosomally women (XX) but due to a genetic defect that affects an enzyme involved in the synthesis of corticosteroid hormones, the adrenal gland produces an enormous amount of androgens that provoke a partial virilization in the individual. Normally CAH is detected at birth or soon after due to an enlargement of the clitoris and swelling of the labia, and they begin to administer corticosteroids to reduce the excess androgens reaffirming the person's sex as female. Nevertheless, in medical literature there are documented extreme untreated cases in which the masculinization was exacerbated and they became XX men with ambiguous genitals.

In some cases of CAH, the virilized genitals are surgically corrected during infancy, but there is some controversy because the effects of high levels of testosterone in the prenatal stage can affect parts of the brain related to sexual identity, and there are people with CAH that grow up feeling mentally male.

German-born researcher Heino Meyer-Bahlburg is one of the preeminent world experts on intersexuality. When I visited him at his center at Columbia University he explained that, traditionally, when there were ambiguous cases they would always opt for feminization, but that that's changing and—except when there are physical problems—now they prefer to wait. It is complicated because, on one hand, Heino Meyer-Bahlburg says that "the gender assigned at birth should be chosen to minimize risk of later gender change, and the genital surgery should wait until a stable gender identity is observed," but at the same time he is of the opinion that "when there is solid evidence, gender reassignment should be done relatively early to avoid more complicated later medical treatments and psychological trauma."

Heino keeps insisting that every intersex condition is different, that each individual must follow a personalized diagnosis, and that it is incredibly important to carry out a rigorous scientific investigation that reports on how patients with each type of syndrome evolve. He has long been studying congenital adrenal hyperplasia and has proven that there is a higher rate of homosexuality among women with CAH, and that around 5% of them have gender dysphoria and as adults feel that they are male. Those aren't really high figures, especially when compared to other conditions of intersexuality like 5-alpha-reductase deficiency (5-ADR), in which more than half of patients who have been assigned female gender begin sex change treatments as adults.

5-ADR deficiency affects individuals who are chromosomally XY and who, like those affected by AIS, began to develop testicles and produce androgens, but who due to a genetic mutation lack the enzyme that converts testosterone into dihydrotestosterone (DHT). DHT is a much more powerful androgen than testosterone and has a key role in the development of the genitalia during the embryonic stage, and whose lack due to 5-ADR leads to the formation of ambiguous genitals. It is also usually identified at birth, but there are various subtypes of 5-ADR, and in some cases the genitals are markedly feminized.

An historic example of extreme 5-ADR took place in the Dominican Republic when, in the 1960s, a local doctor discovered that at adolescence, some girls in a town called Las Salinas started to become men. During their infancy all of them had female genitals and typical feminine behaviors, but when they reached puberty their voices changed, they gained muscle, got body hair and their clitorises grew into penises. The phenomenon repeated for generations and it was well known among the local population, and when the doctor discussed it internationally it immediately attracted the attention of American scientists.

After several studies, the researchers published an article in *Science* describing 24 cases of adolescents with XY chromosomes, who had been born with female genitals due to severe 5-ADR, had been raised as girls in their town, but who masculinized in puberty due to the radical increase in testosterone. Their testicles grew and finally descended, and the same happened with their clitorises, which were actually small penises that had yet to develop. Not all the cases were equally clear, but most of the girl-boys also acquired male personalities and became conventional men. This was tremendously noteworthy, because no matter how much testosterone we inject in a regular teenager we are never going to modify their sexual identity. Perhaps they would begin to act more aggressive, get more body hair or higher libido, but they would still feel like women. But these girl-boys from Las Salinas did become men. The explanation was that their prenatal levels of dihydrotestosterone greatly affected genital differentiation, but that testosterone was still present and determinant in the construction of a brain predisposed to be male. That was interpreted as evidence of the importance of prenatal exposure to androgens in the configuration of the brain, and highlighted the need to wait—in some cases of intersexuality—for sex assignment until sexual identity is well established.

Additionally, Heino Meyer-Bahlburg insists that "the goal of treatments or surgeries should not be tranquilizing the parents, but maximizing each patient's quality of life, emotional wellbeing, gender development and sexual function." And for that it is essential to carry out scientific and medical monitoring of the quality of life, sexual function and gender identity of intersexuals. In 2011, Heino Meyer-Bahlburg published a detailed review of the medical literature on severe gender dysphoria and sex change treatments in intersexuals. According to the data of the sample analyzed, no XY woman with complete androgen insensitivity syndrome said she felt male as an adult. Yet in cases of partial insensitivity, 7% of patients initially assigned as women had gender dysphoria toward the male gender, and 14% of those assigned as men had it toward the female. In the 46XX congenital adrenal hyperplasia syndrome (CAH), 5% of those assigned as women felt they were men in adulthood and 12% of those assigned as men felt they were women. And the proportion shot up in syndromes like 46XY 5-alpha-reductase deficiency (the one the Dominicans had) or 46XY 17-beta-hydroxysteriod dehydrogenase 3 deficiency, in both of which approximately 65% of the patients assigned as women suffered gender dysphoria and began gender reassignment treatments.

It is a polemical subject on ethical and social levels. The number of intersexuals upset because doctors assigned them a gender too quickly is considerable. There are physiopathological situations in which operating on children with ambiguous genitals is necessary in order to avoid different types of problems during development. And at the same time, when the situation is very clear, the sooner the sex is assigned, the better for minimizing future traumas. The big problem are the cases when they get it wrong, since a gender reassignment is more complex if there was previous hormonal or surgical treatments based on an incorrectly assigned gender. It is extremely delicate, because feeling you are a man or a woman does not depend merely on hormones, and much less on whether you have a penis or a vagina. Gender identity is between our ears, not between our legs.

Transsexuality: The Mind Is in Control

One of the people I was most pleased to have met while researching this book is a transsexual woman who asked for complete anonymity, so I won't even give clues that will reveal her case. I know that she will recognize herself immediately in these lines, and I can't help but mention what a positive impact our meetings and conversations had on me, being able to get to know her experience in such depth, discovering her great sensitivity, and learn of the difficulties that she is still overcoming as society and her environment adapt to her new sexual role. Because that is how it should be: it is not her mind or behavior that should adopt to her body and society, but the other way around.

In fact she asked me to maintain her anonymity because part of her most private environment still has not accepted her operation and change of sexual role that she'd done just months before we met. Not being understood is painful, and it is one of the big differences between transsexuality and intersexuality.

Of course they can both be devastating on different levels and circumstances, but intersexuality is associated with clinical causes that society accepts, tolerates and understands better: intersexuals can be operated on while they are still minors, they are covered by public health care, and their families and those around them can point to a gene, enzyme or cellular receptor as the one responsible. On the other hand, even though transsexuality can also have a biological determinant, part of its stigma comes from the fact that more retrograde and less educated sectors still consider it a choice, a decision, or a mental disorder.

"There is nothing wrong with my mind," insists my friend. And that is backed up by her doctorate, impressive self-possession, intellectually demanding profession, emotional wellbeing and personal satisfaction after having overcome such a delicate situation.

Suggesting that my friend has a mental disorder is absurd and offensive. There is absolutely nothing wrong with her mind and brain. They are as "normal" as the brain and mind of any other woman, with the only peculiarity that the rest of her body has male genes, hormones and appearance.

In fact, a profoundly important milestone in the history of transsexuality is that the DSM-V manual of mental disorders, published in 2013, eliminated the "gender identity disorder" (the medical and academic term for transsexuality) from their catalogue of mental disorders. Beginning in mid-2013, transsexuality is no longer classified as a mental illness. It is an enormous achievement, comparable to the removal of homosexuality from the DSM-III in 1973.

However, transsexuality can still be considered a developmental disorder that generates a discord between mind and body, that for some may not be a problem, but in others can provoke a gender dysphoria that does require psychological, hormonal and even surgical support. In fact, transsexuals have no problem with the characteristics of their minds, but many do have a problem with the characteristics of their bodies, and it is precisely that intriguing dissonance between phenotypic sex and sexual identity that science is trying to comprehend.

Trying to Comprehend the Causes of Transsexuality

The etiology of transsexuality is not well known. Always from a biopsychosocial perspective, several studies have identified traumatic events during infancy and certain patterns of family interaction as risk factors, but the fact that the sexual identity of most transsexuals is so consistent from mid-infancy makes us think that there have to be biological determinants directly involved.

Starting with the genes, a bibliographic review of studies on twins and transsexuality, published by Belgian researchers in 2012, observed that the correlation of transsexuals was more frequent among identical twins than fraternal ones—both transsexual men and women[1]—concluding that there could be a certain genetic component involved. Some environmental factors between identical and fraternal twins are different, and the fact that in the large majority of transsexuals who have a twin, that twin is not transsexual indicates that genetics do not play a large role. But it is not far-fetched to think that genes associated with enzymes, cellular receptors or androgen levels could be partially involved in the predisposition toward transsexuality. In fact, an Australian study published in 2009 that compared 112 transsexual women to 258 control men found a significant link between transsexuality and a genetic polymorphism related to androgen receptors.

On a cerebral neuroanatomy level, the research approach is very simple: there are concrete structures in the brains of men and women that are dimorphic, for example the hypothalamus, or very localized aspects of neuronal connectivity. Knowing this, neuroscientists have been analyzing transsexuals' brains to see if these structures are more similar to their chromosomal sex or to their sexual identity.

The first positive result arrived in 1995 when a team headed up by Dutchman Dick Swaab published that a dimorphic cerebral area in the surface of the thalamus called BSTc was indeed more similar to women in female transsexuals, and in male transsexuals to men. Later, similar correlations were observed in the INAH3 area of the hypothalamus, in neuronal connectivity, and even in cognitive aspects whose interpretation was more ambiguous. The last relevant discovery to date was published in 2011 by a Spanish group headed by Dr. Antonio Guillamón, of the UNED, that analyzed the microstructure of some nerve fibers called white matter, which are dimorphic in certain areas of the brains of men and women. In an extensive project, the researchers proved that these cerebral areas were masculinized in transsexual men and feminized in transsexual women. And in another study also published by Dr. Guillamón in 2012, it was seen that concrete areas of the left part of the cerebral cortex in transsexuals were more similar to the gender they identified with than to their biological sex.

---

[1] When we write transsexual women here we are referring to transsexuals born with male genitals but who feel they are women and reassign their sex. And when we discuss transsexual men we are referring to those who were born with female bodies but male sexual identities. We also avoid using the term transgender, which includes not only transsexuals who have hormonal and surgical treatments to reassign their bodily sex but also everyone who feels temporary or permanent discord between their biological gender and social role. The term transgender is widely used, but it is too broad and generic, and comprises so many different identities that it can be confusing. Transsexuals would be a subgroup within transgender people.

In this type of studies with adults there is always the doubt whether the changes in brain structures participate directly in the development of transsexuality or if they are a later consequence of it. But there are many suspicions that these differences would already be present in embryonic stages and be a direct consequence of different androgen exposure in prenatal development.

In fact, the fluctuations in hormone levels during pregnancy, whether due to genetic causes or diverse alterations, is the biological hypothesis with the most weight to justify the predisposition to transsexuality. Accordingly, some interpret transsexuality as a certain type of mental intersexuality, in which differences in androgen levels or cell receptors could have affected brain development differently than the development of the rest of the body.

According to this hypothesis, a transsexual with XX chromosomes could have begun her embryonic development as a woman, generating ovaries, fallopian tubes, uterus and vagina from the sixth week of pregnancy, but weeks later when the brain starts to develop the embryo could have been exposed to more than the normal amount of testosterone and the brain gotten partially masculinized. That could explain the neuroanatomical differences observed in studies like those done by Dick Swaab and Dr. Guillamón, perhaps predispose a sexual attraction toward women, or even a male sexual identity.

It is fundamental to avoid determinism on this point. Thousands of women are exposed to high levels of androgens during pregnancy and still end up being conventional heterosexual women. Brain development is an ongoing process; we know of its vast neuroplasticity, especially during the first stages of life, and we know that both environment and experience constantly modify its configuration. This is to say that exposure to androgens during pregnancy and possible changes in cerebral structures can have an influence, but never a determining one. And it's in this respect that transsexuals can be born with a brain not entirely in agreement with their chromosomal sex, and depending on other physical and environmental conditions they can end up developing a masculine sexual identity or a feminine one. The Canadian Kenneth Zucker, one of the top experts in transsexuality and editor of the journal *Archives of Sexual Behavior*, has conducted many studies that follow children with gender dysphoria and he has proven that as childhood progresses they can opt for one gender or the other.

As for sexual orientation, it is worth mentioning that the large majority of transsexual men feel attracted to women (homosexuality with respect to their gonadal sex and heterosexual with respect to their sexual identity), while there is more flexibility with transsexual women, although attraction to men is also more frequently found.

But returning to hormones, analogously, transsexual women will begin their embryonic development by following the instructions of their Y chromosome, developing testicles in the sixth week of pregnancy that soon start to secrete testosterone. But if from the 13th week on, when the brain begins to differentiate, for whatever reason, those levels are lower or some cellular receptors don't properly identify androgens, the brain can be left more feminized and subject to later influences that end up configuring a female sexual identity.

It's obvious that these biological causes don't capture all the diversity of the community of transsexuals and transgendered people, but in the most common case of a transsexual man or woman who from their infancy feels they are a boy or a girl despite their entire surroundings conditioning them differently, and who continues developing with marked gender dysphoria until one day they realize that their mind has a clearly different gender from their body, it does seem logical to think that some biological causes predisposed that brain to be in discord with the rest of their body from the embryonic stage.

When this happens, there are many transsexuals who take all the steps to make their hormones, genitals and external appearance agree with their sexual identity. There are also those who reject the surgery and accept their body as it is, and have no problem at all in feeling equally men or women despite having the opposite genitals. And even though they are very few—because, as we already mentioned, in adults sexual identity is generally tremendously solid and stable—there are yet other people with more ambiguous feelings who reject this dichotomic social construction between male and female sex, and who defend belonging to a neutral sex that could equal the concepts of bi- or asexuality. This situation is much more frequent in intersexuals, but there are also individuals with well established gonadal sex who alternate cycles in which they feel more male or female and who want to exercise those roles at different moments. It wouldn't be exactly transvestism, but a situation of bigenderism (more extreme than in Martin's case), which has hardly been studied by science, but which does give rise to very interesting hypotheses.

Bigenderism: Being Man and Woman at the Same Time

At the start of 2012, the great neuroscientist Vilayanur Ramachandran published, in the journal *Medical Hypotheses*, possibly the most detailed scientific study on bigenderism to date, in which he analyzed various physical and behavioral characteristics of 32 bigender people who periodically alternated between male and female sexual identities.

For Ramachandran, this ambivalence is strange, but he considers it a subcategory of transgenderism that is part of the wide spectrum of our sexuality, and which from the scientific perspective represents a fascinating condition for investigating the development of gender identity and how body image is internalized.

In his article, Ramachandran describes that of the 32 bigender people, 11 were anatomically women and 21 anatomical men, although five in this latter group were taking estrogens and anti-androgens. As for the frequency of their gender shifts, 23 of the 32 individuals alternated male and female identity several times in a week, 14 of them daily, six at least once a month, and three a few times a year. Most of them defined these changes as involuntary, but ten of them said that they were predictable. Ramachandran observed that there was no delusion in the sense that even when they were in a moment of well-defined sexual identity, they were still fully conscious that they alternated. But he found a curious phenomenon: often, when they were in the identity opposed to their anatomy, 21 of them felt "phantom" breasts or penises, corresponding to their desired gender. According to Ramachandran, this would suggest that they actually had two body maps in their brains, and it strengthened the hypothesis that bigenderism has a certain neurological origin. A curious fact was that their sexual orientation sometimes changed with their alternating identity, but other times it didn't.

On a neurological level, he observed that among those bigender people there was a higher rate of lefties and ambidextrous people than in the general population, a factor associated with alterations in the embryonic development of the brain. He also found a significantly higher index of bipolar disorder that, according to the article, could stem from structures in the parietal lobe involved in body image and connected to the insula and hypothalamus. This all leads Ramachandran to believe that bigenderism has a biological cause, and to propose a new neuropsychiatric condition called alternating gender incongruity or AGI. With the growing tendency to avoid the creation of more mental disorders that pathologize the diversity of our behavior, it doesn't seem that AGI is going to catch on, but what is surely going to stick is the concept that just as sexual orientation and gonadal sex are not static, gender identity itself could be more ambiguous than we have imagined.

Ramachandran cites a survey conducted in San Francisco in which 3% of anatomically male transgendered people and 8% of transgender women identify as bigender, and concludes by reflecting on whether on a neurological level we should consider male and female brains so radically different.

There is still much to learn about bigender people and how biology and sociocultural forces influence their sexual identity not being as strongly marked as it is in the immense majority of the population. It is a tremendously interesting phenomenon, but it should in no way suggest ambiguity in transsexuality. Transsexuals have a highly solid and stable sexual identity, it just doesn't correspond with the external appearance of their bodies. And in those cases, as we've already said, the adapting should be done by society and, if they wish, their bodies.

Sex Change Operations, and the Phantom Penis

I will never forget what I witnessed on November 14th, 2012, at the Diagonal clinic in Barcelona, where I watched an entire sex change operation from inside the operating room. It was done by Dr. Iván Mañero on a 32-year-old transsexual woman, who at ten in the morning was anesthetized on the table with her legs separated and in the air, revealing a completely regular penis and testicles, and four and a half hours later after an extremely delicate artistic surgical job she woke up with a perfect vagina, labia and clitoris. According to what one of the nurses told me, "once the stitches fall out and it heals it's usually impossible to tell." It was impressive on many levels, and truly a fabulous experience.

The process began with a vertical excision in the skin just below the testicles. From there Dr. Mañero gradually cut the internal fibers and tissue with an electric scalpel, while painstakingly opening up a hole in the pubic musculature with the help of some sort of metallic speculum. Surprisingly, with very little blood shed, the future vaginal space grew wider and deeper until it reached the size of a conventional vagina. Once the orifice was finished, Doctor Mañero filled it with some gauze and proceeded with the rest of the operation.

The next step was to cut the skin from the base of the testicles to the start of the penis, and gradually detach it from the internal structures. Once the skin was separated from the testicles, from that same incision he proceeded to disconnect the penis from the inside, so that the internal body of the penis would be loose and the skin could be conserved intact. That skin is what will later be introduced into the orifice to complete the vagina, and with the excess skin of the scrotum they would model the vaginal labia. Doctor Mañero explains to me that when the penis is small and its skin cannot cover the entire interior of the vagina they use a fragment of thin intestine in its place, with the advantage that it is a type of tissue that can lubricate, but the disadvantage that it doesn't have nerve endings and therefore has less sensitivity.

Observing the genital area without skin, I clearly see that the entire structure of the penis is larger than its external part. Actually, the penis begins below the testicles, and from there ascends right up against the body until it folds and extends outside. The complete penis is like a boomerang. In fact, when Doctor Mañero cuts the fibers that hold the base of the penis to the abdomen, it falls down and hangs below the testicles. In that exact moment, the image vividly reminds me of a diagram representing the female reproductive system, where the vagina is in the center and above that the fallopian tubes ascend to two ovaries located on the left and right sides. It is fascinating. The penis is clearly equivalent to the vagina and clitoris, except that it is outside instead of inside, and the testicles are identical to the female gonads, except larger and external. I start to see what I already knew, that male and female genitals unequivocally have the same origin and structure and only differ in size, shape and location.

We must have been an hour and a half into the operation, and obviously I've left out the description of important steps to protect tissues, nerve fibers and suture wounds. It is an artisanal job of great precision, as reflected by the complex play of forceps, incisions and stitches that removed the testicles and left them on the table beside the surgical instruments. Again, it was impressive, in several ways.

Then it comes time to deal with the penis, which is even more delicate work. He has to retain the urinary conduit of the urethra intact, through which he had earlier introduced a yellow plastic tube, and preserve the central corpus spongiosum (spongy body) of the penis, which is attached to the glans. The glans will become the clitoris. With extreme finesse and precision in order to maintain the nerve endings intact and retain the greatest possible sensitivity, he makes an incision in the upper part of the penis from which he meticulously removes the two corpora cavernosa on the left and right sides. The corpora cavernosa are what fill with blood to facilitate erection, and are unnecessary in female physiology. After removing them, the penis begins to be hard to recognize as a penis, and even less so when the urethra is separated from the spongy body and glans. At that point, the urethra and spongy body/glans are already two different entities that Dr. Mañero puts to one side in order to begin sewing the skin on the other side. Then he pulls the skin down and introduces what is left of the penis into the vaginal orifice. He is only measuring right then. With one side partially stitched and the skin placed inside the vagina, Dr. Mañero uses a marker to designate the two places where he will make the incisions through which he will pass the urethra and the glans/clitoris from the inside toward the outside.

In the lower part of both incisions he places the urethra with the yellow tube, which is obviously longer than necessary since it comes from a penis. He cuts off the extra tissue and sews it perfectly to the skin leaving the urinary outlet of the female genitals in place. He does the same in the upper incision, introducing (from below) the spongy body and the glans/clitoris. Both are placed right above the vaginal space, part of the spongy body remains inside the body, the glans is outside, and then he begins an impressive series of stitches, folds and more stitches until the tissue takes on the unmistakable shape of a clitoris with a hood and glans included, right in the space where the labia minora will come together.

With the clitoris already formed, the skin of the former penis is closed off on the end and it is fit into the inner part of the vagina. All the skin of the scrotum is left outside on both sides, and he starts to cut and sew until he has created the labia majora and minora, which after the scarring and recovery time will complete the absolutely functional female genitals. Doctor Mañero steps away from the operating table and I have to keep myself from applauding.

There are many studies following the evolution of transsexuals, both in terms of psychological satisfaction and sexual function. In the case of transsexual women who go from having male genitals to having female ones, during the hormonal treatment there is a substantial drop in libido but, after the operation and a period of recovery, genital sensitivity is very positive, and the large majority can have orgasms through stimulation of their new clitoris, which used to be the penis glans. Also, from the vagina they can reach the prostate, so that penetration is also pleasurable. There are problems of pain and vaginismus, but in general the evaluation is very satisfactory, especially when you add in the psychological wellbeing of seeing their bodies match their minds.

In transsexual men it is a bit more delicate. "The transition from female to male genitalia is more complicated, but it can also achieve good functioning," says Dr. Iván Mañero, when we were relaxing afterward in the clinic's cafeteria.

I didn't get to watch a penis construction live, but I did see a highly detailed presentation on phalloplasties by the British surgeon David Ralph during the conference of the European Federation of Sexology held in Madrid in September 2012. With incredibly explicit images, David Ralph went over the different options and explained that before they used to construct the penis using epithelial tissue and fat from the abdomen, but that now they prefer to take a part of forearm because it has more sensitivity. The hole left in the forearm is covered with tissue taken from beneath the buttocks of the patient.

The testicles are simply two pieces of silicone inserted in the vaginal lips, the urinary conduit sometimes only reaches to half of the new penis, erection is attained with an implant similar to the one used by men with permanent erectile dysfunction, and the clitoral tissue is usually distributed along the base of the penis. This is the most critical process for maintaining sensitivity, and Dr. Ralph showed images of patients who preferred to leave the clitoris intact above the testicles and construct a penis above it so that the clitoris can still be stimulated during penetration.

Genital sensitivity fluctuates in each case. Most can feel contact on the penis shaft, although only a few associate it with an erotic response. But, unlike transsexual women, in transsexual men the administering of androgens usually significantly increases libido, and if the nerve endings of the clitoris remain intact, the sexual response can be even better than before the operation. In the only scientific study published on the systematic monitoring of transsexual men, a Dutch team showed that years after the sex change, 93% of the 49 transsexuals who participated in the study could reach orgasm by masturbation, and almost 80% during intercourse with their partner (not all the time). It was interesting that 65% declared that their orgasms had changed, becoming shorter and more intense, in fact, most reported higher arousal and sexual activity after the hormonal and surgical treatment than before. Even though in the sex change from woman to man there are usually more problems, 92% said they were very satisfied, 4% satisfied and the other 4% neutral, but none of them said they were dissatisfied.

"Never in my entire career have I had a patient who regretted the operation," Dr. Iván Mañero tells me, insisting that, beyond genital sensitivity, what's important is that personal satisfaction is usually 100%. His extensive experience also makes him completely convinced that when a patient decides to make the step, their sexual identity is totally solid and has been since late childhood. This is a very difficult point. Iván explains that there are boys and teens who have striking gender dysphoria, feel that they are girls in every aspect and it is clear that they will be transsexuals as adults, but who have to wait until they are 18 to have the operation because of the laws in effect. This may seem logical, especially because there are cases in which the dysphoria is temporary, but when there are cases that are abundantly clear an operation prior to the pubescent changes of voice, physical constitution and particularly the virilization caused by testosterone would avoid problems and be more effective. "In those cases what is normally done is to hormonally delay puberty until they reach adulthood and the patient can give completely convinced consent for the sex change," Iván Mañero explains.

In any case, this should not be interpreted as an ode to sex change surgery. There is a movement of transsexuals who advocate for accepting their bodies the way they are, accepting the lack of agreement between their minds and their genitals, and feeling perfectly male or female even if they have a vagina or a penis, respectively. For them, the operation is unnecessary, and they defend being fully accepted as transsexuals. Their gender is in their minds, not in their genitals, as is reflected in a curious aspect of sex change operations: the very low incidence of phantom penises.

Phantom Penises

Phantom limbs are one of the most peculiar medical phenomena that exist. Among 60–80 % of people who suffer an amputation of a leg or arm still have the physical sensation that their limb is still there. Some of them feel that they can move it, that it itches, is hot or cold, and the most common is having pain in an extremity they no longer have.

Described for the first time in 1871, for a long time the scientific interpretation of this syndrome was that the nerve endings in the area of the amputation remained intact and that, if activated, they could still transmit information to the brain. The hypothesis seemed logical, but it was ruled out when it was seen that operations to eliminate sensitivity in the amputated area caused no improvement. Later the prevailing hypothesis was that the area of the somatosensory cortex where the brain represented the lost arm or leg was still operative and could activate, creating sensations characteristic of the phantom limb. Following this idea, some therapies use virtual reality to construct the visual illusion that the limb does in fact exist, and that the patient can move it to positions that don't cause pain. This type of therapy does manage to reduce or even eliminate the phantom limb, reinforcing the hypothesis that the amputated leg or arm is still felt because the brain cells that codified its information are slow to readapt.

Possibly because it happens less frequently, and also because of the embarrassment science has toward sexual matters, there is little known about one of the most curious cases of phantom limbs: some transsexual women still feel their penis after being operated.

In the article "Phantom Erectile Penis after Sex Reassignment Surgery," Japanese surgeons describe a series of cases between 2001 and 2007 in which, after amputation, some transsexuals still occasionally had the impression that their penises existed. The effect usually disappeared after a few weeks, but there was one patient in particular who felt the effect for more than 6 months.

The neuroscientist Vilayanur Ramachandran, at the University of San Diego, is perhaps the world expert in phantom limbs. He was the first to design therapies with mirrors used to trick the somatosensory cortex, which later gave rise to virtual reality therapies to eliminate pain. In an article on phantom genitals, Ramachandran proposes an interesting hypothesis that has yet to be experimentally proven: if the brain of a transsexual with a male body really has a female identity, then after the operation they should feel the phantom limb with less frequency than people whose penis has been amputated by an accident, cancer or an illness. And in much the same way, people with female bodies who feel male should also have less of a feeling of phantom breasts than those who have had mastectomies due to breast cancer. The hypothesis is that the unwanted penis or breasts of transsexuals are less represented in their brains. Ramachandran says that he has provisional results that support this hypothesis, since approximately half of the transsexuals operated from male genitals to female ones that he has surveyed admitted having experienced the physical sensation of having a penis at some point. Yet in people who have had their penises removed for other reasons unrelated to transsexuality, the figure reaches 70–80% and the feeling of phantom penis is more intense. If his results are confirmed, according to Ramachandran it will be very solid evidence that our brain has, from birth, a body image that in the case of transsexuals may not correspond to their genital sex.

Why Are Men Attracted to Transsexuals?

Continuing with oddities, one of the most singular is the fascination that some heterosexual men feel for transsexuals.

During the preparation of this book I remember meeting a transsexual who used to be a prostitute, and who still worked as an erotic performer. She was a truly spectacular woman who could certainly dazzle any man who didn't know she had a penis. But she told me something that left me thinking: "I haven't had the operation because I make a lot more money this way." Both when she worked as a prostitute and now as a performer, she says she has many more clients thanks to her penis. All her clients are straight guys—gay men like men, not women—who, when given the choice, prefer to pay for a woman with a penis than another equally attractive one without. Why? At first glance this seems illogical both from an evolutionary biology perspective and because of media influence, since it isn't something encouraged by society. But if it turns out to be as common as I've heard, then there must be some reasons behind it.

I remember that Ogi Ogas, author of the book *A Billion Wicked Thoughts* in which he analyzes the contents of millions of searches on the main erotic websites, told me that one of his primary observations was that porn videos of women with penises are massively searched on the Internet. Ogi defends that we men are fascinated by penises and have genuine interest in transsexuals. This is contentious for several reasons: because his work is not published in academic journals; because the interest could also be due to the growing curiosity of regular porn watchers; and finally because one thing is looking and fantasizing and another thing is doing. But it is true that Ogi Ogas's work on the online world reflected the same offline experienced described by transsexual women: many heterosexual men feel sexually attracted to them. But is it a real, consistent preference or a passing fancy? Is it specifically about the penis or not? Does it cover for some repressed bisexuality or homosexuality? Those are the questions that everyone I asked answered with their own opinions and personal experiences. But that's not good enough data for this book. Had anyone done a systematic study on the subject? Bingo!

In 2010, sociologists Martin Weinberg and Colin Williams published a detailed work in which they reviewed all the academic literature on heterosexual preference for transsexuals. They also supported their research results with 2 months of detailed interviews in a transsexual bar, both with the trans women and the men who courted them.

Weinberg and Williams explain that they ruled out men who had a transsexual woman as their romantic partner, since there could be various factors involved in that and they wanted to specifically analyze men who felt erotic desire and sought out casual sex with transsexuals, preferentially. In particular, they wanted to find out what it was that attracted them, and if the type of relationship established some indication of their sexual orientation.

The bar frequented by transsexuals was in a working-class suburb, in an area with motels in an unspecified American city. The study was based on some initial weeks of observation and a later questionnaire with 25 questions distributed in the same bar to men who came there to be with transsexuals. Among all those interviewed, the sociologists selected the subgroup who truly identified as sexually interested in transsexuals. Of those, 50 % defined themselves as heterosexuals and the other 50 % as bisexuals. Sixty-eight percent had gone to college and 31 % were in a stable relationship with a woman. Of the bisexual group, 46 % went to that bar more than once a week, as opposed to only 14 % of the heterosexuals. Of those who eventually had sexual relations with transsexuals, 54 % paid for them. The most common activity, on 62 % of occasions, was receiving oral sex, while 38 % practiced anal sex, 31 % were masturbated and 7 % masturbated the transsexual. When they were asked about their general preferences, 82% of the clients declared that their first preference was a conventional woman, and only 16% said they were indifferent. This was quite surprising to the researchers, since they were expecting more of an explicit predilection for the transsexuals. When they dug deeper into the motives for frequenting the bar, they found that many responded that it was because the transsexuals were so attentive, affectionate, inviting and kind, and it was so easy and fun to interact with them. The authors interpreted that this type of client liked the "sexual sociability" and personalities of the transsexuals more than other women who devoted themselves to prostitution.

Another aspect of the reasoning had to do with the transsexuals' exaggerated femininity in their behavior, expression, dress and sensuality, which various men defined as very exciting and generally superior to conventional women. Some also said that the trans women were very good at sex and really applied themselves.

But, when analyzing the importance of the penis, there were big differences between the men who had defined themselves as straight and the bisexuals. The latter group clearly felt an attraction to a body with breasts and a penis at the same time, while the heteros didn't mention the penis and usually ignored it. In fact, sometimes some of the bisexuals practiced fellatio on the transsexuals, but none of the straight guys did. The heteros seemed to have certain cognitive dissonance about the penis. They acted as if they were with a spectacular, sensual, exciting and affectionate woman, with a penis that neither attracted them nor bothered them. Some of them admitted that they avoided thinking about it, avoided all sort of contact with it and just received fellatio. Only 23% of the heterosexual group said they felt interest in anal sex, compared to 62% of the bisexuals.

The final conclusions were that there was a very well-defined pattern of men with some bisexuality who were specifically excited by the possibility of being with a woman with a penis. And, on the other hand, other regular clients simply saw the transsexuals as tremendously attractive, welcoming, fun women with a very tempting and very feminine attitude, with whom they had a good time. It wasn't the fact that they had a penis that attracted them and, in fact, they neutralized it.

The authors recognize the many limitations of their study, starting with the small sample size and its being located in a particular setting. They admit that it is impossible to draw general conclusions based on their results, but they cite other studies, conducted in 2000 in Holland and 2005 in San Francisco, with similar proportions of bisexual and heterosexual clients. Of course it is not a general pattern, but it does reflect that many men, when they are with a transsexual, see her as a woman, feel attracted to her and prefer her without hesitation if she is attractive, sensual and welcoming. One of the few cases in which appearances are not deceiving.

18

Marrying Social and Sexual Monogamy in Swingers' Clubs

I'm sitting with María on a red sofa in, supposedly, one of the finest swingers clubs in New York City. When we went in and told them that we were both new, they explained that the most important thing was respect, and told us not to worry, that no one was going to pressure us in any way. Couples come there to have sex with other couples, to have sex with each other but surrounded by other people, to touch and be touched by strangers, or simply to watch. We could do what we wanted. Each swinger establishes his or her own boundaries and strictly respects the limits of others. The club had a common area with a dance floor, sofas and a couple of bars serving drinks. They don't sell alcohol, but they let people bring their own. In that area we could be dressed and chat with other visitors, and even have dinner if we were hungry. That night the buffet included salmon, some kind of stew, salad and vegetables. Everything was included in the 125 dollars per couple entrance price. It was a Saturday and unaccompanied men were not allowed, but single girls could come in for 20 bucks.

In the club there were private rooms for those who wanted some alone time. And if we were up for it, we could head down a long hallway to the various group sex areas. The only condition was that there we could only wear—at most—a towel wrapped around our waists. At the end of the hall on the right side were the dressing rooms and showers.

As we have already discussed in this book, psychologists who investigate the role of emotions in our behavior are constantly proving how bad we are at predicting how we would react in future situations of great emotional intensity. When we imagine the future, we do so depending on our present emotional state, and we are usually wrong. Neither María nor I ever would have thought that 20 minutes after walking into a swingers club we would be naked under a towel sitting on a giant mattress surrounded by couples having all possible sorts of sex. Honestly, don't put much stock in what you believe you would or wouldn't be capable of in a situation like this one. It's better to just give it a try if you have enough interest.

It doesn't matter what María and I ended up doing or not, together or separately, in one room or another. Nor what we saw that night, no matter how explicit it seemed to us at the time. Talking to the swingers, we learned that each club had a different atmosphere and that some nights are wilder than others. "Saturdays usually have a more festive vibe," said the doorman, "Because there are more first time couples who are curious, a lot of people who are in New York as tourists and trying something new, and groups who come to watch and are unlikely to get involved. But with so much going on, the party usually really gets going, like tonight, where things are particularly electrifying."

Joe must have been about 55 years old and looked like… well, he was naked. But a few details seemed to suggest that he had a high cultural and socioeconomic level, and that he and his wife were completely on the same page. He defended group sex as "perfectly natural," since our species isn't monogamous, and social restrictions created unhappiness and cheating. He assured me that his behavior was in line with natural selection and that our ancestors were "swingers." "Maybe," I replied, "but, with all due respect, natural selection would also lead you to attack that guy who is touching your wife…" "Jealousy is no good, and has to be corrected!" snapped Joe.

I'd like to stop and analyze Joe's arguments because, beyond justifications and evolutionary pretexts, the subject of jealousy is very interesting. Evolutionary psychologist David Buss defends that jealousy creates a sexual tension in couples that could even reinforce their bond. It's as if it lit up the first alarm instinct and forced us to invest more energy into our relationship and compete with the other person to maintain our partner's attention and interest. In line with this hypothesis, neuroendocrinologist Sari Van Anders, who works at the University of Michigan studying specifically how hormones fluctuate depending on social context, published in 2007 that levels of testosterone both in swinger and polyamorous men and women were significantly higher than those in married and single folk, and that this reflected the physiological response to the visceral call of competition.

To be exact, the testosterone levels in women rose radically in polyamorous practitioners (people who have various simultaneous romantic relationships with full consent of all parties involved), and less so in swingers (people who generally only have one romantic partner but practice group sex or swapping recreationally without involving love). In the case of men it was already well documented that married men usually have less testosterone than single men, but both swingers and the polyamorous had significantly higher levels than single guys. This could be interpreted as a predisposition, but Van Anders correlates it with the higher tension and sexual desire that having several partners brings about.

Swingers usually agree. One of the many I spoke with told me that he got very excited watching his wife make love to another man. He swore that it viscerally heightened his desire and that for days afterward his relationship was hotter and more intense. In fact, they all told me that it's very common to get home after a night of group sex and make passionate, private love with your partner. Really, many scientific studies have documented that suspicions of infidelity can increase sexual desire toward one's partner,[1] and particularly in the case of swingers, a British study entitled "Management of Jealousy in Heterosexual Swinging Couples" offered data confirming that many of them used the swapping as a way to generate more passion within the couple.

Swingers are fully convinced that their activity strengthens their relationship rather than weakens it. But there is little data on the subject. One of the first empirical studies was carried out in 1988 by the sociologist Richard Jenks, who gathered a large amount of information and opinions during a massive convention of swingers held in the United States. His initial results indicated that the vast majority of swingers say they feel completely satisfied with their lifestyle, but when he made a more detailed analysis, problems showed up. Jenks conducted a specific monitoring of couples who were going through problems, and while in 14% of them the practice did seem to reinforce harmony in the relationship, in 16% of cases the swingers themselves admitted that the partner swapping had contributed to their separation.

Richard Jenk continued to study the swinger lifestyle, and in a review of all the studies published up until 1998 he concluded that swinging did present a challenge for the relationship, especially in couples that had unresolved conflicts, but that in most cases it was usually enriching. He also observed that, unlike what most of society imagines, the most common swinger profile isn't that of a couple who's in crisis or unhappy, who are lacking something that their lover can't give them, but rather it is a clear rebellion against

---

[1] Even though it's an anecdotal case, I can't help but mention the comical situation a friend confessed to me. He was on a long car trip with the woman who was his girlfriend again after a year and a half of separation, and they were using the travel time to tell each other about their various affairs and relationships they'd had during that time. I vividly remember my friend telling me with a surprised expression: "I don't know what was going on with me. I was driving and every time she told me about some affair she had with another guy I started to get hard." Which is perfectly in keeping with studies that link jealousy with heightened sexual desire.

conventionality and a search for new sensations within the context of the couple itself. This "in the context of the couple" is a very important point.

Another anthropologist, whom we'll discuss shortly, William Jankowiak, of the University of Nevada, also conducted studies with swinger communities to analyze whether this activity is a response to dissatisfaction in the relationship, and his conclusion was that their levels of satisfaction were no different than that of conventional couples, and that the practice isn't about "a 5-hour pause to have sex with other people," or an "open relationship that allows for sporadic flings" but rather a way of conceiving of the relationship and having group sex together, not separately. Perhaps the most delicate factor of the swinger lifestyle is sexually transmitted diseases, which despite meticulous hygiene practices usually being promoted, have been documented as proportionately higher among members of this community.

But getting back to the evolutionary arguments, I have repeatedly heard swingers and polyamorous people defend their behavior as "more natural." And I am very interested in critiquing this argument. First of all, because it is only partly true, but mostly because without realizing it they are falling into a rudimentary naturalistic fallacy. In philosophical terms, the naturalist fallacy is associating "natural" with "good" or "justifiable" only when it is in your interests. In discussions about sex it is overused, and Joe was a clear example. He decided that the sexual monogamy constructed by society is something to be corrected, since it is deeply anti-natural, and that what's natural is having a stable partner with whom to share child-rearing duties but also have sporadic sex with others. According to him, this is the instinct that our species has inherited from our evolutionary past, and perhaps he's right. But, at the same time, Joe believes that jealousy and possessiveness should be overcome in order to offer freedom and respect to your partner, when they are also completely natural. Joe and many authors who recur to nature for evidence to support polyamory are guilty of the same contradiction as a clergy person defending celibacy but criticizing the supposed anti-naturalism of homosexuality. The naturalistic argument is truly a fallacy, philosophically it doesn't hold water, and if I'm harping on this it's because it is all too frequently cited in our society.

It makes the strange implication that what's "normal" is morally more permissible or justifiable. Which is patently false. The moral norms we establish must necessarily be above any evolutionary determinant, or otherwise we would be dangerously close to justifying xenophobia, sexist roles or academic postures that argue that pedophilia should not be classified as a mental disorder.

Knowing the genetic instructions we are born with is overwhelmingly positive, not to use as morality criteria but so our cultural growth can allow us to maximize some and struggle against others. Really, quit using the theory of evolution and the naturalist fallacy as arguments. They aren't valid. And, what's more, in the case of defending polyamory, they could lose all meaning.

Polyamory with Emotional Monogamy

The concept of polyamory is wider than the concept of swingers. The polyamorous not only have sex with people outside of their couples, but they also defend the idea of being in love with more than person at the same time, and having simultaneous, consensual romantic relationships with total honesty on all sides. Some also argue that their lifestyle is more coherent, and undoubtedly many of them achieve this balance and are very happy with it. But the data seems to contradict the supposed naturalness of polyamory: it isn't so common to be passionately in love with two people simultaneously, nor to be able to be happy for very long accepting that our beloved is sharing more lust and intimacy with their other partner. Again, it is natural to feel sexual desire and sincere love for various people at the same time, and anyone who has had an affair can attest to that, but accepting that your partner also feels that for others doesn't form part of the internal logic of our brains. And, ultimately, as hard as it is for the polyamorous to admit, this situation usually brings unhappiness.

For years, anthropologist William Jankowiak has been studying polygamous Mormons in Utah, polygamous communities in Mongolia and the characteristics of sexual behavior and prostitution in China. My conversations with him were very revealing. In his work with polygamous Mormons in Utah he observed something curious: when he directly asked the husband and his wives, the man always responded that he had no preference among his wives, and the women all assured him that they accepted the relationship model and felt totally happy with it. However, when Jankowiak followed their behavior over several days, and consulted with neighbors and acquaintances, it always turned out that there was a favorite wife, and that some of the others were deeply unhappy. The equality was false, there were different roles. Jankowiak clearly observed that with some there was more passion, with others there was more companionship, with others more comfort and some showed significant prostration. Identical patterns were observed in other polygamous cultures, and even in normalized relationships that married men, in some parts of China, established with prostitutes and madams. Jankowiak says that these men feel fraternal love for their wives, passionate love for the prostitutes, and that they very often fall romantically in love with the brothels' madams. In all of the polygamous settings he has studied there is usually a complementary relationship. In order words, "you can love various people at the same time, but rarely in the same way. Or at least in a stable way, since sometimes the roles are swapped," Jankowiak told me.

With similar methodology, Jankowiak began to study polyamorous communities and—with the obvious caveat of the enormous distance created by the sexism in polygamous relationships—he is observing some analogous patterns. While most polyamorous people, when asked directly, say they are perfectly happy and love all their partners equally, those around them say there are obvious preferences, differentiated roles, and unhappiness and anxiety in those who feel they aren't receiving enough attention. The preferred man or woman is usually satisfied, but the one who feels they are second or third shows clear signs of despondency in the medium term.

Jankowiak recognizes that he still has not gathered sufficient data to confirm these tendencies, but he points out something very interesting: the very few empirical studies conducted on polyamory have been done by academics who are part of those communities, which Jankowiak feels means they cannot be objective. I was able to see that for myself during the meeting of the International Academy of Sex Research held in June 2012 in Portugal. There was an entire session devoted exclusively to polyamory, where in theory the latest data on the subject was presented. In addition to Sari Van Anders, who detailed the different levels of testosterone in polyamorous men and women as opposed to conventional couples, the rest of the presentations were by defenders of polyamory who offered, more than empirical data, definitions and lectures on the advantages of the lifestyle. Speaking with one of them, researcher Alex Iantaffi, after the session, she admitted that the complementary effect did exist in many polyamorous relationships. Even a passionate supporter of polyamory like filmmaker Tristan Taormino, when we spoke in New York, ended up arguing that many people start polyamory to make up for something that is lacking in their partnership, and that it is very difficult to maintain stable relationships.

Even on a neurochemical level, anthropologist Helen Fisher in her book *Why We Love: The Nature and Chemistry of Romantic Love* defends the idea that there are several types of love. Fraternal love increases oxytocin levels and would be easier to share, but passionate love diminishes serotonin and increases dopamine creating that unhealthy obsession focused on one person that we call falling in love. Many of us have been romantically in love with two people at the same time, but the effect doesn't last long and usually ends in tragedy. Because, in reality, we are emotional monogamists.

As we saw in Chap. 8, there are various levels of monogamy. It is clear that sexual faithfulness is a social construct without the slightest evolutionary meaning, that instinctively both men and women feel desire for many people at the same time and that our brains aren't programmed to act in a sexually monogamous way. Social monogamy—a nuclear family formed by two individuals—is more favored in our species, but when looking at our ancestors it also seems that we, especially men, can feel love and family commitment to various people at the same time. On the other hand, we are emotionally monogamous, we only fall passionately and addictively in love with one person at a time, and we experience possessive feelings toward our partners. Jealousy forms part of our deepest nature. Actually we are suspicious and promiscuous, and what's truly natural is for us to want to have several relationships with people who are faithful to us. And therein lies the paradox. Polyamorous people tackle it by fighting against jealousy, conventional couples with an agreement of social monogamy that in theory is worth it for them, and swingers by opening themselves up to the possibility of sporadic sex, which in strict terms of our primate nature is the most coherent.[2]

Partner Issues When Desire Wanes

There is a question, almost of a philosophical nature, that profoundly affects some of our everyday concerns and dissatisfactions: can we desire what we already have? Take some time to think about it. It's a question I find deeply disconcerting.

Obviously we can appreciate, admire, enjoy, care for, miss, protect, take pleasure in, prefer, love, want… but desire with all our might something that we already have no risk of losing? Of course we *can*. But it is more complicated because the neurological circuit of search and motivation doesn't get activated, so there is no dopamine and serotonin fluctuating between various parts of our brain setting off emotional charges of anxiousness, reward and satisfaction.

---

[2] This text should in no way be interpreted as a criticism of polyamory. I have met many people who practice this lifestyle and I admire their laudable efforts to maximize their positive feelings of love toward various people at the same time. Their honesty and ability to communicate are impressive, and many of them live full lives. But psychology does not agree with their vision that polyamory is a more natural state than social monogamy, nor that you can be passionately in love with two people at the same time, nor that jealousy in a couple is something negative to be avoided. Many of us would not tolerate being with someone who didn't feel any jealousy over us.

It is true that the dichotomy between desiring and having doesn't apply equally to all stimuli, habits and individual personalities, but when referring to sex, all the sexual therapists I've consulted assure me that the most frequent problem they deal with is the loss of desire and sexual interest in couples. Helping couples overcome infidelities is also a big one, but sex in the marriage is definitely more of a problem than extramarital sex. And it's not only a question of the relationship being in crisis. There are many couples who love and adore each other, but who have lost their passion, attraction and sexual chemistry.

Here I again want to stress a fundamental nuance: lack of sex is not necessarily a problem and there are no quotas or external pressures that should be paid attention to. What is important is satisfaction. There are countless couples for whom sexuality has passed into the background of their relationship, but who are perfectly happy enjoying other aspects of life, and neither of them feels they or their relationship are missing anything. But there are also many others who used to enjoy a terrific sex life, still feel full of life and have fantasies, but for whom the loss of desire for their partner creates tension, apathy and is negatively affecting their relationship.

Unfortunately it is such a frequent and important problem for our wellbeing both as individuals and as couples, and so difficult to tackle scientifically. In this book we have talked about attraction, falling in love, biopsychosociological keys to desire, physiology of male and female sexual response, various dysfunctions of both genders, learned sexual behaviors, genetic determinants and the vast diversity of human sexuality. But when we add in thousands of other family, work, and conjugal factors, coming up with a diagnosis and therapy for a couple who has lost their desire for each other is—at this point in time—still more art than science.

I remember discussing the subject with Julia Heiman, an eminent sexologist and researcher who was at the time the director of the Kinsey Institute. We talked about the Masters and Johnson models, Helen Kaplan's approach, the overwhelming variety of techniques, books and sexual therapies now on offer, and she told me: "There are too many therapists proposing all sorts of solutions to sexual problems. Some of them are good, others more or less appropriate to each case, and others that are truly aberrant. We need professionals with more solid formation, and better scientific knowledge of the effectiveness of each therapeutic approach."

Heiman maintains that the problems of sexuality in couples will always continue to be analyzed in a personalized way, but she asks that the people who do it be psychologists and specialized therapists who keep scientific criteria in mind. She cites, for example, a therapy based on using laughter to relieve tension and facilitate sexual predisposition and says: "It's logical in theory, and I'm sure it can stimulate a conventional couple, but I don't know if those who lead it have enough training to be able to distinguish hidden problems that are not at all helped by it, and it could even be counterproductive because it will keep them from seeing better professionals." With her extensive experience, Julia Heiman sees that the approach to treating lack of desire in couples has evolved and diversified over time, but that little scientific attention has been paid to analyzing what models are most effective.

Apart from medical problems—which should always be considered—and focusing just on cognitive and behavioral therapies, there are those who put more weight on psychotherapy for exploring conscious and unconscious conflicts, searching for the origin of the lack of desire, or trying to control emotional reactions, while others use more behavioral approaches to promote changes in the couple's attitude.

Accordingly, sexual exercises are very common for couples who, under the umbrella term of "sensate focus" developed by Masters and Johnson, are designed to touch, feel and explore both their own and their partner's body with the goal of learning about their sensual reactions and increasing arousal. There are a vast number of books on sensate focus exercises, and therapists who design individualized programs and guide couples through periodic sessions. The basic idea is forcing sexuality a bit to see if it generates desire. There are all sort of techniques, including tantra or meditation, incorporating erotic toys and pornographic films, or planning specific sexual activities that the couple might not have dared to propose before. In fact, one of the most frequent comments of the sexologists I spoke with was how surprisingly common it was to find couples who are resistant to talking to each other about their fantasies and sexual preferences. Improving explicit communication is a general need.

Since Masters and Johnson, therapies based on exercises and direct communication have increased, but so has a growing tendency to work harder on eroticism and sensuality than on sexual function and practice itself. Sex motivated by desire is much better than sex for pleasure's sake. Marta Meana, a sexologist and researcher at the University of Nevada, expressed it perfectly: "In therapy what is difficult is not getting a couple to have more sex, but getting them to want to do it." I also met Marta Meana at the International Society for Sexual Medicine conference in Chicago, where she talked about her studies showing that most sexual problems between couples are not due to functional or physiological aspects, but rather to the loss of desire, and that eroticism and satisfaction deserve more attention than functionality. Meana stresses that they have to be adapted to the expectations of each couple's age and that there has to be a multidisciplinary vision, but she warns of the risk that the growing fragmentation of unproven therapies leads to a devaluation of sexual therapy.

An interesting example of the controversy on how to confront the loss of passion in a marriage revolves around the question "can you want what you already have?" Some therapists believe that the trust, proximity and security of love in a couple can, in the medium term, come into conflict with the adventure, mystery and novelty needed for the emergence of sexual desire; accordingly, increasing distance in the couple and fostering individual activities has positive effects because it staves off apathy, increases insecurity and pursuit, and awakens desire. On the other hand, many therapists advocate for the opposite, promoting greater unity and shared activities. It is a paradigmatic example of a situation in which there is no empirical data to evaluate what strategy works better and under which circumstances, and each therapist has his or her own preferred method, to which they cling.

In any case, psychologists have for years been investigating and comparing the effectiveness of different treatments, and a curious conclusion the studies come to is that the patient's attitude and connection with the therapist usually has more of an influence than the concrete method followed. I haven't found specific studies on lack of desire, but it is very probable that this effect is also true in sex therapy. Generally, doing something with good intentions and trusting in the person guiding us will bear good results, definitely better than ignoring the problem and letting it grow into a trauma. Resignation is the worst.

I don't want to give the impression that there are no research projects and scientific studies on couple relationships. Of course there are (few specifically on sexuality), and to cite one of the most paradigmatic examples we could mention the work of psychologist John Gottmann, who, after analyzing the interaction within middle-aged couples, developed various models to predict which married couples would stay together and which would separate in 4–6 years. His results, published in various scientific articles during the 1990s, showed a higher than 80 % accuracy rate. Of course many factors had a simultaneous influence and there was never a single determining one, but in addition to signs of conflict, boredom and individual personality, Gottmann established that the four most relevant factors for predicting a divorce were: constant criticism, staying always on the defensive, showing scorn and systematically refusing communication.

In a more neurobiological version that is reminiscent of Gottman's work, researchers from the State University of New York at Stony Brook in September 2012 published a curious study that identified different patterns of brain activity in couples that were going to stay together from those who were going to separate. Some members of the psychology department of that university have long been studying the different phases of romantic love using fMRI brain scanners. Among them are Arthur Aron, Lucy Brown and Bianca Acevedo, whom I met along with Helen Fisher in 2008, when they were starting to lead the neuroscan love study. The peculiar thing about the 2012 work is that they compiled various brain scans from their first studies with couples in love, and they contacted them again 3 and a half years later to ask them if they were still together or had separated. They analyzed the initial brain activity, and they saw slight differences years later between those who broke up and those who didn't, particularly in areas of the orbitofrontal cortex and right parts of the nucleus accumbens and the cingulate cortex involved in long-term love and relationship satisfaction. We are all aware of the big limitations of fMRI in interpreting our behavior, but the researchers believe that observing cerebral activity could predict certain stability in couples. The authors don't suggest that it should be used in therapy, but John Gottman has developed methods of marriage therapy that he classifies as scientific.

Somewhere in between we can find studies such as Brent Atkinson's, "Emotional Intelligence in Couples Therapy," whose main idea is that knowing the brain's neurobiology could allow us to better predict our reactions and develop an emotional intelligence of great therapeutic value. For example, in an experiment it was observed that when in-love volunteers saw images of their partners tense and stressed, they also felt sudden, immediate stress. This showed the intimate connection and empathy that exists in romantic couples, but it also explained the paradoxical "little patience" that some of us have when our partner is tense over some problem, just when we should be supporting instead of reacting. Atkinson argues that by accepting the imperfections in our brain we can deduce that it is more important to respond well when our partner has an unfair or inconsiderate reaction, than to try to avoid at all costs being unfair or inconsiderate ourselves. We should use our mind to control the muscular tension, heart palpitations, shrinking stomach, and contain our stress and aggressiveness. Just as we saw that accepting and naturally controlling unwanted sexual fantasies is better than trying to suppress them, since doing so generates a counterproductive rebound effect that reinforces them, being well acquainted with our emotional brain will lead us to conclude that we shouldn't struggle to inhibit our emotions, but rather understand them and redirect them.

In any case, many still disagree with the idea of applying scientific solutions to marriage therapy. I remember specifically talking about sexual crises in couples with Stephen B. Levine, a very famous sexologist, researcher, author of sexuality textbooks and experienced therapist, and that he was firmly convinced when he said, "Look, when a couple with lack of desire comes into my office for therapy, it's not one case, but three. I have to analyze each of their circumstances and those of the relationship between them. It's difficult to analyze individual people, so imagine what it's like to combine all those factors. Any attempt to understand that scientifically is futile." Levin is categorical: "Here in the United States there are couples who stay together for financial reasons. They commit infidelities, they ignore each other and they aren't in love at all but they stay together out of fear and convenience. Then when you see them in therapy they tell you whatever story they can come up with as if they didn't want to face up to the reality. You don't know if they are trying to lie to you, or if they are deceiving themselves. Often it's the latter. And they complain of a lack of desire between them as if that were the problem… do you think there's a model that could capture that?"

Stephen Levin is convincing and makes one see that it would be difficult for science to tackle the great diversity of factors that intervene in every relationship, but that doesn't mean that scientific investigation should give up on trying to offer keys. For example, Levin has written extensively about infidelity and is one of the many therapists who relativize its importance and believe we stigmatize it excessively. They say that sometimes the feelings of guilt can turn out to be positive, even though they admit that there is a great risk of having a relapse after a first slip. And that it depends on each case, since there are external influences and people who are more inclined toward infidelity. Here is where science has something to contribute.

Genes Don't Justify Infidelity

Throughout this book we have seen that more than 90 % of men and 80 % of women have sexual fantasies about someone who is not their partner, that in nature there are socially monogamous species but not sexually monogamous ones, and, clearly, having extramarital temptations is absolutely natural. But do some have more than others? Undoubtedly distance, how close the couple is, their previous experiences and their need for reaffirmation, the surroundings, the available incitements to cheating, and other countless factors contribute to having more or less socioculturally illicit thoughts. But are there people who are more prone to cheating than others? The answer seems to be a clear yes, and they have even identified specific genes associated with it.

It's not that they're exactly "infidelity genes" but they are genetic alterations, some associated with a lack of attachment and others with restless personalities, which can lead to a higher predisposition to infidelity.

With regard to the less attachment, research goes back to that famous species of prairie voles that form monogamous couples and permanent family ties, unlike other species of practically identical voles that, like most other rodents, are promiscuous and never form stable partnerships. Intrigued by this clearly evident difference, scientists searched in the brains of the voles to find that the monogamous ones had many, many more oxytocin hormone receptors in their limbic systems and that their vasopressin gene had various repeated sequences as compared to other rodents.

It was already known that oxytocin and vasopressin were hormones involved in the feeling of attachment between mothers and children, and several investigations confirmed that they were also involved in the affection and closeness in couples. In fact, injecting or inhibiting these hormones and genes in the brains of voles could convert monogamous species into polygamous ones and vice versa. That was when oxytocin started to be called the love hormone, and the suspicion emerged that it could have the same function in humans. That was indeed confirmed but, in addition, various studies revealed that among us there are also people who have some polymorphisms or others in their vasopressin receptor gene. Swedish researchers did a long study that concluded that men who have repetitions in the R2 allele of the AVPR1A gene—related to the vasopressin receptors in the brain—usually felt less satisfied in their relationships, had a lower estimation of commitment and, as a result, had a greater predisposition to infidelity.

As for the restless nature, studies are focused on the metabolism of dopamine (the hormone involved in motivation and pleasure), whose differences in levels and number of neuronal receptors have been associated with compulsive personalities and what is called novelty seeking, which among other character traits also entails a greater possibility of infidelity.

Specifically, Justin Garcia, of the Kinsey Institute, whom we spoke with in Chap. 9, in 2010 published an association between infidelity and polymorphisms in the dopamine D4 receptor gene. Of the 181 individuals analyzed by Justin, those who had a repetition in the 7 (7R+) allele showed a 50% higher probability of committing an infidelity than those who didn't. This genetic polymorphism was already associated with the search for sensations and impulsivity, and is more frequent in immigrants, innovators and extreme sports enthusiasts, but also in people who suffer from attention deficit and hyperactivity, addiction to drugs and alcoholism. In other words, this gene would provoke the constant searching for new sensations and increased dopamine. The metabolism of dopamine is really a key factor, and having less D2 dopamine receptors in the brain's striatum—which communicates the pleasure center to the cerebral cortex—has also been related to drug addiction and with breaking social norms, like faithfulness.

There are countless other studies of relative importance on cheating behavior. It has been observed that women with higher estradiol levels and men with higher testosterone are more unfaithful, that the higher one is on the business hierarchy the more chances for adultery, that women perceive deep voices as less faithful, and that men who are more economically dependent on their wives have more possibilities of infidelity, while in women the opposite is true. Who knows. A serious study published in 2011 in *Archives of Sexual Behavior* concluded that personality has more of an affect than demographic data (education or socioeconomic position) on infidelity. The study also reinforced the stereotypes that in men infidelity is associated with sexual anxiety and excitability, and in women, to unhappiness and the state of their relationships.

In a sane brain, behavior is not only dictated by desire but also by self-control, but yes, we are hard-wired to feel the call of infidelity and this enters into clear conflict with the relationship model that Western society has constructed. That is why there are such high rates of cheating, frustration, and different models like swingers or the polyamorous who seek to "marry" the powerful but sporadic instinct of the sexual affair with the marvelous, constant and no less powerful impulse of romantic love.

Love Addicts

Better alone than in bad company, that's for sure. But all studies indicate that the state of highest happiness, emotional wellbeing and physical health is located within a satisfying romantic relationship. Of course there are exceptions and ups and downs, but the tendency is unequivocal: solitude ends up being harmful, and our impulse toward romantic love responds more to an instinctive need than to an emotion or feeling that comforts us. We don't chase after it because it makes us feel good, but because we need it. And when we find it, we become addicted to—much more than to sex—that love in particular and none other.

It all starts in the brain's pleasure center, also called the reward system. Located in the limbic system (our emotional headquarters), this is what compensates us with pleasurable dopamine when we do something good for our bodies. When we eat, have sex, exercise or overcome a challenge, the tegmental ventral area rewards us with bursts of dopamine that reach our nucleus accumbens along mesolimbic channels, generating a wonderful feeling of euphoria and wellbeing that we want to repeat as soon as possible. The dependency is so great that if something like drugs manages to increase dopamine levels radically and consistently in the reward system, it can become an addiction.

Studies with brain scans have revealed that when someone is enjoying being with the person they love, the pleasure center activates and starts to secrete intoxicating dopamine. But not just that. Other areas of the brain also activate, related to perception, memory and learning. We are learning that it is this person in particular and not someone else who generates pleasure in us, and every time we have a good time with them the neurochemical bond is strengthened. That is key in love. We are addicted to food, but not to a specific dish. On the other hand, during a period of falling in love that is the only person who is activating our reward system and reducing the connection with our frontal cortex. And if the sex is overwhelmingly satisfying too, the dopamine levels are even higher and the bond is even greater.

The bond with that person who has been giving us all sorts of pleasure over weeks or months is so strong that the attraction is not only manifested because being with them drives us wild, but because just thinking about them, seeing a photo, getting a message or imagining a shared future also activates the dopaminergic channels, generating wellbeing and turning on the pursuit circuit that makes us desire them even more. The information between the pleasure center and the outer layers of the brain is traveling in both directions, as if we were addicted to heroin or tobacco, and the slightest abstinence produces anxiety. We are hooked on that person and only them, because they are the only one who gets us to secrete dopamine just through our memory and imagination. And like any addict, when we don't have them by our side we not only miss them, but we need them. They are the person we have associated with pleasure, desire and the future, and in these circumstances, the satisfaction we could get from sex with a stranger is laughable. Our imaginations already secrete more arousal and dopamine than any other stimulus.

Let's put aside for a moment the loving oxytocin that offers us a feeling of wellbeing and attachment after orgasm. It relaxes us and makes us feel affectionate, but it doesn't make us fall in love. It calms us, but it doesn't arouse us. Oxytocin is the hormone of lovey-dovey love, but not the hormone of addictive love that lights up our dopaminergic pursuit circuit at the mere thought of seeing our beloved again. When we do and the search is over, our cortisol drops with each hug and stress dissipates. We are no longer anxious and we feel calm and relaxed. Soon, familiar oxytocin will make us feel connected to that person and we will start to express our affection. But be on the lookout for marital boredom, which affects so many couples who love each other but don't desire each other. Celebrate your successes together and search out novelty hand in hand. Our brain must always keep strong the link between that person and pleasure, because if it starts to associate pleasure with other activities or stimuli, that's when marital boredom begins.

Avoiding conflicts isn't enough. In fact, it is better to face up to them with fights or sex than hide them and enter into a state of mediocre tranquility. If there is no emotion, romantic love dies and all that's left is companionship.

Fight with all your emotional intelligence and the strength of your frontal lobes against that. Know how your brain works, and make sure you figure out what romantic love needs to stay alive before it's too late. It is the best of all addictions. Feed it constantly by creating mutual shared pleasures that make your irrational brain secrete dopamine just at the thought of that person you love so passionately, and who will alter all your possible hormonal levels from now until happily ever after.

ERRATUM

S=EX²

© Springer International Publishing Switzerland 2017
P. Estupinyà, *S=EX²*, DOI 10.1007/978-3-319-31726-7

DOI 10.1007/978-3-319-31726-7_19

An error in the production process unfortunately led to publication of this book prematurely, before incorporation of the final corrections. The version supplied here has been corrected and approved.

The updated online version of the original book can be found at
http://dx.doi.org/10.1007/978-3-319-31726-7

Epilogue: Sex and Science Don't End at Orgasm

I remember the night I was introduced to the former professional cyclist Peio Ruiz Cabestany. It was summer and I was spending a few days in Zahara de los Atunes, trying to disconnect from the process of writing this book. But it was no use, my mind was possessed—academically—by sex, and I soon found myself excitedly explaining to Peio a scientific study about women who had orgasms while exercising, which said that it often happened when riding a bicycle. Peio was disconcerted by my unexpected comment, and I told myself that I should tone down my scientific sexual asides, but then all of a sudden he answered in a faint voice, "well… that's happened to me several times, especially going uphill."

We've come to the end of the book and I still feel I have so many curious things to tell you about this peculiar intersection between science and sex. How could I leave out the orgasms induced by exercise! The study I cited to Peio was published in late 2011 by Debby Herbenick of Indiana University, whom I interviewed during my visit to the Kinsey Institute. Conducted with 246 women who sometimes felt intense sexual arousal when practicing sports, and another 124 who had even had orgasms, the first thing that Debby pointed out to me was that "it doesn't seem to be anecdotal, since when we started looking for women to take part in the study we found many, in just a few weeks." What's more, at first the researchers thought that it was an almost exclusively female phenomenon, but "when we published the study, men started to get in touch with us to say that it happened to them too, and we are investigating that now," explained Debby, citing the case of a college student on the track team who always did his abdominal exercises at home because he was occasionally surprised by an embarrassing ejaculation.

In fact, doing abdominals was the most frequent physical activity that women said gave them spontaneous orgasms, followed by climbing, lifting weights and riding a bicycle. Most women assured researchers that they were not having any erotic fantasies or feeling any arousal when the genital stimulation appeared, and they simply had the sensation that if they continued the repetitions of the exercise they were doing they would reach orgasm. They explained that it could be embarrassing if they were with other people, but that it was very pleasurable. The average age of the first "orgasm induced by exercise" was 19, and for many women it only happened a few times, but in almost half it had happened more than ten times. A few even said that they could make it happen if they tried to. For Debby it is an extremely interesting phenomenon because it represents a new example of spontaneous orgasmic reflex unrelated to desire or sexual activity, such as orgasms produced by electrical stimulations on specific points of the spine, those that happen while sleeping and those induced after taking certain drugs. The team at Indiana University is investigating which physiological mechanisms could be involved, and while the most logical assumption is that muscular tension in the pelvic region could press on the internal part of the clitoris and be the origin of the arousal, Debby suggests that the activation of the sympathetic nervous system due to physical stress could also be the orgasm trigger.

Her idea coincides with Peio's experience. He explained that erections on the bicycle were relatively frequent but that the "orgasms came in extreme situations of almost agonizing suffering." Joking about how much fun he had on his bike, Peio told me: "I remember one quite intense orgasm I had on a steep slope in Lasatte heading toward Donosti, while making an enormous effort to stay drafted."

The study of orgasms induced by exercise may seem fairly anecdotal, but Debby reflects on a very interesting point: if it is found that the muscular mechanisms involved are the same as the typical pelvic movements carried out during the sexual act, this could indicate that, in addition to clitoral stimulation, for some women the abrupt, almost spasmodic swaying increases arousal and is necessary for reaching orgasm. "And this could lead to the design of new exercises—in addition to kegels—in order to facilitate orgasmic response," Debby tells me.

After consulting more people about this issue, I realized that—especially in women—genital arousal during intense physical activity is something relatively common. So is arousal during sleep, even though on many occasions we don't notice it. Another study that I can't leave out is the fascinating review of scientific bibliography on abnormal sexual behavior while sleeping, published in 2007 by specialists in sleep disorders. Among the numerous cases discussed we find a 27-year-old woman who several times a week would start moaning with pleasure after 15–20 minutes of sleeping, another—26 years old—who, in the middle of the night, would take off her clothes and start masturbating and scream angrily without being aware of it, and yet another woman who sometimes stroked and played with her boyfriend's genitals, and when he responded and she woke up she would accuse him of taking her by force while she was sleeping.

In fact, there were countless cases of somnambulistic sexual activity: men who touch the breasts and buttocks of their partners while sleeping, others who moan and make pelvic movements, and still others who turn aggressively toward their partners and try to penetrate them. There are curious testimonials, such as the woman who tells of how her husband was much more affectionate and tender with her while he was sleeping than awake, a man who always avoided having sex with his wife when she was menstruating but didn't seem to mind that when he was asleep, and even a woman who swore that her husband snored while they were having sex.

There are also unpleasant situations documented, of forced contact, women shouting obscenities they would never dare to say while awake, and very legally problematic cases of sleeping men who have tried to abuse their partners' teenage daughters. It seems there is an absolute loss of control, as reflected in the case of a 27-year-old man who broke two fingers when, while sleeping, he tried to free himself from the knots he himself had tied to keep him from the masturbation and ejaculation that had woken him up almost daily over the prior 5 years.

The scientific information on sexuality is limitless, and this book would never end if we kept adding more and more studies to the many already cited. We could discuss a scientific article that describes a woman who was capable of having more than a hundred orgasms in each sexual session, or reports of people who go to hospitals with amnesia after a sexual encounter. It is more frequent than we imagine and, in fact, in 2006, doctors from Zamora and Salamanca described in the *Revista de Neurología* [*Neurology Journal*] the case of a 57-year-old woman who was brought to the emergency room by her husband after becoming disoriented and losing her memory for an hour and a half after having sexual relations. The woman could only remember that she had started intercourse, but not whether she had washed up after, or how she had gotten to the hospital. Episodes of transient global amnesia can appear after intense physical exercise, situations of emotional stress, metabolic alterations and the consumption of toxic substances, but occasionally also immediately after sex. They usually last for a few hours and are believed to perhaps be due to small ischemias in the thalamus or hippocampus during an abrupt rise in blood pressure, as happens during orgasm and extreme physical activity. Who knows, maybe when we wake up in the morning in someone else's house and don't remember how we got there it could be due to something more—this is just gratuitous speculation—than the alcohol we consumed or a defense mechanism that forces us to forget what happened.

In medical literature we can find more dramatic clinical cases of intense pain or depression following orgasm (post-orgasmic illness), articles on allergic reactions to semen, and see images ranging from men with two penises and women with two vaginas to clitoral atrophy or boys with aphallia, where you see two testicles with no penis above them. I'm not trying to overwhelm you with half-told curiosities, but rather to show you that science is an inexhaustible source of novel, interesting, singular and thought-provoking information in all realms of knowledge, including, of course, human sexuality. Actually, scientists' real secret is—besides their methodology—in something very simple that I recommend you take to heart: being more interested in the unfamiliar than in what we already know. In that way we will constant experience the intellectual stimulation of new knowledge, be encouraged to learn continuously, assuming that doubts are more interesting than certainties, and we will be predisposed to our ideas evolving instead of remaining stagnant.

With regard to the constant renewal of preconceived ideas, allow me to mention a recent study whose design I found fascinating. Researchers at the University of Texas recruited a wide number of students and asked them all—male and female—to read the same erotic short story. The story was 60 lines long and described a heterosexual affair that began with a romantic dinner, progressed to the early stages of seduction and ended with different sexual practices and intercourse. The text was filled with details about the food, the conversation the couple had, their thoughts and explicit descriptions of their bodies and intimate contact. After reading the story, the students were given a distracting task for 30 minutes, and then they were given a questionnaire with very specific questions in order to see what type of information they had retained from the text. As was to be expected—and therefore the study was not particularly valuable—in general the males remembered much better the descriptions of the female anatomy and the explicit erotic details, while the females recalled the emotional information and the protagonists' personalities, or the interaction in the seduction phase.

I cite this study because of its peculiarity (and again to stick it in before concluding), but also to contrast it against another study and reinforce one of the messages conveyed in this book: in reality, men and women aren't as different as we think we are, and very often we fall into the "trap" of using expressions such as "generally" or "on average." For example, take a look at this other investigation where they also asked almost 300 student—cheap guinea pigs but not very representative ones (and yes, that is a criticism)—to note down for a week every moment when they thought about sex, food and sleeping. The study has methodological limitations but contributes a very interesting result: the range of frequencies was overwhelming. The young women thought about sex an average of 18 times a day and the young men, 34, but the responses ranged from 1 to 140 in females and from 1 to 388 in males. The researchers compared the relative frequency of sexual thoughts to those of food and sleep, and they discuss the results in terms of erotophilia and social desirability, but there are two very obvious conclusions to be drawn: the first is that young men do think more about sex than young women, but not really that much more; and the second is that the differences within the male and female groups are impressively higher than the differences between the genders. This reflects the distortion that some researchers and media outlets are fostering when they speak generically about differences between men and women. As we've seen in various parts of this book, we really aren't as different as we've always been told. It is obvious that in the development of our behavior biological determinants have an early influence and then sociocultural ones come in, but the differences are so generic and form part of such a wide diversity that we should begin to change the discourse and be more rigorous when referring to things that are actually more typical of one gender or the other.

Really, it's more interesting to compare sexual and romantic behavior between homosexuals and heterosexuals; and while we're on the subject, I'll confess that I would have also liked to go further in depth in this book on aspects such as sexuality in minorities, characteristics of different sexual orientations, comparisons between cultures, the impact of the Internet and the development of sexuality in childhood. I didn't, in part, because of the scant empirical studies that exist on some of those topics. Sex is also a taboo for science, and several researchers have acknowledged to me the difficulties of financing studies, for example, on the sexual behavior of preadolescent children. This reticence affects the scientific community itself, and it is not demagogical to say that it is simpler to investigate animal sexual behavior than human. To give one example, in early 2013 a Spanish newspaper published a list of the top ten news items on sexuality in 2012, and seven of them referred to studies done with animals (squid, flies, frogs, rams and gay fish), one was about Neanderthals and only two about human sexuality. It is telling, and a shame because the medicine, psychology and sociology of sex are a gold mine of highly interesting investigations and they are not being sufficiently explored by science.

Of course I could have also talked more about sex in nature and about our evolutionary past, but it seems to me that there's a disproportionately large number of books about "science and sex" that, in order to simply the discussion, take refuge in logical stories of evolutionary psychology that are not experimentally confirmable. It is usually a very limited vision, scientific but not journalistic, simplistic, not at all multidisciplinary, and—I would add—not very rigorous when it neglects the importance of reinforcements conditioned by experience, learning and the sociocultural environment of our sexual development. Evolution by natural selection undoubtedly represents an essential conceptual framework, but it is absolutely inadequate to the task of tackling the complexity of human sexuality. One of the changes on the horizon is the burgeoning of more and better scientific research on sex.

Concluding a book by talking about the future is always a tempting recourse, and particularly when it seems that the online world could affect a very wide range of aspects of our sexuality. Accordingly, one line of thinking holds that the Internet will transform everything and that we are at the start of a new sexual revolution, and another perspective maintains that our brain and instincts are the same as ever and there won't really be any seismic shift. In most aspects of this debate, I lean more toward the latter viewpoint, especially after doing the following exercise, which is another attempt at stalling before saying farewell.

One of the books I consulted during my research phase of this project is *On Sex and Human Loving*, published in 1985 by William Masters and Virginia Johnson. In that period Masters and Johnson had already described their sexual response models, accumulated vast therapeutic and research experience since the late 1950s, and had become the primary point of reference in the scientific study of human sexuality. And as such, in the last chapter of their book, they ventured to predict how our sexuality was going to change in the following 25 years, which is to say, up until 2010. I set about to check whether they'd gotten it right.

Masters and Johnson introduced that chapter by discussing how the sexual revolution of the 1960s and 1970s was impacting on that moment in the mid-1980s, the advances in reproductive medicine and the novelty of mothers-for-hire, before, on page 558, literally saying that "there is a strong possibility that a substantial number of women will opt to have children without experiencing a pregnancy." They were referring to the possibility of a woman fertilizing her ovum with her partner's semen and hiring a surrogate to continue the pregnancy. Masters and Johnson even stated that with the development of artificial placentas, "it is quite possible that within the next decade the surrogate mother can be circumvented entirely by the development of an artificial uterus" and that, according to their data, 40% of women in their 20s would prefer that option. This erroneous prediction is a clear example of how we usually overestimate the power and speed of technological and cultural change, and undervalue the solidity of the most basic human nature, which in this case tells us that the instinct to want to become a mother is much more powerful than all the varicose veins, pains, disadvantages and risks that could be avoided by using a womb-for-hire. This is the same reason why all the futuristic projections about cybersex scenarios replacing conventional intercourse could fall flat. There are some things that really are very slow to change.

Masters and Johnson made a similar mistake when predicting that in only 10 years sex selection would be very common, that 75% of first children would be male, and that 60–65% of the second children would be female. They also foresaw that there would be almost perfect birth control options for both sexes and that, as a result, the number of abortions would be cut in half. They were wrong about the first part but right about the second, since according to data from the American Center for Disease Control (CDC), in 1984–1985 there were 24 abortions performed for every thousand women, 23 in 1986, 20 in 1995, and 15 in 2009. Masters and Johnson accurately estimated "a continuation and accentuation of the present trend of young adults to postpone marriage until later ages, so that by 2010 the average age of first marriage in America will be closer to twenty-five than to twenty-one," but they fell short, since currently in the United States it is 29 for men and 27 for women, and in Spain, 32 and 30, respectively.

Something similar happened when they predicted that sex in adolescence wouldn't change substantially and that in 2010 half of teenagers would have lost their virginity at 16, and 75% at the age of 18. According to data from 2005, 37% of American boys and 40% of girls had their first sexual encounter before 16, and 62 and 70% respectively before the age of 18. Curiously, they foresaw that after the period of sexual liberation there would be some sort of backlash where the values of monogamy and love would resurge, there would be "less emphasis on sex for the sake of sex," but at the same time new types of relationships would appear and—especially among the middle and upper classes—there would be more tolerance of extramarital sexuality. These are ambiguous forecasts and, depending on one's interpretation, could be considered to have come true.

AIDS had just broken out and the effects it would have were still unclear, but they spoke of vaccines and the appearance of new sexual transmitted diseases, such as the cancers provoked by some strains of the human papilloma virus. They were right about the increased tolerance of homosexuality, the free access to explicit sexual material, to a certain extent that sexual activity in senior citizens would be more accepted, and that some stereotypes—like that men have to "make sexual conquests to 'prove' their masculinity" and that women have much less sexual desire—would gradually start to fall away. Now we mock macho attitudes and are sometimes overwhelmed by female sexual ability.

There are more aspects to their analysis, but a significant one is that "over the next 25 years there will be a gradual but definite increase in recognizing the importance of sex education for children. This will eventually lead to the routine inclusion of sex education in public schools [...] It is also possible, but somewhat less likely, that such a widespread prevalence of childhood and adolescent sex education will in time contribute to a reduced rate of adult sexual problems, particularly those that stem from misinformation and inhibitions about sex." The educational system's progress has not come anywhere close to their prognosis but, on the other hand, the Internet has made it possible for quality information to be available in the West for everyone who really wants it.

If there is one thing that's clear after reading the final chapter of this Masters and Johnson book it is how little attention we should pay to predictions, and that if we are destined to err, we may as well make our forecasts with our heads but also with our hearts. I do believe that the differences in sexual patterns between men and women will narrow, that homosexuality will be completely accepted in future generations, that there will be more communication but that sex will continued to be tied to our deepest intimacy, that pornography is here to stay but much more sensual and stimulating erotic alternatives will arise, that not even the virtual subworld or sad sex dolls will ever satisfactorily substitute conventional sex, that women will continue to liberate themselves and that casual sex will be increasingly more frequent, that older adults will be the sector of the population that will most benefit from online dating sites, that sexual monogamy will be a growing challenge for couples, that sexual medicine will establish itself as a discipline in clinical practice, that sexologists will continue to treat cases in an individualized way but with more and more of a basis in better scientific information, and that therapies will not only seek to cure dysfunctions but also to increase sexual function and pleasure. The mass media will talk about sex in a much more rigorous way, we will lose our hesitation to consult doctors and therapists, and I'd like to think that science will deepen this biopsychosociological perspective of sex until it becomes the meeting point of all multidisciplinary knowledge of sexuality.

There will be much more, or much less, but I'm firmly convinced that this is an absolutely captivating subject. While writing this book I have learned from academics but also from asexuals, the polyamorous, transsexuals, practitioners of Tantra, the disabled, sadomasochists and so many others who have been willing to share their personal experiences with me, proving how wide the range of sexual expressions is and how narrow my own view of sexuality was. Among the most fun moments, I remember having been confused for several minutes at one of the first conferences I attended, trying to figure out why the slides were talking about Artificial Intelligence and Polyvinyl Chloride, until I realized that AI was Anal Intercourse and PVC was Penile-Vaginal Coitus. I swear I was really puzzled, and I interpreted it as an example of our brains' resistance to change when conditioned to see, think and believe based on our pasts, and how we tend to reaffirm our old knowledge rather than soak up the new. The adventure of researching and writing this book has given me a much more complete view of the study of human nature and behavior; it has been enriching both intellectually and personally, and I would even go so far as to say that I now know myself better. I hope that reading it has been stimulating for you as well and that you are finishing it with more curiosity and eagerness to learn than when you began. Knowing how much is left to know is spellbinding. We are lucky to live in a fascinating moment, when science is contributing the most revolutionary knowledge in the history of human thought. As I said at the end of *The Brain Snatcher*, let's take advantage of that. Let's enjoy science the way we do art, music and literature (or sex). Let's not take off our scientific goggles, let's scratch where we have no itch and let's search for new destinations in the fascinating ocean that is science!

Acknowledgments

In the academic world, my main thanks go to Jim Pfaus and Barry Komisaruk, for opening up the doors of their friendship to me, as well as those of their laboratories.

Many other researchers agreed to let me rob their time and pick their brains. I have fond memories of meetings with Roy Levin, Mayte Parada, Julia Heiman, Jennifer Bass, Justin Garcia, Lori Brotto, Helen Fisher, Irv Binik, Meredith Chivers, Iván Mañero, Gonzalo Giribet, Laura Sánchez, Francisco Cabello, Joan Vidal, Pedro Nobre, Debbie Herbenick, Meg Kaplan, Richard Krueger, John Bancroft, Irwin Goldstein, Heino Meyer Bahlburg, Anke Ehrhardt, Sasha Stulhofer, Carlos Beyer, Frédérique Courtois, Fernando Bianco, Emmanuele Jannini, Erwin Haeberle, Janniko Georgiadis, Raymond Rosen, Erick Janssenn, and the long conversations with Beverly Whipple, Genaro Coria, Stephen Levine, Ray Blanchard, Juan Carlos Sierra, Siri Lekness and William Jankowiak. I would like to thank them all here.

From the most everyday experiences to the testimonies of asexuals, sado-masochists, tantrics and polyamorous people, this book was enriched by the contributions of countless people who generously agreed to share their sexual universes with me. Their names are cited in the different chapters. And for all those who preferred to us pseudonyms, I did want to recognize them, even if just by their initials: C. A., A. S., P. C., E., E. H., C. S., M. P., M. S., M. B., A. D., N. J., C. L., N., R. K., J. V., D. M., B. S., B. A., A. C. Huge thanks for your confessions and inspiration.

A very special thank you to Mikel Urmeneta for the cover design and for being a fantastic partner in adventure in the New York jungle.

Thanks to my editors Miguel Aguilar and Xisca Mas, for trusting in this project despite my insisting that science was more interesting than sex and for their valuable contributions to the final text.

Thanks to the readers of *The Brain Snatcher*, who have been showing me their fondness for and interest in science over the past 2 years. Unbeknownst to them, they have been pushing this book forward each and every day.

Thanks to enthusiasm, because in important moments that is what guides our decisions.

Even though I still can't talk about sex with them, my final and most heartfelt thanks will always be for the constant support and affection I get from my fabulous parents. Kisses and hugs to everyone.

About the Author

Pere Estupinyà is a trained chemist and biochemist and a science communicator by vocation who, after a short time as a PhD researcher, switched to science journalism. After serving as the editor of a leading Spanish television program on science, he is currently the director and presenter of a new show on Spanish public television named after his book "*The Brain Snatcher.*"

Pere Estupinyà has lectured on Science, Technology and Society at the Ramon Llull University in Barcelona and spent an academic year at the prestigious Massachusetts Institute of Technology (MIT) and Harvard University as a Knight Science Journalism Fellow. He has been a consultant for the Inter-American Development Bank (IADB) and the Organization Of American States (OAS) on topics concerning the improvement of science communications in Latin America. His remarkably transparent and clever writing led him to write about science in the main Spanish newspapers.

In November 2010 Pere Estupinyà published his first book on science for the general public: "El Ladrón de cerebros" (*The Brain Snatcher*), which is now in its fifth reprinting. In summer 2011 he released the eBook "Rascar donde no pica" (*To Scratch Where It Doesn't Itch*), and in 2013 he published the Spanish and Catalan editions of $S=EX^2$. In 2016, Pere Estupinyà published his most recent book "Comer cerezas con los ojos a ciegas" (*Eating Cherries with Blind Eyes*).

Pere Estupinyà lives in Washington, DC and Barcelona, and defines himself as a scientific omnivore who writes about science as an excuse to learn more about, and enjoy, its wonders.

Bibliography

Chapter 1: Sex in Our Cells

Bancroft, J., Graham, C. A., Janssen, E., & Sanders, S. A. (2009). The dual control model: Current status and future directions. *Journal of Sex Research, 46*(2–3), 121–142.

Brody, S., & Krüger, T. H. C. (2006). The post-orgasmic prolactin increase following intercourse is greater than following masturbation and suggests greater satiety. *Biological Psychology, 71*, 312–315.

Georgiadis, J. R., Kringelbach, M. L., & Pfaus, J. G. (2012). Sex for fun: A synthesis of human and animal neurobiology. *Nature Reviews Urology, 9*(9), 486–498.

Hyde, J. S. (2005). The gender similarities hypothesis. *American Psychologist, 60*(6), 581–592.

Komisaruk, B. R., Beyer-Flores, C., & Whipple, B. (2006). *The science of orgasm*. Baltimore: Johns Hopkins University Press.

Parada, M., Chamas, L., Censi, S., Coria-Ávila, G., & Pfaus, J. G. (2010). Clitoral stimulation induces conditioned place preference and Fos activation in the rat. *Hormones and Behavior, 57*(2), 112–118.

Parada, M., Vargas, E. B., Kyres, M., Burnside, K., & Pfaus, J. G. (2012). The role of ovarian hormones in sexual reward states of the female rat. *Hormones and Behavior, 62*(4), 442–447.

Petersen, J. L., & Hyde, J. S. (2010). A meta-analytic review of research on gender differences in sexuality, 1993-2007. *Psychological Bulletin, 136*(1), 21–38.

Petersen, J. L., & Hyde, J. S. (2011). Gender differences in sexual attitudes and behaviors: A review of meta-analytic results and large datasets. *Journal of Sex Research, 48*(2–3), 149–165.

© Springer International Publishing Switzerland 2016
P. Estupinyà, *S=EX²*, DOI 10.1007/978-3-319-31726-7

Pfaff, D. W. (1999). *Drive: Neurobiological and molecular mechanisms of sexual motivation*. Cambridge, MA: MIT Press.

Pfaff, D. W. (2010). *Man and woman: An inside story*. Oxford: Oxford University Press.

Pfaus, J. G. (2009). Pathways of sexual desire. *Journal of Sex Medicine, 6*, 1506–1533.

Pfaus, J. G., Kippin, T. E., & Coria-Ávila, G. A. (2003). What can animal models tell us about human sexual response? *Annual Review Sex Research, 14*, 1–63.

Pfaus, J. G., Kippin, T. E., Coria-Ávila, G. A., Gelez, H., Afonso, V. M., Ismail, N., et al. (2012). Who, what, where, when (and maybe even why)? How the experience of sexual reward connects sexual desire, preference, and performance. *Archives of Sexual Behavior, 41*(1), 31–62

Salonia, A., Giraldi, A., Chivers, M. L., Georgiadis, J. R., Levin, R., Maravilla, K. R., et al. (2010). Physiology of women's sexual function: Basic knowledge and new findings. *Journal of Sex Medicine, 7*(8), 2637–2660.

Stone, C. P. (1939). Copulatory activity in adult male rats following castration and injections of testosterone propionate. *Endocrinology, 24*, 165–174.

Chapter 2: Sex in Our Genitals

Bancroft, J., & Graham, C. A. (2011). The varied nature of women's sexuality: Unresolved issues and a theoretical approach. *Hormones and Behavior, 59*(5), 717–729.

Buisson, O., Foldes, P., Jannini, E., & Mimoun, S. (2010). Coitus as revealed by ultrasound in one volunteer couple. *The Journal of Sexual Medicine, 7*(8), 2750–2754.

Burri, A. V., Cherkas, L. M., & Spector, T. D. (2009). The genetics and epidemiology of female sexual dysfunction (FSD): A review. *Journal of Sexual Medicine, 6*(3), 646–657.

Dunn, K. M., Cherkas, L. F., & Spector, T. D. (2005). Genetic on variation in female orgasmic function: A twin study. *Biology Letters, 1*(3), 260–263.

Foldes, P., & Buisson, O. (2009). The clitoral complex: A dynamic sonographic study. *The Journal of Sexual Medicine, 6*(5), 1223–1231.

Friedman, D. M. (2001). *A mind of its own: A cultural history of the penis*. New York: Free Press.

Herbenick, D., & Schick, V. (2011). *Read my lips: A complete guide to the vagina and vulva*. Lanhan: Rowman and Littlefield.

Jannini, E. A., Whipple, B., Kingsberg, S. A., Buisson, O., Foldès, P., & Vardi, Y. (2010). Who's afraid of the G-spot? *The Journal of Sexual Medicine, 7*(1 Pt 1), 25–34.

Janssen, E. (2011). Sexual arousal in men: A review and conceptual analysis. *Hormones and Behavior, 59*, 708–716.

Laumann, E. O., Paik, A., & Rosen, R. C. (1999). Sexual dysfunction in the United States prevalence and predictors. *JAMA, 281*(6), 537–544.

Lloyd, E. (2005). *The case of the female orgasm bias in science of evolution.* Cambridge: Harvard University Press.

Meston, C. M., Levin, R. J., Sipski, M. L., Hull, E. M., & Heiman, J. R. (2004). Women's orgasm. *Annual Review of Sex Research, 15,* 173–257.

Narjani, A. (1924). Considerations sur les causes anatomiques de frigidité chez la femme. *Bruxelles Médical, 42*(27), 768–778.

Shifren, J. L., Monz, B. U., Russo, P. A., Segreti, A., & Johannes, C. (2008). Sexual problems and distress in United States women: Prevalence and correlates. *Obstetrics and Gynecology, 112*(5), 970–978.

Wallen, K., & Lloyd, E. A. (2011). Female sexual arousal: Genital anatomy and orgasm in intercourse. *Hormones and Behavior, 59*(5), 780–792.

Yang, C. C., & Jiang, X. (2009). Clinical autonomic neurophysiology and the male sexual response: An overview. *Journal of Sexual Medicine, 6*(Suppl. 3), 221–228.

Chapter 3: Sex in Our Brains

Berglund, H., Lindstrom, P., & Savic, I. (2006). Brain response to putative pheromones in lesbian women. *Proceedings of the National Academy of Sciences of the United States of America, 103,* 8269–8274.

Childress, A. R., Ehrman, R. N., Wang, Z., Li, Y., Sciortino, N., Hakun, J., et al. (2008). Prelude to passion: Limbic activation by unseen drug and sexual cues. *PLoS One, 3,* e1506.

Georgiadis, J. R., & Kringelbach, M. L. (2012). The human sexual response cycle: Brain imaging evidence linking sex to other pleasures. *Progress in Neurobiology, 98*(1), 49–81.

Jiang, Y., Costello, P., Fang, F., Huang, M., & He, S. (2006). A gender- and sexual orientation-dependent spatial attentional effect of invisible images. *Proceedings of the National Academy of Sciences of the United States of America, 103,* 17048–17052.

Komisaruk, B. R. (2012). A scientist's dilemma: Follow my hypothesis or my findings? *Behavioral Brain Research, 231*(2), 262–265.

Komisaruk, B. R., Beyer-Flores, C., & Whipple, B. (2006). *The science of orgasm.* Baltimore: Johns Hopkins University Press.

Komisaruk, B. R., & Whipple, B. (2005). Functional MRI of the brain during orgasm in women. *Annual Review of Sex Research, 16,* 62–86.

Komisaruk, B. R., Whipple, B., Crawford, A., Liu, W. C., Kalnin, A., & Mosier, K. (2004). Brain activation during vaginocervical self-stimulation and orgasm in women with complete spinal cord injury: fMRI evidence of mediation by the vagus nerves. *Brain Research, 1024*(1–2), 77–88.

Komisaruk, B. R., Wise, N., Frangos, E., Liu, W. C., Allen, K., & Brody, S. (2011). Women's clitoris, vagina, and cervix mapped on the sensory cortex: fMRI evidence. *The Journal of Sexual Medicine, 8*(10), 2822–2830.

Loken, L. S., Wessberg, J., Morrison, I., McGlone, F., & Olausson, H. (2009). Coding of pleasant touch by unmyelinated afferents in humans. *Nature Neuroscience, 12*, 547–548.

Ponseti, J., Bosinski, H. A., Wolff, S., Peller, M., Jansen, O., Mehdorn, H. M., et al. (2006). A functional endophenotype for sexual orientation in humans. *Neuroimage, 33*(3), 825–833.

Zhou, W., & Chen, D. (2008). Encoding human sexual chemosensory cues in the orbito-frontal and fusiform cortices. *The Journal of Neuroscience, 28*, 14416–14421.

Chapter 4: Sex in Our Minds

Basson, R. (2000). The female sexual response: A different model. *Journal of Sex and Marital Therapy, 26*, 51–65.

Chivers, M. L., Rieger, G., Latty, E., & Bailey, J. M. (2004). A sex difference in the specificity of sexual arousal. *Psychological Science, 15*, 736–744.

Chivers, M. L., Seto, M. C., & Blanchard, R. (2007). Gender and sexual orientation differences in sexual response to the sexual activities versus the gender of actors in sexual films. *Journal of Personality and Social Psychology, 93*, 1108–1121.

Goldey, K. L., & Van Anders, S. M. (2011). Sexy thoughts: Effects of sexual cognitions on testosterone, cortisol, and arousal in women. *Hormones and Behavior, 59*(5), 754–764.

Haeberle, E. J. Online archive for sexology. http://www2.hu-berlin.de/sexology/

Janus, S., & Janus, C. (1993). *The Janus report on sexual behavior*. New York: Wiley.

Kaplan, H. (1974). *The new sex therapy*. New York: Brunner Mazel.

Kinsey, A. C., Pomeroy, W. B., & Martin, C. E. (1948). *Sexual behavior in the human male*. Philadelphia: Saunders.

Kinsey, A. C., Pomeroy, W. B., Martin, C. E., & Gebhard, P. H. (1953). *Sexual behavior in the human female*. Philadelphia: Saunders.

Laumann, E. O., Gagnon, J. H., Michael, R. T., & Michaels, S. (1994). *The social organization of sexuality: Sexual practices in the United States*. Chicago: University of Chicago Press.

Laumann, E. O., Nicolosi, A., Glasser, D. B., Paik, A., Gingell, C., Moreira, E., et al. (2005). Sexual problems among women and men aged 40 to 80 years: Prevalence and correlates identified in the Global Study of Sexual Attitudes and Behaviors. *International Journal of Impotence Research, 17*, 39–57.

Masters, W. H., & Johnson, V. E. (1966). *Human sexual response*. Boston: Little, Brown.

Masters, W. H., & Johnson, V. E. (1970). *Human sexual inadequacy*. Boston: Little, Brown.

Mosher, W. D., Chandra, A., & Jones, J. (2004). Sexual behavior and selected health measures: Men and women 15-44 years of age, United States, 2002. *Advance Data from Vital and Health Statistics*, n.° 362.

National Survey of Sexual Health and Behavior (NSSHB). (2010). Findings from the National Survey of Sexual Health and Behavior, Centre for Sexual Health Promotion, Indiana University. *Journal of Sexual Medicine, 7*(Suppl. 5), 255–265.

Ponseti, J., Granert, O., Jansen, O., Wolff, S., Beier, K., Neutze, J., et al. (2012). Assessment of pedophilia using hemodynamic brain response to sexual stimuli. *Archive of General Psychiatry, 69*(2), 187–194.

Richters, J., Grulich, A. E., De Visser, R. O., Smith, A. M., & Rissel, C. E. (2003). Sex in Australia: Attitudes towards sex in a representative sample of adults. *Australia and New Zealand Journal of Public Health, 27*(2), 118–123.

Stulhofer, A., Soh, D., Jelaska, N., Baćak, V., & Landripet, I. (2011). Religiosity and sexual risk behavior among Croatian college students, 1998-2008. *Journal of Sex Research, 48*(4), 360–371.

Suschinsky, K. D., & Lalumière, M. L. (2012). Is sexual concordance related to awareness of physiological states? *Archives of Sexual Behavior, 41*, 199–208.

The Kinsey Institute—Facts and statistics. http://www.kinseyinstitute.org/resources/

Chapter 5: Sex in Our Beds

Addiego, F., Belzer, E. G., Comolli, J., Moger, W., Perry, J. D., & Whipple, B. (1981). Female ejaculation: A case study. *Journal of Sex Research, 17*, 13–21.

Balayssac, S., Gilard, V., Zedde, C., Martino, R., & Malet-Martino, M. (2012). Analysis of herbal dietary supplements for sexual performance enhancement: First characterization of propoxyphenyl-thiohydroxyhomosildenafil and identification of sildenafil, thiosildenafil, phentolamine and tetrahydropalmatine as adulterants. *Journal of Pharmaceutical and Biomedical Analysis, 63*, 135–150.

Bogaert, A. F., & Hershberger, S. (1999). The relation between sexual orientation and penile size. *Archives of Sexual Behavior, 28*(3), 213–221.

Brody, S. (2010). The relative health benefits of different sexual activities. *Journal of Sexual Medicine, 7*, 1336–1361.

Brody, S., & Costa, R. M. (2011). Vaginal orgasm is more prevalent among women with a prominent tubercle of the upper lip. *Journal of Sexual Medicine, 8*, 2793–2799.

Brody, S., & Krüger, T. H. C. (2006). The post-orgasmic prolactin increase following intercourse is greater than following masturbation and suggests greater satiety. *Biological Psychology, 71*, 312–315.

Burleson, M. H., Trevathan, W. R., & Gregory, W. L. (2002). Sexual behavior in lesbian and heterosexual women: Relations with menstrual cycle phase and partner availability. *Psychoneuroendocrinology, 27*(4), 489–503.

Burri, A. V., Cherkas, L. M., & Spector, T. D. (2009). Emotional intelligence and its association with orgasmic frequency in women. *Journal of Sexual Medicine, 6*(7), 1930–1937.

Chia, M. (1997). *El hombre multiorgásmico: Secretos sexuales que todo hombre debería conocer*. Buenos Aires: Neo-person.

Costa, R. M., & Brody, S. (2012). Greater resting heart rate variability is associated with orgasms through penile-vaginal intercourse, but not with orgasms from other sources. *Journal of Sexual Medicine, 9*, 188–197.

Darling, C. A., Davidson, J. K., Sr., & Conway-Welch, C. (1990). Female ejaculation: Perceived origins, the Grafenberg spot/area, and sexual responsiveness. *Archives of Sexual Behavior, 19*, 29–47.

Dillon, B. E., Chama, N. B., & Honig, S. C. (2008). Penile size and penile enlargement surgery. A review. *International Journal of Impotence Research, 20*(6), 519–529.

Dunn, K. M., Cherkas, L. F., & Spector, T. D. (2005). Genetic influences on variation in female orgasmic function: A twin study. *Biology Letters, 1*(3), 260–263.

Eisenman, R. (2001). Penis size: Survey of female perceptions of sexual satisfaction. *BMC Womens Health, 1*(1), 1.

Ellison, C. R. (2000). *Women's sexualities*. Oakland, CA: New Harbinger.

Fiorino, D. F., Coury, A., & Phillips, A. G. (1997). Dynamic changes in nucleus accumbens dopamine efflux during the Coolidge effect in male rats. *Journal of Neuroscience, 17*(12), 4849–4855.

Fisher, D. G., Reynolds, G. L., Ware, M. R., & Napper, L. E. (2011). Methamphetamine and Viagra use: Relationship to sexual risk behaviors. *Archives of Sexual Behavior, 40*(2), 273–279.

Francken, A. B., Van de Wiel, H. B., Van Driel, M. F., & Weijmar Schultz, W. C. (2002). What importance do women attribute to the size of the penis? *European Urology, 42*(5), 426–431.

Friedman, D. M. (2001). *A mind of its own: A cultural history of the penis*. New York: Free Press.

Fugl-Meyer, K. S., Oberg, K., Lundberg, P. O., Lewin, B., & Fugl-Meyer, A. (2006). On orgasm, sexual techniques, and erotic perceptions in 18- to 74-year-old Swedish women. *Journal of Sexual Medicine, 3*(1), 56–68.

George, W. H., Davis, K., Heiman, J., Norris, J. R., Stoner, S. A., Schacht, R. L., et al. (2011). Women's sexual arousal: Effects of high alcohol dosages and self-control instructions. *Hormones and Behavior, 59*, 730–738.

George, W. H., Davis, K. C., Norris, J., Heiman, J. R., Schacht, R. L., Stoner, S. A., et al. (2006). Alcohol and erectile response: The effects of high dosage in the context of demands to maximize sexual arousal. *Experimental and Clinical Psychopharmacology, 14*(4), 461–470.

Gerressu, M., Mercer, C. H., Graham, C. A., Wellings, K., & Johnson, A. M. (2008). Prevalence of masturbation and associated factors from a British national probability survey. *Archives of Sexual Behaviour, 37*, 266–278.

Goodman, M. P. (2011). Female genital cosmetic and plastic surgery: A review. *Journal of Sexual Medicine, 8*(6), 1813–1825.

Grafenberg, E. (1950). The role of the urethra in female orgasm. *International Journal of Sexology, 3*, 145–148.

Gravina, G. L., Brandetti, F., Martini, P., Carosa, E., Di Stasi, S. M., Morano, S., et al. (2008). Measurement of the thickness of the urethrovaginal space in women with or without vaginal orgasm. *Journal of Sexual Medicine, 5*(3), 610–618.

Halsey, L. G., Huber, J. W., Bufton, R. D., & Little, A. C. (2010). An explanation for enhanced perceptions of attractiveness after alcohol consumption. *Alcohol, 44*(4), 307–313.

Harris, J. M., Cherkas, L. F., Kato, B. S., Heiman, J. R., & Spector, T. D. (2008). Normal variations in personality are associated with coital orgasmic infrequency in heterosexual women: A population-based study. *Journal of Sexual Medicine, 5*(5), 1177–1183.

Herbenick, D., Reece, M., Hensel, D., Sanders, S., Jozkowski, K., & Fortenberry, J. D. (2011). Association of lubricant use with women's sexual pleasure, sexual satisfaction, and genital symptoms: A prospective daily diary study. *Journal of Sexual Medicine, 8*(1), 202–212.

Herbenick, D., Reece, M., Sanders, S., Dodge, B., Ghassemi, A., & Fortenberry, J. D. (2009). Prevalence and characteristics of vibrator use by women in the United States: Results from a nationally representative study. *Journal of Sexual Medicine, 6*(7), 1857–1866.

Herbenick, D., Schick, V., Reece, M., Sanders, S., & Fortenberry, J. D. (2010). Pubic hair removal among women in the United States, prevalence, methods, and characteristics. *Journal of Sexual Medicine, 7*(10), 3322–3330.

Hines, T. (2001). The G-spot: A modern gynecologic myth. *American Journal of Obstetrics and Gynecology, 185*(2), 359–362.

Humphries, A. K., & Cioe, J. (2009). Reconsidering the refractory period: An exploratory study of women's post-orgasmic experiences. *Canadian Journal of Human Sexuality, 18*(3), 127.

Janet, L., Peplau, L. A., & Frederick, D. A. (2006). Does size matter? Men's and women's views on penis size across the lifespan. *Psychology of Men and Masculinity, 7*(3), 129–143.

Jannini, E. A., Whipple, B., Kingsberg, S. A., Buisson, O., Foldès, P., & Vardi, Y. (2010). Who's afraid of the G-spot? *Journal of Sexual Medicine, 7*, 25–34.

Jern, P., Westberg, L., Johansson, A., Gunst, A., Eriksson, E., Sandnabba, K., et al. (2012). A study of possible associations between single nucleotide polymorphisms in the serotonin receptor 1A, 1B, and 2C genes and self-reported ejaculation latency time. *Journal of Sexual Medicine, 9*(3), 866–872.

Kamel, I., Gadalla, A., Ghanem, H., & Oraby, M. (2009). Comparing penile measurements in normal and erectile dysfunction subjects. *Journal of Sexual Medicine, 6*(8), 2305–2310.

Kilchevsky, A., Vardi, Y., Lowenstein, L., & Gruenwald, I. (2012). Is the female G-spot truly a distinct anatomic entity? *Journal of Sexual Medicine, 9*(3), 719–726.

Krüger, T. H., Haake, P., Hartmann, U., Schedlowski, M., & Exton, M. S. (2002). Orgasm-induced prolactin secretion: Feedback control of sexual drive? *Neuroscience Biobehavioral Reviews, 26*(1), 31–44.

Levin, R. J. (2009). Revisiting post-ejaculation refractory time-what we know and what we do not know in males and in females. *Journal of Sexual Medicine, 6*(9), 2376–2389.

Levin, R., & Meston, C. (2006). Nipple/breast stimulation and sexual arousal in young men and women. *Journal of Sexual Medicine, 3*, 450–454.

Levitas, E., Lunenfeld, E., Weiss, N., Friger, M., Har-Vardi, I., Koifman, A., et al. (2005). Relationship between the duration of sexual abstinence and semen quality: Analysis of 9,489 semen samples. *Fertility and Sterility, 83*(6), 1680–1686.

Lloyd, J., Crouch, N. S., Minto, C. L., Liao, L. M., & Creighton, S. M. (2005). Female genital appearance: "Normality" unfolds. *BJOG, 112*(5), 643–646.

McBride, K. R., & Fortenberry, J. D. (2010). Heterosexual anal sexuality and anal sex behaviors: A review. *Journal of Sex Research, 47*(2), 123–136.

Meston, C. M., Rellini, A. H., & Telch, M. J. (2008). Short- and long-term effects of Ginkgo biloba extract on sexual dysfunction in women. *Archives of Sexual Behavior, 37*, 530–547.

Meston, C. M., & Worcel, M. (2002). The effects of yohimbine plus L-arginine glutamate on sexual arousal in postmenopausal women with sexual arousal disorder. *Archives of Sexual Behavior, 31*, 323–332.

Mondaini, N., Ponchietti, R., Gontero, P., Muir, G. H., Natali, A., Caldarera, E., et al. (2002). Penile length is normal in most men seeking penile lengthening procedures. *International Journal of Impotence Research, 14*(4), 283–286.

Nicholas, A., Brody, S., De Sutter, P., & De Carufel, F. (2008). A woman's history of vaginal orgasm is discernible from her walk. *Journal of Sexual Medicine, 5*, 2119–2124.

Parker, L. L., Penton-Voak, I. S., Attwood, A. S., & Munafò, M. R. (2008). Effects of acute alcohol consumption on ratings of attractiveness of facial stimuli: Evidence of long-term encoding. *Alcohol, 43*(6), 636–640.

Patel, D. K., Kumar, R., Prasad, S. K., & Hemalatha, S. (2011). Pharmacologically screened aphrodisiac plant: A review of current scientific literature. *Asian Pacific Journal of Tropical Biomedicine,* S131–S138.

Petersen, J. L., & Hyde, J. S. (2010). Meta-analytic review of research on gender differences in sexuality, 1993-2007. *Psychological Bulletin, 136*, 21–38.

Ponchietti, R., Mondaini, N., Bonafè, M., Di Loro, F., Biscioni, S., Masieri, L., et al. (2001). Penile length and circumference: A study on 3,300 young Italian males. *European Urology, 39*(2), 183–186.

Promodu, K., Shanmughadas, K. V., Bhat, S., & Nair, K. R. (2007). Penile-length and circumference: An Indian study. *International Journal of Impotence Research, 19*(6), 558–563.

Rubio-Casillas, A., & Jannini, E. A. (2011). New insights from one case of female ejaculation. *Journal of Sexual Medicine, 8*(12), 3500–3504.

Shah, J., & Christopher, N. (2002). Can shoe size predict penile length? *BJU International, 90*(6), 586–587.

Söylemez, H., Atar, M., Sancaktutar, A. A., Penbegül, N., Bozkurt, Y., & Onem, K. (2012). Relationship between penile size and somatometric parameters in 2276 healthy young men. *International Journal of Impotence Research, 24*(3), 126–129.

Spyropoulos, E., Borousas, D., Mavrikos, S., Dellis, A., Bourounis, M., & Athanasiadis, S. (2002). Size of external genital organs and somatometric parameters among physically normal men younger than 40 years old. *Urology, 60*(3), 485–489. Discussion, pp. 490–491.

Stulhofer, A. (forthcoming). It made me realize what I like and dislike in sex: A mixed method exploration of women's experiences of and meanings attached to pain at anal intercourse, International Academy of Sex Research annual conference, Estoril, 9 July 2012.

Stulhofer, A., & Ajduković, D. (2011). Should we take anodyspareunia seriously? A descriptive analysis of pain during receptive anal intercourse in young heterosexual women. *Journal of Sex and Marital Therapy, 37*(5), 346–358.

Wallen, K., & Lloyd, E. A. (2011). Female sexual arousal: Genital anatomy and orgasm in intercourse. *Hormones and Behavior, 59*(5), 780–792.

Waterman, J. M. (2010). The adaptive function of masturbation in a promiscuous African ground squirrel. *PLoS One, 5*(9), pii: e13060.

Wylie, K. R., & Eardley, I. (2007). Penile size and the "small penis syndrome". *BJU International, 99*(6), 1449–1455.

Yang, M. L., Fullwood, E., Goldstein, J., & Mink, J. W. (2005). Masturbation in infancy and early childhood presenting as a movement disorder. 12 cases and a review of the literature. *Pediatrics, 116*(6), 1427–1432.

Zietsch, B. P., Miller, G. F., Bailey, J. M., & Martin, N. G. (2011). Female orgasm rates are largely independent of other traits: Implications for "female orgasmic disorder" and evolutionary theories of orgasm. *Journal of Sexual Medicine, 8*(8), 2305–2316.

Chapter 6: Sex in the Doctor's Office

Basson, R. (2000). The female sexual response: A different model. *Journal of Sex and Marital Therapy, 26*, 51–65.

Ben-Zion, I., Rothschild, S., Chudakov, B., & Aloni, R. (2007). Surrogate versus couple therapy in vaginismus. *Journal of Sexual Medicine, 4*(3), 728–733.

Both, S., Laan, E., & Schultz, W. W. (2010). Disorders in sexual desire and sexual arousal in women, a 2010 state of the art. *Journal of Psychosomatic Obstetrics and Gynecology, 31*(4), 207–218.

Brindley, G. S. (1983). Cavernosal alpha-blockade: A new technique for investigating and treating erectile impotence. *The British Journal of Psychiatry, 143*, 332–337.

Burrows, L. J., Basha, M., & Goldstein, A. T. (2012). The effects of hormonal contraceptives on female sexuality: A review. *Journal of Sexual Medicine, 9*(9), 2213–2223.

Davis, S. R., Moreau, M., Kroll, R., Bouchard, C., Panay, N., Gass, M., et al. (2008). Testosterone for low libido in postmenopausal women not taking estrogen. *New England Journal of Medicine, 359*(19), 2005–2017.

Gajer, P., Brotman, R. M., Bai, G., Sakamoto, J., Schütte, U. M., Zhong, X., et al. (2012). Temporal dynamics of the human vaginal microbiota. *Science Translational Medicine, 2*(4), 132ra52.

Haake, P., Exton, M. S., Haverkamp, J., Krämer, M., Leygraf, N., Hartmann, U., et al. (2002). Absence of orgasm-induced prolactin secretion in a healthy multi-orgasmic male subject. *International Journal of Impotence Research, 14*(2), 133–135.

Kaschak, E., & Tiefer, E. (2002). *A new view of women's sexual problems*. New York: Haworth Press.

Klotz, L. (2005). How (not) to communicate new scientific information: A memoir of the famous Brindley lecture. *BJU International, 96*(7), 956–957.

Laumann, E. O., Paik, A., & Rosen, R. C. (1999). Sexual dysfunction in the United States: Prevalence and predictors. *JAMA, 281*(6), 537–544.

Montorsi, F., Adaikan, G., Becher, E., Giuliano, F., Khoury, S., Lue, T. F., et al. (2010). Summary of the recommendations on sexual dysfunctions in men. *Journal of Sexual Medicine, 7*(11), 3572–3588.

Price, L. B., Liu, C. M., Johnson, K. E., Aziz, M., Lau, M. K., Bowers, J., et al. (2010). The effects of circumcision on the penis microbiome. *PLoS One, 5*(1), e8422.

Salonia, A., Giraldi, A., Chivers, M. L., Georgiadis, J. R., Levin, R., Maravilla, K. R., et al. (2010). Physiology of women's sexual function: Basic knowledge and new findings. *Journal of Sexual Medicine, 7*(8), 2637–2660.

Sand, M., & Fisher, W. A. (2007). Women's endorsement of models of female sexual response: The nurses' sexuality study. *Journal of Sexual Medicine, 4*(3), 708–719.

Chapter 7: Sex in Nature

Brennan, P. L., Clark, C. J., & Prum, R. O. (2010). Explosive eversion and functional morphology of the duck penis supports sexual conflict in waterfowl genitalia. *Proceedings of Biological Sciences, 277*(1686), 1309–1314.

Brennan, P. L., Prum, R. O., McCracken, K. G., Sorenson, M. D., Wilson, R. E., & Birkhead, T. R. (2007). Coevolution of male and female genital morphology in waterfowl. *PLoS One, 2*(5), e418.

Judson, O. (2002). *Dr Tatiana's sex advice to all creation.* New York: Metropolitan Books.
Pillsworth, E. G., Haselton, M. G., & Buss, D. M. (2004). Ovulatory shifts in female sexual desire. *Journal of Sex Research, 41*(1), 55–65.
Tan, M., Jones, G., Zhu, G., Ye, J., Hong, T., Zhou, S., et al. (2009). Fellatio by fruit bats prolongs copulation time. *PLoS One, 4*(10), e7595.

Chapter 8: Sex in Evolution

Durrleman, S., Pennec, X., Trouvé, A., Ayache, N., & Braga, J. (2012). Comparison of the endocranial ontogenies between chimpanzees and bonobos via temporal regression and spatiotemporal registration. *Journal of Human Evolution, 62*(1), 74–88.
Haun, D. B., Nawroth, C., & Call, J. (2011). Great apes' risk-taking strategies in a decision making task. *PLoS One, 6*(12), e28801.
Herrmann, E., Hare, B., Cissewski, J., & Tomasello, M. (2011). A comparison of temperament in nonhuman apes and human infants. *Developmental Science, 14*(6), 1393–1405.
Ridley, M. (2003). *The red queen: Sex and the evolution of human nature.* New York: Harper Perennial.
Rilling, J. K., Scholz, J., Preuss, T. M., Glasser, M. F., Errangi, B. K., & Behrens, T. E. (2012). Differences between chimpanzees and bonobos in neural systems supporting social cognition. *Social and Cognitive Affective Neuroscience, 7*(4), 369–379.
Schaller, F., Fernandes, A. M., Hodler, C., Münch, C., Pasantes, J. J., et al. (2010). Y chromosomal variation tracks the evolution of mating systems in chimpanzee and bonobo. *PLoS One, 5*(9), e12482.
Wobber, V., Hare, B., Maboto, J., Lipson, S., Wrangham, R., & Ellison, P. T. (2010). Differential changes in steroid hormones before competition in bonobos and chimpanzees. *Proceedings of National Academy of Sciences of the United States of America, 107*(28), 12457–12462.
Woods, V., & Hare, B. (2011). Bonobo but not chimpanzee infants use socio-sexual contact with peers. *Primates, 52*(2), 111–116.

Chapter 9: Sex in Bars

Barrett, D., Greenwood, J. G., & McCullagh, J. F. (2006). Kissing laterality and handedness. *Laterality, 11*(6), 573–579.
Clark, R. D., & Hatfield, E. (1989). Gender differences in receptivity to sexual offers. *Journal of Psychology and Human Sexuality, 2*, 39–55.

Coetzee, V., Re, D., Perrett, D. I., Tiddeman, B. P., & Xiao, D. (2011). Judging the health and attractiveness of female faces: Is the most attractive level of facial adiposity also considered the healthiest? *Body Image, 8*(2), 190–193.

DeBruine, L. M., Jones, B. C., Watkins, C. D., Roberts, S. C., Little, A. C., Smith, F. G., et al. (2011). Opposite-sex siblings decrease attraction, but not prosocial attributions, to self-resembling opposite-sex faces. *Proceedings of the National Academy of Sciences of the United States of America, 108*(28), 11710–11714.

Derti, A., Cenik, C., Kraft, P., & Roth, F. P. (2010). Absence of evidence for MHC-dependent mate selection within HapMap populations. *PLoS Genetics, 6*(4), e1000925.

Finkel, E. J., Eastwick, P. W., Karney, B. R., Reis, H. T., & Sprecher, S. (2012). Online dating: A critical analysis from the perspective of psychological science. *Psychological Science in the Public Interest, 13*(1), 3–66.

Fisak, B., Jr., Tantleff-Dunn, S., & Peterson, R. D. (2007). Personality information: Does it influence attractiveness ratings of various body sizes? *Body Image, 4*(2), 213–217.

Fisher, H. E. (2011). Serial monogamy and clandestine adultery: Evolution and consequences of the dual human reproductive strategy. In S. C. Roberts (Ed.), *Applied evolutionary psychology*. New York: Oxford University Press.

Fisher, M. L., Worth, K., Garcia, J. R., & Meredith, T. (2012). Feelings of regret following uncommitted sexual encounters in Canadian university students. *Culture, Health and Sexuality, 14*, 45–57.

Garcia, J. R., & Reiber, C. (2008). Hook-up behavior: A biopsychosocial perspective. *The Journal of Social, Evolutionary, and Cultural Psychology, 2*, 192–208.

Garcia, J. R., Reiber, C., Massey, S. G., & Merriwether, A. M. (2012). Sexual hookup culture: A review. *Review of General Psychology, 16*(2), 161–176.

Givens, D. B. (1978). The nonverbal basis of attraction: Flirtation, courtship, and seduction. *Psychiatry, 41*(4), 346–359.

Glassenberg, A. N., Feinberg, D. R., Jones, B. C., Little, A. C., & Debruine, L. M. (2010). Sex-dimorphic face shape preference in heterosexual and homosexual men and women. *Archives of Sexual Behavior, 39*(6), 1289–1296.

Gueguen, N. (2007). Women's bust size and men's courtship solicitation. *Body Image, 4*(4), 386–390.

Gueguen, N. (2008). The effect of a woman's smile on men's courtship behavior. *Social Behavior and Personality, 36*, 1233–1236.

Gueguen, N. (2011). The effect of women's suggestive clothing on men's behavior and judgment: A field study. *Psychological Reports, 109*(2), 635–638.

Havlicek, J., & Roberts, S. C. (2009). MHC-correlated mate choice in humans: A review. *Psychoneuroendocrinology, 34*(4), 497–512.

Hill, S. E., & Buss, D. M. (2008). The mere presence of opposite-sex others on judgments of sexual and romantic desirability: Opposite effects for men and women. *Personality and Social Psychology Bulletin, 34*(5), 635–647.

Hughes, S. M., & Kruger, D. J. (2011). Sex differences in post-coital behaviors in long- and short-term mating: An evolutionary perspective. *Journal of Sex Research, 48*(5), 496–505.

Kim, J. L., & Ward, L. M. (2012). Striving for pleasure without fear: Short-term effects of reading a women's magazine on women's sexual attitudes. *Psychology of Women Quarterly, 36*(3), 326–336.

Kirshenbaum, S. (2011). *The science of kissing*. New York: Hachette.

Law Smith, M. J., Deady, D. K., Moore, F. R., Jones, B. C., Cornwell, R. E., Stirrat, M., et al. (2012). Maternal tendencies in women are associated with estrogen levels and facial femininity. *Hormones and Behavior, 61*(1), 12–16.

Lee, L., Loewenstein, G., Ariely, D., Hong, J., & Young, J. (2008). If I'm not hot, are you hot or not? Physical-attractiveness evaluations and dating preferences as a function of one's own attractiveness. *Psychological Science, 19*(7), 669–677.

Moore, F. R., Al Dujaili, E. A., Cornwell, R. E., Smith, M. J., Lawson, J. F., Sharp, M., et al. (2011). Cues to sex- and stress-hormones in the human male face: Functions of glucocorticoids in the immunocompetence handicap hypothesis. *Hormones and Behavior, 60*(3), 269–274.

Moore, M. M. (1985). Nonverbal courtship patterns in women: Context and consequences. *Ethology and Sociobiology, 6,* 237–247.

Moore, M. M. (1998). Nonverbal courtship patterns in women: Rejection signaling. An empirical investigation. *Semiotica, 3,* 205–215.

Moore, M. M. (2010). Human nonverbal courtship behavior. A brief historical review. *Journal of Sex Research, 47*(2–3), 171–180.

Owen, J., & Fincham, F. D. (2011). Young adults' emotional reactions after hooking up encounters. *Archives of Sexual Behavior, 40,* 321–330.

Paul, E. L., & Hayes, K. A. (2002). The casualties of "casual" sex: A qualitative exploration of the phenomenology of college students' hook-ups. *Journal of Social and Personal Relationships, 19,* 639–661.

Paul, E. L., McManus, B., & Hayes, A. (2000). "Hook-ups": Characteristics and correlates of college students' spontaneous and anonymous sexual experiences. *Journal of Sex Research, 37,* 76–88.

Perret, D. (2010). *In your face: The new science of human attraction*. New York: Palgrave Macmillan.

Re, D. E., Whitehead, R. D., Xiao, D., & Perrett, D. I. (2011). Oxygenated blood colour change thresholds for perceived facial redness, health, and attractiveness. *PLoS One, 6*(3), e17859.

Swami, V. (2009). An examination of the love-is-blind bias among gay men and lesbians. *Body Image, 6*(2), 149–151.

Swami, V., & Allum, L. (2012). Perceptions of the physical attractiveness of the self, current romantic partners, and former partners. *Scandinavian Journal of Psychology, 53*(1), 89–95.

Swami, V., & Barrett, S. (2011). British men's hair color preferences: An assessment of courtship solicitation and stimulus ratings. *Scandinavian Journal of Psychology, 52*(6), 595–600.

Swami, V., Frederick, D. A., Aavik, T., Alcalay, L., Allik, J., Anderson, D., et al. (2010). The attractive female body weight and female body dissatisfaction in 26 countries across 10 world regions: Results of the International Body Project I. *Personality and Social Psychology Bulletin, 36*, 309.

Swami, V., & Furnham, A. (2007). *The psychology of physical attraction*. New York: Routledge.

Swami, V., Furnham, A., Chamorro-Premuzic, T., Akbar, K., Gordon, N., Harris, T., et al. (2010). More than just skin deep? Personality information influences men's ratings of the attractiveness of women's body sizes. *Journal of Social Psychology, 150*(6), 628–647.

Swami, V., & Tovée, M. J. (2006). Does hunger influence judgments of female physical attractiveness? *British Journal of Psychology, 97*(Pt 3), 353–363.

Swami, V., & Tovée, M. J. (2012). The impact of psychological stress on men's judgments of female body size. *PLoS One, 7*(8), e42593.

Wedekind, C., Seebeck, T., Bettens, F., & Paepke, A. J. (1995). MHC-dependent mate preferences in humans. *Proceedings of Biological Sciences, 260*(1359), 245–249.

Whitehead, R. D., Re, D., Xiao, D., Ozakinci, G., & Perrett, D. I. (2012). You are what you eat: Within-subject increases in fruit and vegetable consumption confer beneficial skin-color changes. *PLoS One, 7*(3), e32988.

Willis, F., & Briggs, L. (1992). Relationship and touch in public settings. *Journal of Nonverbal Behavior, 16*, 55–63.

Chapter 10: Having an Orgasm with the Power of the Mind

Botto, L. A., Klein, C., & Gorzalka, B. B. (2009). Laboratory-induced hyperventilation differentiates female sexual arousal disorder subtypes. *Archives of Sexual Behavior, 38*, 463–475.

Beck, J. G., & Baldwin, L. E. (1994). Instructional control of female sexual responding. *Archives of Sexual Behavior, 23*, 665–684.

Bradford, A., & Meston, C. M. (2006). The impact of anxiety on sexual arousal in women. *Behaviour Research and Therapy, 44*, 1067–1077.

Bradford, A., & Meston, C. M. (2011). Behavior and symptom change among women treated with placebo for sexual dysfunction. *Journal of Sexual Medicine, 8*(1), 191–201.

Brotto, L. A., Basson, R., & Luria, M. (2008). A mindfulness-based group psycho-educational intervention targeting sexual arousal disorder in women. *Journal of Sexual Medicine, 5*(7), 1646–1659.

Brotto, L. A., Erskine, Y., Carey, M., Ehlen, T., Finlayson, S., Heywood, M., et al. (2012). A brief mindfulness-based cognitive behavioral intervention improves sexual functioning versus wait-list control in women treated for gynecologic cancer. *Gynecologic Oncology, 125*(2), 320–325.

Brotto, L. A., Klein, C., & Gorzalka, B. B. (2002). Genital and subjective sexual arousal in postmenopausal women: Influence of laboratory-induced hyperventilation. *Journal of Sex and Marital Therapy, 28*(Suppl. 1), 39–53.

Brotto, L. A., Krychman, M., & Jacobson, P. (2008). Eastern approaches for enhancing women's sexuality: Mindfulness, acupuncture, and yoga (CME). *Journal of Sexual Medicine, 5*(12), 2741–2748.

Brotto, L. A., Seal, B. N., & Rellini, A. (2012). Pilot study of a brief cognitive behavioral versus mindfulness-based intervention for women with sexual distress and a history of childhood sexual abuse. *Journal of Sex and Marital Therapy, 38*(1), 1–27.

Dhikav, V., Karmarkar, G., Gupta, R., Verma, M., Gupta, R., Gupta, S., et al. (2010). Yoga in female sexual functions. *Journal of Sexual Medicine, 7*(2 Pt 2), 964–970.

Dhikav, V., Karmarkar, G., Verma, M., Gupta, R., Gupta, S., Mittal, D., et al. (2010). Yoga in male sexual functioning: A noncomparative pilot study. *Journal of Sexual Medicine, 7*(10), 3460–3466.

George, W. H., Davis, K. C., Heiman, J. R., Norris, J., Stoner, S. A., Schacht, R. L., et al. (2011). Women's sexual arousal: Effects of high alcohol dosages and self-control instructions. *Hormones and Behavior, 59*, 730–738.

Chapter 11: Pornography: From Distortion to Education

Chivers, M. L., Seto, M. C., Lalumière, M. L., Laan, E., & Grimbos, T. (2010). Agreement of self-reported and genital measures of sexual arousal in men and women: A meta-analysis. *Archives of Sexual Behavior, 39*(1), 5–56.

Diamond, M., Jozifkova, E., & Weiss, P. (2010). Pornography and sex crimes in the Czech Republic. *Archives of Sexual Behavior, 40*, 1037–1043.

Griffith, J. D., Hayworth, M., Adams, L. T., Mitchell, S., & Hart, C. (2013). Characteristics of pornography film actors: Self-report versus perceptions of college students. *Archives of Sexual Behavior, 42*(4), 637–647 [Epub 2012 Nov 22].

Griffith, J. D., Mitchell, S., Hart, C. L., Adams, L. T., & Gu, L. L. (2013). Pornography actresses: An assessment of the damaged goods hypothesis. *Journal of Sex Research, 50*(7), 621–632 [Epub 2012 Nov 20].

Hald, G. M. (2006). Gender differences in pornography consumption among young heterosexual Danish adults. *Archives of Sexual Behavior, 35*(5), 577–585.

Hald, G. M., & Malamuth, N. M. (2008). Self-perceived effects of pornography consumption. *Archives of Sexual Behavior, 37*(4), 614–625.

Hald, G. M., Malamuth, N. M., & Yuen, C. (2010). Pornography and attitudes supporting violence against women: Revisiting the relationship in nonexperimental studies. *Aggressive Behavior, 36*(1), 14–20.

Hamann, S., Herman, R. A., Nolan, C. L., & Wallen, K. (2004). Men and women differ in amygdala response to visual sexual stimuli. *Nature Neuroscience, 7*(4), 411–416.

Hilton, D. L., Jr., & Watts, C. (2011). Pornography addiction: A neuroscience perspective. *Surgical Neurology International, 2*, 19.

Peter, J., & Valkenburg, P. M. (2011). The influence of sexually explicit Internet material and peers on stereotypical beliefs about women's sexual roles: Similarities and differences between adolescents and adults. *Cyberpsychology, Behaviour, and Social Networking, 14*(9), 511–517.

Reid, R. C., Carpenter, B. N., & Fong, T. W. (2011). Neuroscience research fails to support claims that excessive pornography consumption causes brain damage. *Surgical Neurology International, 2*, 64.

Short, M. B., Black, L., Smith, A. H., Wetterneck, C. T., & Wells, D. E. (2012). A review of internet pornography use research: Methodology and content from the past 10 years. *Cyberpsychology, Behaviour, and Social Networking, 15*(1), 13–23.

Sinković, M., Stulhofer, A., & Božić, J. (2013). Revisiting the association between pornography use and risky sexual behaviors: The role of early exposure to pornography and sexual sensation seeking. *Journal of Sex Research, 50*(7), 633–641.

Wright, P. J. (2013). U.S. males and pornography, 1973-2010: Consumption, predictors, correlates. *Journal of Sex Research, 50*(1), 60–71.

Chapter 12: Let's Do It Tonight, Dear, I Have a Headache

Anand, K. S., & Dhikav, V. (2009). Primary headache associated with sexual activity. *Singapore Medical Journal, 50*(5), e176–e177.

Beckman, N., Waern, M., Gustafson, D., & Skoog, I. (2008). Secular trends in self reported sexual activity and satisfaction in Swedish 70 year olds: Cross sectional survey of four populations, 1971-2001. *British Medical Journal, 337*, A279.

Brody, S. (2006). Blood pressure reactivity to stress is better for people who recently had penile–vaginal intercourse than for people who had other or no sexual activity. *Biological Psychology, 71*(2), 214–222.

Charnetski, C. J., & Brennan, F. X. (2004). Sexual frequency and salivary immunoglobulin A (IgA). *Psychological Reports, 94*(3 Pt 1), 839–844.

Chen, X., Zhang, Q., & Tan, X. (2009). Cardiovascular effects of sexual activity. *Indian Journal of Medical Research, 130*, 681–688.

Costa, R. M. (2012). Greater resting heart rate variability is associated with orgasms through penile-vaginal intercourse, but not with orgasms from other sources. *Journal of Sexual Medicine, 9*(1), 188–197.

DeLamater, J. (2012). Sexual expression in later life: A review and synthesis. *Journal of Sex Research, 49*(2–3), 125–141.
DeLamater, J., Hyde, J. S., & Fong, M. C. (2008). Sexual satisfaction in the seventh decade of life. *Journal of Sex and Marital Therapy, 34*(5), 439–454.
Derby, C. A., Mohr, B. A., Goldstein, I., Feldman, H. A., Johannes, C. B., & McKinlay, J. B. (2000). Modifiable risk factors and erectile dysfunction: Can lifestyle changes modify risk? *Urology, 56*, 302–306.
Hamilton, L. D., Fogle, E. A., & Meston, C. M. (2008). The roles of testosterone and alpha-amylase in exercise-induced sexual arousal in women. *Journal of Sexual Medicine, 5*(4), 845–853.
Hsiao, W., Shrewsberry, A. B., Moses, K. A., Johnson, T. V., Cai, A. W., Stuhldreher, P., et al. (2012). Exercise is associated with better erectile function in men under 40 as evaluated by the international index of erectile function. *Journal of Sexual Medicine, 9*, 524–530.
Hu, C. M., Lin, Y. J., Fan, Y. K., Chen, S. P., & Lai, T. H. (2010). Isolated thunderclap headache during sex: Orgasmic headache or reversible cerebral vasoconstriction syndrome? *Journal of Clinical Neuroscience, 17*(10), 1349–1351.
Hyde, Z., Flicker, L., Hankey, G. J., Almeida, O. P., McCaul, K. A., Chubb, S. A., et al. (2012). Prevalence and predictors of sexual problems in men aged 75-95 years: A population-based study. *Journal of Sexual Medicine, 9*(2), 442–453.
Lamina, S., Okoye, C. G., & Dagogo, T. T. (2009). Therapeutic effect of an interval exercise training program in the management of erectile dysfunction in hypertensive patients. *The Journal of Clinical Hypertension, 11*, 125–129.
Leitzmann, M. F., Platz, E. A., Stampfer, M. J., Willett, W. C., & Giovannucci, E. (2004). Ejaculation frequency and subsequent risk of prostate cancer. *JAMA, 291*(13), 1578–1586.
Levine, G. N., Steinke, E. E., Bakaeen, F. G., Bozkurt, B., Cheitlin, M. D., Conti, J. B., et al. (2012). Sexual activity and cardiovascular disease: A scientific statement from the American Heart Association. *Circulation, 125*(8), 1058–1072.
Maines, R. P. (1999). *The technology of orgasm: Hysteria, vibrators and women's sexual satisfaction.* Baltimore: Johns Hopkins University Press.
Menezes, R. G., Shetty, A. J., Lobo, S. W., Kanchan, T., Suresh Kumar Shetty, B., Sreeramareddy, C. T., et al. (2008). Does increased sexual activity increase body weight gain? *Medical Hypotheses, 71*(4), 601–602.
Meston, C. M. (2000). Sympathetic nervous system activity and female sexual arousal. *American Journal of Cardiology, 86*(2A), 30F–34F.
Palmeri, S. T., Kostis, J. B., Casazza, L., Sleeper, L. A., Lu, M., Nezgoda, J., et al. (2007). Heart rate and blood pressure response in adult men and women during exercise and sexual activity. *American Journal of Cardiology, 100*(12), 1795–1801.
Pascual, J., González-Mandly, A., Martín, R., & Oterino, A. (2008). Headaches precipitated by cough, prolonged exercise or sexual activity: A prospective etiological and clinical study. *Journal Headache Pain, 9*(5), 259–266.

Poryazova, R., Khatami, R., Werth, E., Claudio, L., & Bassetti, M. (2009). Weak with sex: Sexual intercourse as a trigger for cataplexy. *Journal of Sexual Medicine, 6*, 2271–2277.

Smith, L. J., Mulhall, J. P., Deveci, S., Monaghan, N., & Reid, M. C. (2007). Sex after seventy: A pilot study of sexual function in older persons. *Journal of Sexual Medicine, 4*(5), 1247–1253.

Tessler Lindau, S., Philip Schumm, L., Laumann, E. O., Levinson, W., O'Muircheartaigh, C. A., & Waite, L. J. (2007). A study of sexuality and health among older adults in the United States. *New England Journal of Medicine, 357*, 762–774.

Trompeter, S. E., Bettencourt, R., & Barrett-Connor, E. (2012). Sexual activity and satisfaction in healthy community-dwelling older women. *The American Journal of Medicine, 125*(1), 37.

Valença, M. M., Valença, L. P., Bordini, C. A., Da Silva, W. F., Leite, J. P., Antunes-Rodrigues, J., et al. (2004). Cerebral vasospasm and headache during sexual intercourse and masturbatory orgasms. *Headache, 44*(3), 244–248.

Weaver, A. D., & Byers, E. S. (2006). The relationships among body image, body mass index, exercise, and sexual functioning in heterosexual women. *Psychology of Women Quarterly, 30*(4), 333–339.

Xue-Rui, T., Ying, L., Da-Zhong, Y., & Xiao-Jun, C. (2008). Changes of blood pressure and heart rate during sexual activity in healthy adults. *Blood Press Monitoring, 13*(4), 211–217.

Chapter 13: Sex in a Wheelchair, for Love and Pleasure

Anderson, K. D. (2004). Targeting recovery, priorities of the spinal cord-injured population. *Journal of Neurotrauma, 21*(10), 1371–1383.

Courtois, F., Charvier, K., Leriche, A., Vézina, J. G., Côté, I., Raymond, D., et al. (2008). Perceived physiological and orgasmic sensations at ejaculation in spinal cord injured men. *Journal of Sexual Medicine, 5*(10), 2419–2430.

Courtois, F., Charvier, K., Vézina, J. G., Journel, N. M., Carrier, S., Jacquemin, G., et al. (2011). Assessing and conceptualizing orgasm after a spinal cord injury. *BJU International, 108*(10), 1624–1633.

Everaert, K., de Waard, W. I., Van Hoof, T., Kiekens, C., Mulliez, T., & D'herde, C. (2010). Neuroanatomy and neurophysiology related to sexual dysfunction in male neurogenic patients with lesions to the spinal cord or peripheral nerves. *Spinal Cord, 48*, 182–191.

Overgoor, M. L., de Jong, T. P., Cohen-Kettenis, P. T., Edens, M. A., & Kon, M. (2013). Increased sexual health after restored genital sensation in male patients with spina bifida or a spinal cord injury: The TOMAX procedure. *Journal of Urology, 189*(2), 626–632 [Epub 2012 Oct 16].

Overgoor, M. L., Kon, M., Cohen-Kettenis, P. T., Strijbos, S. A., De Boer, N., & De Jong, T. P. (2006). Neurological bypass for sensory innervation of the penis in patients with spina bifida. *Journal of Urology, 176*(3), 1086–1090.

Sønksen, J., & Ohl, D. A. (2002). Penile vibratory stimulation and electroejaculation in the treatment of ejaculatory dysfunction. *International Journal of Andrology, 25*(6), 324–332.

Chapter 14: Science in Sexual Orientation

Allen, L. S., & Gorski, R. A. (1992). Sexual orientation and the size of the anterior commissure in the human brain. *Proceedings of the National Academy of Sciences of the United States of America, 89,* 7199–7202.

Bagemihl, B. (1999). *Biological exuberance: Animal homosexuality and natural diversity.* New York: St. Martin's Press.

Bancroft, J. (2003). Can sexual orientation change? A long-running saga. *Archives of Sexual Behavior, 32*(5), 419–468.

Bancroft, J., & Marks, I. (1968). Electric aversion therapy of sexual deviations. *Proceedings of the Royal Society of Medicine, 61,* 796–799.

Blanchard, R. (2007). Older-sibling and younger-sibling sex ratios in Frisch and Hviid's (2006) National Cohort Study of two million Danes. *Archives of Sexual Behavior, 36,* 860–863.

Blanchard, R., Cantor, J. M., Bogaert, A. F., Breedlove, S. M., & Ellis, L. (2000). Interaction of fraternal birth order and handedness in the development of male homosexuality. *Hormones and Behavior, 49*(3), 405–414.

Blanchard, R., & Lippa, R. A. (2007). Birth order, sibling ratio, handedness and sexual orientation of male and female participants in a BBC Internet research project. *Archives of Sexual Behavior, 36,* 163–176.

Bogaert, A. F. (2004). Asexuality: Prevalence and associated factors in a national probability sample. *Journal of Sex Research, 41,* 279–287.

Bogaert, A. F. (2006). Biological versus nonbiological older brothers and men's sexual orientation. *Proceedings of the National Academy of Sciences of the United States of America, 103*(28), 10771–10774.

Brotto, L. A., Knudson, G., Inskip, J., Rhodes, K., & Erskine, Y. (2010). Asexuality: A mixed-methods approach. *Archives of Sexual Behavior, 39,* 599–618.

Brotto, L. A., & Yule, M. A. (2011). Physiological and subjective sexual arousal in self-identified asexual women. *Archives of Sexual Behavior, 40*(4), 699–712.

Diamond, L. M. (2008a). Female bisexuality from adolescence to adulthood: Results from a 10-year longitudinal study. *Developmental Psychology, 44*(1), 5–14.

Diamond, L. M. (2008b). *Sexual fluidity: Understanding women's love and desire.* Cambridge, MA: Harvard University Press.

Green, R. (1987). *The "sissy boy" syndrome and the development of homosexuality.* New Haven, CT: Yale University Press.

Kallmann, F. J. (1952). Twin and sibship study of overt male homosexuality. *The American Journal of Human Genetics, 4*, 136–146.

Lalumière, M. L., Blanchard, R., & Zucker, K. J. (2000). Sexual orientation and handedness in men and women: A meta-analysis. *Psychological Bulletin, 126*(4), 575–592.

LeVay, S. (1991). A difference in hypothalamic structure between heterosexual and homosexual men. *Science, 253*, 1034–1037.

LeVay, S. (2010). *Gay, straight, and the reason why: The science of sexual orientation.* New York: Oxford University Press.

Marshal, M. P., Dietz, L. J., Friedman, M. S., Stall, R., Smith, H. A., McGinley, J., et al. (2011). Suicidality and depression disparities between sexual minority and heterosexual youth: A meta-analytic review. *Journal of Adolescent Health, 49*(2), 115–123.

Paul, J. P., Catania, J., Pollack, L., Moskowitz, J., Canchola, J., Mills, T., et al. (2002). Suicide attempts among gay and bisexual men: Lifetime prevalence and antecedents. *American Journal of Public Health, 92*, 1338–1345.

Ponseti, J., Granert, O., Jansen, O., Wolff, S., Mehdorn, H., Bosinski, H., et al. (2009). Assessment of sexual orientation using the hemodynamic brain response to visual sexual stimuli. *The Journal of Sexual Medicine, 6*, 1628–1634.

Richters, J., Butler, T., Schneider, K., Yap, L., Kirkwood, K., Grant, L., et al. (2012). Consensual sex between men and sexual violence in Australian prisons. *Archives of Sexual Behavior, 41*(2), 517–524.

Rieger, G., Chivers, M. L., & Bailey, J. M. (2005). Sexual arousal patterns of bisexual men. *Psychological Science, 16*(8), 579–584.

Rosenthal, A. M., Sylva, D., Safron, A., & Bailey, J. M. (2012). The male bisexuality debate revisited: Some bisexual men have bisexual arousal patterns. *Archives of Sexual Behavior, 41*(1), 135–147.

Savic, I., Berglund, H., & Lindstrom, P. (2005). Brain response to putative pheromones in homosexual men. *Proceedings of the National Academy of Sciences of the United States of America, 102*, 7356–7361.

Swaab, D. F., & Hofman, M. A. (1990). An enlarged suprachiasmatic nucleus in homosexual men. *Brain Research, 537*, 141–148.

Vrangalova, Z., & Savin-Williams, R. C. (2012). Mostly heterosexual and mostly gay/lesbian: Evidence for new sexual orientation identities. *Archives of Sexual Behavior, 41*(1), 85–101 [Epub 11 Feb 2012].

Yule, M. A., Brotto, L. A., & Gorzalka, B. B. (2014). Biological markers of asexuality: Finger length ratios, handedness, and birth order in self-identified asexual men and women. *Archives of Sexual Behavior, 43*(2), 299–310.

Chapter 15: Learning from S&M Clubs

Bastian, B., Jetten, J., & Fasoli, F. (2011). Cleansing the soul by hurting the flesh: The guilt reducing effect of pain. *Psychological Science, 22*, 334–335.

Bastian, B., Jetten, J., & Stewart, E. (2013). Physical pain and guilty pleasures. *Social Psychological and Personality Science, 4*, 215–219.
Benamou, P. H. (2006). Erotic and sadomasochistic foot and shoe. *Medicine et Chirurgie du Pied, 22*, 43–64.
Berridge, K. (2007). The debate over dopamine's role in reward: The case for incentive salience. *Psychopharmacology, 191*, 391–431.
Bhugra, D., Popelyuk, D., & McMullen, I. (2010). Paraphilias across cultures: Contexts and controversies. *Journal of Sex Research, 47*(2–3), 242–256.
Bivona, J., & Critelli, J. (2009). The nature of women's rape fantasies: An analysis of prevalence, frequency, and contents. *Journal of Sex Research, 46*(1), 33–45.
Byers, E. S., Purdon, C., & Clark, D. A. (1998). Sexual intrusive thoughts of college students. *Journal of Sex Research, 35*, 359–369.
Coria-Ávila, G. A., Ouimet, A. J., Pacheco, P., Manzo, J., & Pfaus, J. G. (2005). Olfactory conditioned partner preference in the female rat. *Behavioral Neuroscience, 119*(3), 716–725.
Critelli, J. W., & Bivona, J. M. (2008). Women's erotic rape fantasies: An evaluation of theory and research. *Journal of Sex Research, 45*(1), 57–70.
Dawson, S. J., Suschinsky, K. D., & Lalumière, M. L. (2012). Sexual fantasies and viewing times across the menstrual cycle: A diary study. *Archives of Sexual Behavior, 41*(1), 173–183.
Franklin, J. C., Hessel, E. T., Aaron, R. V., Arthur, M. S., Heilbron, N., & Prinstein, M. J. (2010). The functions of nonsuicidal self-injury: Support for cognitive-affective regulation and opponent processes from a novel psychophysiological paradigm. *Journal of Abnormal Psychology, 119*(4), 850–862.
Holtzman, D., & Kulish, N. (2012). Female exhibitionism: Identification, competition and camaraderie. *International Journal of Psychoanalysis, 93*(2), 271–292.
Hsu, B., Kling, A., Kessler, C., Knapke, K., Diefenbach, P., & Elias, J. E. (1994). Gender differences in sexual fantasy and behavior in a college population: A ten-year replication. *Journal of Sex and Marital Therapy, 20*(2), 103–118.
Kafka, M. P. (2010a). The DSM diagnostic criteria for fetishism. *Archives of Sexual Behavior, 39*(2), 357–362.
Kafka, M. P. (2010b). The DSM diagnostic criteria for paraphilia not otherwise specified. *Archives of Sexual Behavior, 39*(2), 373–376.
Kantorowitz, D. A. (1978). An experimental investigation of preorgasmic reconditioning and postorgasmic deconditioning. *Journal of Applied Behavioral Analysis, 11*, 23–34.
Kirsch, L. G., & Becker, J. V. (2007). Emotional deficits in psychopathy and sexual sadism: Implications for violent and sadistic behavior. *Clinical Psychological Review, 27*(8), 904–922.
Komisaruk, B. R., & Whipple, B. (1986). Vaginal stimulation-produced analgesia in rats and women. *Annals of New York Academy of Sciences, 467*, 30–39.
Korner, A., Gerull, F., Stevenson, J., & Meares, R. (2007). Harm avoidance, self-harm, psychic pain, and the borderline personality: Life in a "haunted house". *Comprehensive Psychiatry, 48*, 303–308.

Långström, N. (2010). The DSM diagnostic criteria for exhibitionism, voyeurism, and frotteurism. *Archives Sexual Behavior, 39*(2), 317–324.

Leitenberg, H., & Henning, K. (1995). Sexual fantasy. *Psychological Bulletin, 117*, 469–496.

Leknes, S. G., Bantick, S., Willis, C. M., Wilkinson, J. D., Wise, R. G., & Tracey, I. (2007). Itch and motivation to scratch: An investigation of the central and peripheral correlates of allergen- and histamine-induced itch in humans. *Journal of Neurophysiology, 97*, 415–422.

Leknes, S. G., Brooks, J. C. W., Wiech, K., & Tracey, I. (2008). Pain relief as an opponent process: A psychophysical investigation. *European Journal of Neuroscience, 28*, 794–801.

Leknes, S. G., & Tracey, I. (2008). A common neurobiology for pain and pleasure. *National Review of Neuroscience, 9*(4), 314–320.

Moyano, N., & Sierra, J. C. (2012). Adaptación y validación de la versión española del Sexual Cognitions Checklist (SCC). *Anales de Psicología, 28*(3), 904–914.

Nichols, M. (2006). Psychotherapeutic issues with "kinky" clients: Clinical problems, yours and theirs. *Journal of Homosexuality, 50*(2–3), 281–300.

Person, E. S., Terestman, N., Myers, W. A., Goldberg, E. L., & Salvadori, C. (1989). Gender differences in sexual behaviors and fantasies in a college population. *Journal of Sex and Marital Therapy, 15*(3), 187–198.

Pfaus, J. G., Kippin, T. E., & Centeno, S. (2001). Conditioning and sexual behavior: A review. *Hormones and Behavior, 40*(2), 291–321.

Pfaus, J. G., Kippin, T. E., Coria-Ávila, G. A., Gelez, H., Afonso, V. M., Ismail, N., et al. (2012). How the experience of sexual reward connects sexual desire, preference, and performance. *Archives of Sexual Behavior, 41*, 31–62.

Pollok, B., Krause, V., Legrain, V., Ploner, M., Freynhagen, R., Melchior, I., et al. (2010). Differential effects of painful and non-painful stimulation on tactile processing in fibromyalgia syndrome and subjects with masochistic behaviour. *PLoS One, 5*(12), e15804.

Rantala, M. J., Pölkki, M., & Rantala, L. M. (2010). Preference for human male body hair changes across the menstrual cycle and menopause. *Behavioral Ecology, 21*, 419–423.

Richters, J., de Visser, R. O., Rissel, C. E., Grulich, A. E., & Smith, A. M. (2008). Demographic and psychosocial features of participants in bondage and discipline, "sadomasochism" or dominance and submission (BDSM): Data from a national survey. *Journal of Sexual Medicine, 5*(7), 1660–1668.

Sagarin, B. J., Cutler, B., Cutler, N., Lawler-Sagarin, K. A., & Matuszewich, L. (2009). Hormonal changes and couple bonding in consensual sadomasochistic activity. *Archives of Sexual Behavior, 38*, 186–200.

Sánchez-Sánchez, L. C., Luciano Soriano, C., & Barnes-Holmes, D. (2009). The formation of sexual fantasies by means of the rebound effect of suppressed thoughts. *Sexología Integral, 6*(2).

Sierra, J. C., Ortega, V., & Zubeidat, I. (2006). Confirmatory factor analysis of a Spanish version of the sex fantasy questionnaire: Assessing gender differences. *Journal of Sex and Marital Therapy, 32*(2), 137–159.
Silberberg, A., & Adler, N. (1974). Modulation of the copulatory sequence of the male rat by a schedule of reinforcement. *Science, 185,* 374–376.
Stulhofer, A., & Ajduković, D. (2011). Should we take anodyspareunia seriously? A descriptive analysis of pain during receptive anal intercourse in young heterosexual women. *Journal of Sex and Marital Therapy, 37*(5), 346–358.
Wegner, D. M. (1989). *White bears and other unwanted thoughts: Suppression, obsession, and the psychology of mental control.* New York: Viking/Penguin.
Wegner, D. M., Schneider, D. J., Carter, S., & White, T. (1987). Paradoxical effects of thought suppression. *Journal of Personality and Social Psychology, 53,* 5–13.
Whipple, B., & Komisaruk, B. R. (1985). Elevation of pain threshold by vaginal stimulation in women. *Pain, 21,* 357–367.
Wright, S. (2010). Depathologizing consensual sexual sadism, sexual masochism, transvestic fetishism, and fetishism. *Archives of Sexual Behavior, 39,* 1229–1230.
Zubieta, J. K., Ketter, T. A., Bueller, J. A., Xu, Y., Kilbourn, M. R., Young, E. A., et al. (2003). Regulation of human affective responses by anterior cingulate and limbic muopioid neurotransmission. *Archives of General Psychiatry, 60*(11), 1145–1153.

Chapter 16: Disorders of Obsession, Impulsivity, and Lack of Control

Aggarwal, G., Satsangi, B., Raikwar, R., Shukla, S., & Mathur, R. (2011). Unusual rectal foreign body presenting as intestinal obstruction: A case report. *Ulusal Travma ve Acil Cerrahi Dergisi, 17*(4), 374–376.
Barlow, D. H., Sakheim, D. K., & Beck, J. H. (1983). Anxiety increases sexual arousal. *Journal of Abnormal Psychology, 93,* 49–54.
Bhugra, D., Popelyuk, D., & McMullen, I. (2010). Paraphilias across cultures: Contexts and controversies. *Journal of Sex Research, 47*(2–3), 242–256.
Boureghda, S. S., Retz, W., Philipp-Wiegmann, F., & Rösler, M. (2011). A case report of necrophilia-a psychopathological view. *Journal of Forensic and Legal Medicine, 18*(6), 280–284.
Brass, M., & Haggard, P. (2007). To do or not to do: The neural signature of self-control. *The Journal of Neuroscience, 27*(34), 9141–9145.
Chivers, M. L., Rieger, G., Latty, E., & Bailey, J. M. (2004). A sex difference in the specificity of sexual arousal. *Psychological Science, 15,* 736–744.
Chivers, M. L., Seto, M. C., & Blanchard, R. (2007). Gender and sexual orientation differences in sexual response to the sexual activities versus the gender of actors in sexual films. *Journal of Personality and Social Psychology, 93,* 1108–1121.

Davis, J. F., Loos, M., Di Sebastiano, A. R., Brown, J. L., Lehman, M. N., & Coolen, L. M. (2010). Lesions of the medial prefrontal cortex cause maladaptive sexual behavior in male rats. *Biological Psychiatry, 67*(12), 1199–1204.

Figner, B., Knoch, D., Johnson, E. J., Krosch, A. R., Lisanby, S. H., Fehr, E., et al. (2010). Lateral prefrontal cortex and self-control in intertemporal choice. *Nature Neuroscience, 13*, 538–539.

Goldberg, J. E., & Steele, S. R. (2010). Rectal foreign bodies. *Surgical Clinic of North America, 90*(1), 173–184.

Harris, G. T., Lalumière, M. L., Seto, M. C., Rice, M. E., & Chaplin, T. C. (2012). Explaining the erectile responses of rapists to rape stories: The contributions of sexual activity, non-consent, and violence with injury. *Archives of Sexual Behavior, 41*, 221–229.

Heatherton, T. F., & Wagner, D. D. (2011). Cognitive neuroscience of self-regulation failure. *Trends in Cognitive Sciences, 15*(3), 132–139.

Hedgcock, W. M., Vohs, K. D., & Rao, A. R. (2012). Reducing self-control depletion effects through enhanced sensitivity to implementation: Evidence from fMRI and behavioral studies. *Journal of Consumer Psychology, 22*(4), 486–495.

Janssen, E. (2011). Sexual arousal in men: A review and conceptual analysis. *Hormones and Behavior, 59*, 708–716.

Kaplan, M. S., & Krueger, R. B. (2010). Diagnosis, assessment, and treatment of hypersexuality. *Journal of Sex Research, 47*(2), 181–198.

Krueger, R. B., & Kaplan, M. S. (2002). Behavioral and psychopharmacological treatment of the paraphilic and hypersexual disorders. *Journal of Psychiatric Practice, 8*, 21–32 (10131).

Krueger, R. B., & Kaplan, M. S. (2006). Chemical castration. Treatment for pedophilia. *DSM-IV-TR Casebook, 2*, 309–334.

Levin, R. J., & Van Berlo, W. (2004). Sexual arousal and orgasm in subjects who experience forced or non-consensual sexual stimulation. A review. *Journal of Clinical and Forensic Medicine, 11*, 82–88.

Levine, S. B. (2010). What is sexual addiction? *Journal of Sex and Marital Therapy, 36*(3), 261–275.

Mahoney, S., & Zarate, C. (2007). Persistent sexual arousal syndrome: A case report and review of the literature. *Journal of Sex and Marital Therapy, 33*(1), 65–71.

Pfaus, J. G., Kippin, T. E., Coria-Ávila, G. A., Gelez, H., Afonso, V. M., Ismail, N., et al. (2012). Who, what, where, when (and maybe even why)? How the experience of sexual reward connects sexual desire, preference, and performance. *Archives of Sexual Behavior, 41*(1), 31–62.

Rantala, M. J., Polkki, M., & Rantala, L. M. (2010). Preference for human male body hair changes across the menstrual cycle and menopause. *Behavioral Ecology, 21*, 419–423.

Schott, J. C., Davis, G. J., & Hunsaker, J. C., 3rd. (2003). Accidental electrocution during autoeroticism: A shocking case. *American Journal of Forensic Medicine and Pathology, 24*(1), 92–95.

Suschinsky, K. D., & Lalumière, M. L. (2012). Is sexual concordance related to awareness of physiological states? *Archives of Sexual Behavior, 41*, 199–208.

Von Krafft-Ebing, R. (1929). *Psychopathia Sexualis*. New York: Physicians and Surgeons.

Chapter 17: Sexual Identities Beyond XX and XY

Case, L. K., & Ramachandran, V. S. (2012). Alternating gender incongruity: A new neuropsychiatric syndrome providing insight into the dynamic plasticity of brain sex. *Medical Hypotheses, 78*(5), 621–631.

Crone-Munzebrock, A. (1951). A phantom sensation after amputation of the penis. *Z Urology, 41*, 819–822.

Hare, L., Bernard, P., Sánchez, F. J., Baird, P. N., Vilain, E., Kennedy, T., et al. (2009). Androgen receptor repeat length polymorphism associated with male-to-female transsexualism. *Biological Psychiatry, 65*(1), 93–96.

Heylens, G., De Cuypere, G., Zucker, K. J., Schelfaut, C., Elaut, E., Vanden Bossche, H., et al. (2012). Gender identity disorder in twins: A review of the case report literature. *Journal of Sexual Medicine, 9*(3), 751–757.

Hughes, I. A., Houk, C., Ahmed, S. F., & Lee, P. A. (2006). Consensus statement on management of intersex disorders. *Archives of Disease in Childhood, 91*, 554–563.

Imperato-McGinley, J., Guerrero, L., Gautier, T., & Peterson, R. E. (1974). Steroid 5alpha-reductase deficiency in man: An inherited form of male pseudohermaphroditism. *Science, 186*(4170), 1213–1215.

Meyer-Bahlburg, H. F. L. (2011). Gender monitoring and gender re-assignment of children and adolescents with a somatic disorder of sex development. *Child and Adolescent Psychiatric Clinics of North America, 20*(4), 639–649.

Meyer-Bahlburg, H. F. L., Dolezal, C., Baker, S. W., Carlson, A. D., Obeid, J. S., & New, M. I. (2004). Prenatal androgenization affects gender-related behavior but not gender identity in 5-12 year old girls with congenital adrenal hyperplasia. *Archives of Sexual Behavior, 33*, 97–104.

Meyer-Bahlburg, H. F. L., Dolezal, C., Baker, S. W., & New, M. I. (2008). Sexual orientation in women with classical or non-classical congenital adrenal hyperplasia as a function of degree of prenatal androgen excess. *Archives of Sexual Behavior, 37*(1), 85–99.

Mitchell, S. W. (1871). Phantom limbs. *Lippincotts Magazine, 8*, 563–569.

Ramachandran, V. S., & McGeoch, P. D. (2007). Occurrence of phantom genitalia after gender reassignment surgery. *Medical Hypotheses, 69*, 1001–1003.

Ramachandran, V. S., & Rogers-Ramachandran, D. (1996). Synaesthesia in phantom limbs induced with mirrors. *Proceedings of Biological Sciences, 263*, 377–386.

Rametti, G., Carrillo, B., Gómez-Gil, E., Junque, C., Zubiarre-Elorza, L., Segovia, S., et al. (2011). The microstructure of white matter in male to female transsexuals before cross-sex hormonal treatment. A DTI study. *Journal of Psychiatric Research, 45*(7), 949–954.

Weinberg, M. S., & Williams, C. J. (2010). Men sexually interested in transwomen (MSTW): Gendered embodiment and the construction of sexual desire. *Journal of Sex Research, 47*(4), 374–383.

Wierckx, K., Van Caenegem, E., Elaut, E., Dedecker, D., Van de Peer, F., Toye, K., et al. (2011). Quality of life and sexual health after sex reassignment surgery in transsexual men. *Journal of Sexual Medicine, 8*(12), 3379–3388.

Zhou, J. N., Hofman, M. A., Gooren, L. J. G., & Swaab, D. F. (1995). A sex difference in the human brain and its relation to transsexuality. *Nature, 378*(6552), 68–70.

Zubiaurre-Elorza, L., Junque, C., Gómez-Gil, E., Segovia, S., Carrillo, B., Rametti, G., et al. (2013). Cortical thickness in untreated transsexuals. *Cerebral Cortex, 23*(12), 2855–2862 [Epub 31 Aug 2012].

Chapter 18: Marrying Social and Sexual Monogamy in Swingers' Clubs

Acevedo, B. P., Aron, A., Fisher, H. E., & Brown, L. L. (2012). Neural correlates of long-term intense romantic love. *Social Cognitive and Affective Neuroscience, 7*(2), 145.

Bancroft, J., Loftus, J., & Long, J. S. (2003). Distress about sex: A national survey of women in heterosexual relationships. *Archives of Sexual Behavior, 32*, 193–208.

Buss, D. M. (2000). *The dangerous passion: Why jealousy is as necessary as love and sex*. New York: Free Press.

De Visser, R., & McDonald, D. (2007). Swings and roundabouts: Management of jealousy in heterosexual swinging couples. *British Journal of Social Psychology, 46*(Pt 2), 459–476.

Durante, K. M., & Li, N. P. (2009). Oestradiol level and opportunistic mating in women. *Biology Letters, 5*(2), 179–182.

Ember, C., & Ember, M. (2003). Gender and sexuality in Mormon polygamous society. In *Encyclopedia of sex and gender* (pp. 433–442).

Fisher, H. (2004). *Why we love: The nature and chemistry of romantic love*. New York: Henry Holt.

Fisher, H. E., Aron, A., Mashek, D., Li, H., & Brown, L. L. (2002). Defining the brain systems of lust, romantic attraction, and attachment. *Archives of Sexual Behavior, 31*, 413–419.

Garcia, J. R., MacKillop, J., Aller, E. L., Merriwether, A. M., Wilson, D. S., & Lum, J. K. (2010). Associations between dopamine D4 receptor gene variation with both infidelity and sexual promiscuity. *PLoS One, 5*(11), e14162.

High rates of sexually transmitted infections among older swingers. *British Medical Journal*, 28 June 2010.

Jankowiak, W. (2008). Co-wives, desires and conflicts in a USA polygamous community. *Ethnology, 52*(3), 163–180.

Jankowiak, W. (2011). One vision: The making, unmaking and remaking of an American polygamous community. In C. K. Jacobson & L. Burton (Eds.), *Modern polygamy in the United States: Historical, cultural, and legal issues* (pp. 41–76). Oxford: Oxford University Press.

Jankowiak, W., & Gerth, H. (2012). Anthropologica. Can you love more than one person at the same time? A research report. *Anthropologica, 54*, 95–105.

Kazdin, A. E. (2008). Evidence-based treatment and practice: New opportunities to bridge clinical research and practice, enhance the knowledge base, and improve patient care. *American Psychologist, 63*(3), 146–159.

Leavitt, E. (1988). Alternative lifestyle and marital satisfaction: A brief report. *Annals of Sex Research, 1*(3), 455–461.

Levine, S. B. (2006). *Demystifying love: Plain talk for the mental health professional.* New York: Routledge.

Mark, K. P., Janssen, E., & Milhausen, R. R. (2011). Infidelity in heterosexual couples: Demographic, interpersonal, and personality-related predictors of extradyadic sex. *Archives of Sexual Behavior, 40*(5), 971–982.

Meana, M., & Jones, S. (2011). Developments and trends in sex therapy. *Advanced Psychosomatic Medicine, 31*, 57–71.

O'Connor, J. J. M., Re, D. E., & Feinberg, D. R. (2011). Voice pitch influences perceptions of sexual infidelity. *Evolutionary Psychology, 9*(1), 64–78.

Sternberg, R. J., & Weis, K. (2008). *The new psychology of love.* New Haven: Yale University Press.

Van Anders, S. M., Hamilton, L. D., & Watson, N. V. (2007). Multiple partners are associated with higher testosterone in North American men and women. *Hormones and Behavior, 51*(3), 454–459.

Walum, H., Westberg, L., Henningsson, S., Neiderhiser, J. M., Reiss, D., Igl, W., et al. (2008). Genetic variation in the vasopressin receptor 1a gene (AVPR1A) associates with pair-bonding behavior in humans. *Proceedings of the National Academy of Sciences of the United States of America, 105*(37), 14153–14156. Early edition, 2–5 September.

Winslow, J. T., Hastings, N., Carter, C. S., Harbaugh, C. R., & Insel, T. R. (1993). A role for central vasopressin in pair bonding in monogamous prairie voles. *Nature, 365*, 545–548.

Xu, X., Brown, L., Aron, A., Cao, G., Feng, T., Acevedo, B., et al. (2012). Regional brain activity during early-stage intense romantic love predicted relationship outcomes after 40 months: An fMRI assessment. *Neuroscience Letter, 526*(1), 33–38.

Xu, X., Wang, J., Aron, A., Lei, W., Westmaas, J. L., & Weng, X. (2012). Intense passionate love attenuates cigarette cue-reactivity in nicotine-deprived smokers: An fMRI study. *PLoS One, 7*(7), e42235.

Young, L. J., & Wang, Z. (2004). The neurobiology of pair-bonding. *Nature Neuroscience, 7*, 1048–1054.

Epilogue: Sex and Science Don't End at Orgasm

Alonso-Navarro, H., & Jiménez-Jiménez, F. J. (2006). Transient global amnesia during sexual intercourse. *Revista de Neurología, 42*(6), 382–383.

Amsterdam, A., & Krychman, M. (2009). Clitoral atrophy: A case series. *Journal of Sexual Medicine, 6*(2), 584–587.

Bancroft, J., Carnes, L., Janssen, E., Goodrich, D., & Long, J. S. (2005). Erectile and ejaculatory problems in gay and heterosexual men. *Archives of Sexual Behavior, 34*(3), 285–297.

Bohlen, J. G., Held, J. P., Sanderson, M. O., & Ahlgren, A. (1982). The female orgasm: Pelvic contractions. *Archives of Sexual Behavior, 11*, 367–386.

Diez cosas que los científicos nos enseñaron sobre el sexo en 2012. ABC.es, 2 Jan 2013.

Fisher, T. D., Moore, Z. T., & Pittenger, M. J. (2012). Sex on the brain: An examination of frequency of sexual cognitions as a function of gender, erotophilia, and social desirability. *Journal of Sex Research, 49*(1), 69–77.

Herbenick, D., & Fortenberry, J. D. (2011). Exercise-induced orgasm and pleasure among women. *Sexual and Relationship Therapy, 26*(4), 373.

Hernandez, B. C., Schwenke, N. J., & Wilson, C. M. (2011). Spouses in mixed-orientation marriage: A 20-year review of empirical studies. *Journal of Marital and Family Therapy, 37*(3), 307–318.

Masters, W. H., Johnson, V. E., & Kolodny, R. C. (1985). *Masters and Johnson on sex and human loving*. Boston: Little, Brown.

McCall, K. M., Rellini, A. H., Seal, B. N., & Meston, C. M. (2007). Sex differences in memory for sexually-relevant information. *Archives of Sexual Behavior, 36*(4), 508–517 [Epub 21 Dec 2006].

Rattan, K. N., Kajal, P., Pathak, M., Kadian, Y. S., & Gupta, R. (2010). Aphallia: Experience with 3 cases. *Journal of Pediatric Surgery, 45*(1), E13–E16.

Schenck, C. H., Arnulf, I., & Mahowald, M. W. (2007). Sleep and sex: What can go wrong? A review of the literature on sleep related disorders and abnormal sexual behaviors and experiences. *Sleep, 30*(6), 683–702.

Shtarkshall, R. A., & Feldman, B. S. (2008). A woman with a high capacity for multi-orgasms: A non-clinical case-report study. *Sexual and Relationship Therapy, 23*(3), 259–269.

Truitt, W. A., & Coolen, L. M. (2002). Identification of a potential ejaculation generator in the spinal cord. Science, 297, 1566–1569.

GPSR Compliance

The European Union's (EU) General Product Safety Regulation (GPSR) is a set of rules that requires consumer products to be safe and our obligations to ensure this.

If you have any concerns about our products, you can contact us on

ProductSafety@springernature.com

In case Publisher is established outside the EU, the EU authorized representative is:

Springer Nature Customer Service Center GmbH
Europaplatz 3
69115 Heidelberg, Germany

www.ingramcontent.com/pod-product-compliance
Lightning Source LLC
LaVergne TN
LVHW010334260326
834688LV00036B/706